Ulrich Poth

Automotive Coatings Formulation

Chemistry, Physics und Practices

Cover picture: Evonik Tego Chemie Services GmbH, Essen/Germany

Poth, Ulrich:
Automotiv Coatings Formulation
Hannover: Vincentz Network, 2008
(European Coatings Tech Files)
ISBN 3-86630-904-X
ISBN 978-3-86630-904-3

Please ask for our book catalogue
Vincentz Network, Plathnerstr. 4c, 30175 Hanover, Germany
T (202) 684-6630, F (202) 380-9129
E-mail: books@american-coatings.com, www.american-coatings.com

Layout: Maxbauer & Maxbauer, Hanover

ISBN 3-86630-904-X
ISBN 978-3-86630-904-3

European Coatings Tech Files

Ulrich Poth

Automotive Coatings Formulation

Chemistry, Physics und Practices

Preface

Very few objects capture the public interest as much as automobiles. In practice, automobiles are means of transporting persons and goods. And therein lies the particularity of this form of transportation - its individuality, which is very much in vogue with current ideas. Individual mobility is often linked to the individual appearance of objects, and automobiles are no different in this regard. Popular tastes in automobiles vary enormously. Some people like to drive very prestigious cars, others prefer sporty cars, or cars which "look different" - and there are many different tastes in between. The whole appearance of automobiles is dictated by the shape of the car body and its accessories, and not least by the surface impression created by the coatings.

Makers of cars and coating materials spend a lot of effort on generating particularly appealing car surfaces. These include optimising resistance to weathering, corrosion, chemicals and mechanical impact. The most important target is to preserve the appearance as long as possible, ultimately over the vehicle's life-time.

The formulation and composition of coating materials depend on the requirements of processes of different industrial application methods. Therefore, automotive coating systems are a special class of industrial coating systems. That is why some paint producers specialise in the development and production of automotive paints. Although criticism has been levelled at ever increasing production levels (e.g. due to rising emissions of carbon dioxide), it must nonetheless be borne in mind that producers of coating systems for automotive applications pioneered the development and introduction of environmental paint systems. Currently the most important target is to avoid hazardous compounds or to replace them, and to cut emissions of organic compounds during the production and application of paints. This book presents the results of such developments within the coatings industry.

The development and production of automotive coatings is not the preserve of paint producers only. Suppliers of raw materials also participate in the developments and improvements. These include major chemical companies who produce basic raw materials, and producers of specialty chemicals, polymers for coatings, pigments, and additives. Makers of application equipment contribute to the development process, too. The automotive industry itself is also interested in achieving optimum systems and processes. Some companies therefore expend a great deal of effort on researching and developing coating systems.

This book is directed at persons who are involved in those sectors of industry which are related to the development, production, testing, marketing, supply of coatings raw materials, coating systems and application equipment. The information is addressed to chemists, physicists, engineers, and other persons with technical interests who are involved with coatings. Of course, another goal is to educate persons who are working their way into the technology. In addition, the book is directed at persons who are mainly involved in the application processes, but would like to know more about coating systems.

The book mainly contains descriptions of automotive coating formulations. There is a defined correlation between composition and properties. The composition and production of coating systems and application conditions greatly influences the application properties of coating films. By application properties is meant the application behaviour and the properties of the coating film - including those of protection and appearance. Application processes are described in basic terms only with a view to showing their influence, together with the composition of paints, on the entire coating properties.

For a better understanding of the special demands and properties of the various coating systems, the book first deals with general aspects of coating materials, before providing more detail of the chemistry and physics of automotive coating systems, with a view to explaining why their components are selected from the plurality of different ingredients that are available for paints.

Ulrich Poth

Münster, March 2008

Content

1 Introduction

Automobile production is one of the most important economic activities in the world. In 2006, some 50 million passenger cars, and nearly 70 million vehicles including passenger cars, trucks, and buses, were produced [1]. The list of countries that produced the most passenger cars in 2006 is headed by Japan (about 9.8 million), followed by Germany (5.4 million); People's Republic of China (5.2 million) ahead of the USA (4.4 million). The world's ten largest producers of automobiles (passenger cars, trucks, and buses) are as follows: General Motors (8.9 million), Toyota (8.0 million), Ford Motor Corp. (6.3 million), Volkswagen (5.7 million), Honda (3.7 million), PSA (3.6 million), Nissan (3.2 million), Chrysler (2.5 million), Renault (2.5 million), and Hyundai (2.5 million). In 2007, Toyota took over from General Motors, which now occupies second position. Whereas some 20 million cars were registered in Germany in 1975, there were 46,569,657 as of 1 January 2007 [1].

Besides the large car producers, there are small companies that mostly make custom car types. Given that a medium-sized car is coated with about 9 kg paint (solids) [2], automobiles around the world accounted for 630,000 t and, in terms of delivery form, up to 1.6 million t paint materials in 2006.

The production of automotive OEM (original equipment manufacturing) coating systems is dominated by large companies. The most important paint producers for automotive OEM coatings are PPG, DuPont, and BASF followed by Nippon-Paint and Kansai-Paint [3]. However, smaller companies play niche roles in car-paint production. The concentration process undergone by paint automotive OEM paint makers is being driven by demand and the need to expend vast energies on innovation and warranties. Incessant demand for ever-better automotive paints calls for a great deal of development effort that small companies are unable to achieve. This not only involves personnel levels and specialised equipment, but also continuous improvements in production processes and quality control. Optimum quality control requires several analytical measurements and a multitude application tests that are only achieved with expensive equipment. As for warranties, there are more than the material costs to be borne in mind. If, for example, a paint system is applied to car bodies in a thickness of just 40 µm and defects arise, it is possible for the day's entire production of cars (e.g. 800 units) to be faulty. Compensation claims may extend to more than just the deficient car bodies - but may also entail replacement deliveries, compensation for production downtimes, and refurbishments.

Two different strategies are deployed to safeguard against such problems. In most cases, at least two suppliers of coating systems will have been approved for the products. This emphasises competition between the various paint producer as regards quality and cost of the coatings materials. Totally different is the strategy that fully involves the paint producer in the coatings application process. The reason for this approach is mainly that the coating application process is totally different from all other production steps in car making, and - from the viewpoint of the car maker - constitutes a very special type of processing. In the most pronounced case, the paint producer accepts the car body from construction, executes the entire paint application process and then hands over only the well coated car bodies for further production steps. In most of these cases, the maker and deliverer of application equipment are involved in quality control [4].

Although the most important producers of automotive OEM coatings belong to major chemical corporations, the development, production and supply of automotive coating materials is more or less in the hands of medium-sized companies. The reason for this is that the technology is strongly application and customer oriented. The major producers of automotive paint systems have become more vertically integrated and produce their own synthetic resins. Part of this decision is the tradition of the old 'oil varnish-making houses'. However, another part is the plan to offer optimum, custom systems to the market that meet the specific application conditions of the customers very well and offer the possibility of differentiating oneself from the competitors. In addition, if the entire production process is efficient, vertical integration increases the degree of added value (from raw material to finished product).

There are many ancillary companies which produce attached parts that must of course be coated. These are mostly small and medium-sized companies that need coating materials which are easy to apply at a low cost. By contrast, repair coating is very much manual work. However, that is not to say that scientific methods are not employed. The key requirement imposed on repair coatings is faithful reproduction of colours and effects. Makers of repair coatings - which are the large companies and several other small companies - offer not only materials for their customers, but also programs and software for determining colours and effects that lead to individual formulation guidelines for reproducing required colours and effects. The paints are produced with the aid of standardised tinting coatings available in mixing units (for storage of suitable tinting coatings and mixing equipment). Some producers and suppliers also offer training programs for repair shop employees.

On account of the different structure and size of trucks, buses and other large vehicles, these require application methods which differ from those for passenger cars, and they therefore also need different coating materials. In this application field, the most important property is general durability (mostly corrosion resistance). Besides coating materials, other materials (e.g. films) are used for large vehicles.

2 General aspects of coatings

2.1 Composition of paints

All paints, including automotive coatings, contain four types of raw material: resins or binders, pigments or pigment-like substances, solvents or dispersing media, and additives (see Figure 2.1).

2.1.1 Resins

Resins are the most important component as they provide the basis of the film forming layer. They are also called binders because they have to wet and disperse the pigments, and bind them in the paint or film layer. Resins or binders are polymers. Physically, polymers are very high viscous liquids which give the impression of being solid. But, unlike true solids in the physical sense that have crystalline structures, polymers do not consist of a structured arrangement of molecules; like liquids, the distances between the molecules have only average values. Nonetheless, polymer molecules are so large that they are virtually immobile [5]. It is these properties that help polymers meet all the demands imposed on coating layers. They can form thin layers that protect the surface against mechanical impact (bending and shock), weathering (light, humidity, and fall-out), chemicals, and solvents. Such properties result particularly from the formation of crosslinked polymers that are elastic and no longer soluble or capable of melting. Crosslinking of polymers therefore has to take place after application to the substrate. Paint systems which form crosslinked films consist of resins that in certain conditions react by crosslinking. In the applied state, these resins are an intermediate stage in the formation of the final polymer, and they are therefore oligomers. Such paint systems are called "reactive paints". The entire film forming process consists of two steps: evaporation of solvents or dispersing media, followed by chemical reaction to form crosslinked polymer molecules of nearly infinite molecular size. There are also paint formulations which consist of resins that form films merely by evaporation of solvents or dispersing media. As evaporation proceeds merely physically, film formation is said to take place by "physical drying". Physically drying resins are mostly chosen from high polymers, because there is no possibility otherwise of creating the necessary minimum resistance to the various types of impact. High polymers have to be either diluted in solvents to a relatively low solids concentration or converted into dispersions. Both methods have some disadvantages. Some automotive paints consist of combinations of reactive resins (crosslinkable) and physically drying resins.

Paint resins are closely related to the polymers familiar from plastics. They are made by the established methods of polymer chemistry, namely polycondensation, polyaddition, and polymerisation (in the narrow sense) [6]. Modified natural substances also serve as resins or binders for paints [7]. All resins must first be converted into an application form which enables the paint to be produced and applied by the various methods. They then have to form films that have the required properties. The most important

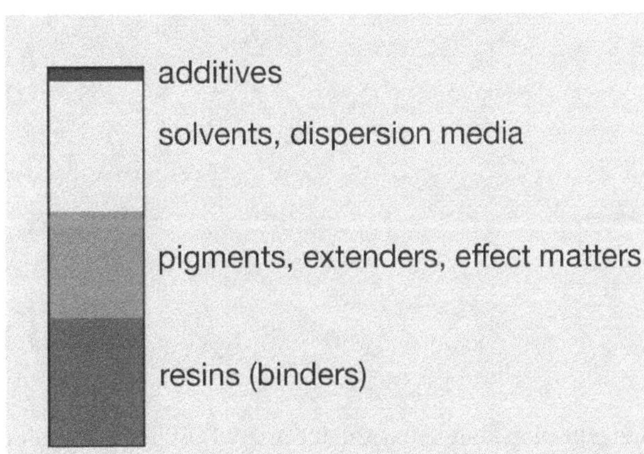

Figure 2.1: Basic composition of paints

resins for automotive paints are polyesters, alkyd resins, acrylic resins, amino resins, polyisocyanates, polyurethanes, epoxy resins, and cellulose esters. These are described in detail in the Chapter on the various paint systems.

2.1.2 Pigments and pigment-like substances

Pigments and pigment-like substances are divided into the following groups: pigments in the narrower sense (coloured pigments), effect substances, and functional pigments, which do not contribute to colour or effect, but support other properties. This definition also includes extenders (fillers). Coloured pigments and functional pigments are introduced into the paint composition by special dispersion methods.

2.1.2.1 Colour generation

Coloured pigments, including white and black pigments, are fine powders of crystalline solids that are insoluble in both water and organic solvents. They have to be distributed efficiently in the film matrix of the coating layer, where they give the coating its colour. The colour stems from the absorption and scattering of visible light. This principle is described in Figure 2.2.

Absorption of light

Light, as the totality of different electromagnetic beams, is absorbed to various extents by different chemical compounds. Absorption takes place when the light energy is consumed due to changes in molecular valences, e.g. valence resonance, electron resonance, and valence angles. In these cases, the light energy is converted into heat energy [8]. The extent of light absorption varies with the thickness of the layer through which the light beams pass. If the light beam enters the layer with an initial intensity I_0, the first layer absorbs a certain fraction of the light. The light, now of reduced intensity, passes through the next layer, and its intensity is again reduced to the same extent, and so on. Thus, the change in the intensity of light dI depends on the intensity I_x at a depth x of the absorbing layer and on the specific fraction K of it absorbed by the layer thickness dx. The total energy absorption is described by the following equation (Lambert's law [9]).

Equation 2.1: Lambert's law

$$dI = -K \cdot I_x \cdot dx$$

$$\frac{dI}{I_x} = -K \cdot dx$$

$$\int_{I_0}^{I_x} \frac{dI}{I_x} = -K \cdot \int_0^x dx$$

$$\ln \frac{I_x}{I_0} = -K \cdot x$$

$$I_{(x)} = I_0 \cdot e^{-K \cdot x}$$

This equation states that the intensity I_0 of light which passes through a layer thickness x of an absorbing medium is reduced to intensity I_x in accordance with an exponential function. The factor K is the specific extinction coefficient for the material through which the light passes.

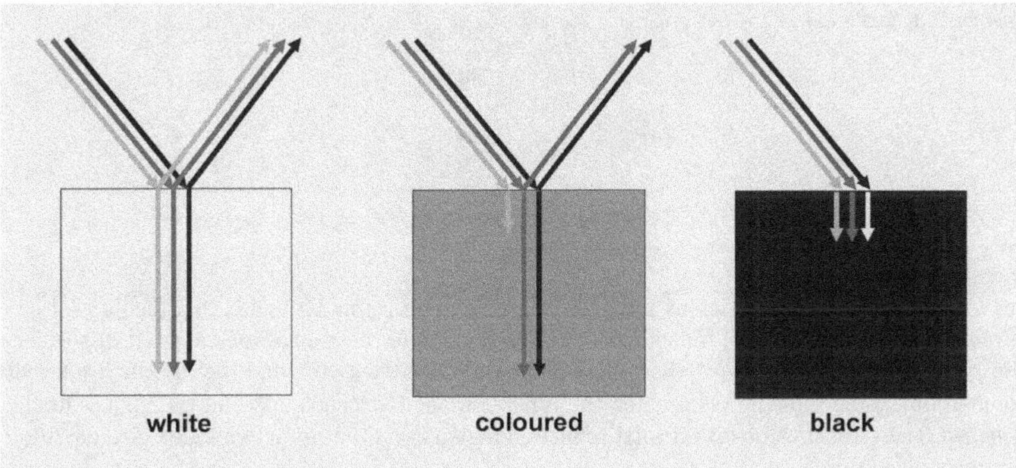

Figure 2.2: Colour effect created by the absorption and scattering of visible light

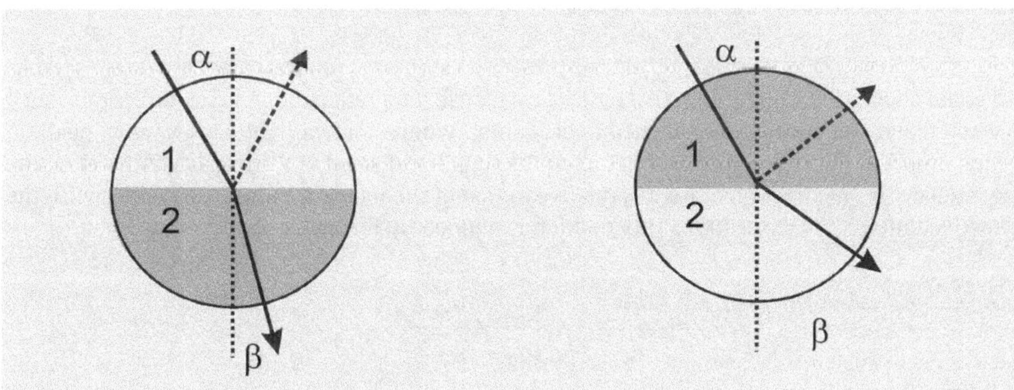

Figure 2.3: Light refraction in media of different optical density

For soluble dyes or coloured ions, this extinction coefficient is directly proportional to the concentration of dyes or ions. This is known as Beer's law [10] and it gave rise to the methods of colorimetric analysis.

Equation 2.2: Beer's law

$$K = K' \cdot c$$

K' is the specific coefficient for the concentration of diluted compound per millimetre per mole. For coloured media, the coefficient varies with the wavelength of the light. After passing through the medium or being reflected from a white surface, the light has the wavelength which is complementary to that of the absorbed light (absorption maximum) or the corresponding wavelength if there is more than one absorption maximum.

Light refraction

When light passes from a medium of low optical density into a medium of high optical density, the beam is refracted towards the axis of incidence (see Figure 2.3). The sine values of the angles of incidence and refraction are inversely proportional to the optical densities of the media (refractive indices, n, Snell's law [11]). Since the refractive index of air is more or less 1, this law enables the refractive indices of different media to be defined.

Equation 2.3: Snell's law

$$\frac{n_2}{n_1} = \frac{\sin \alpha}{\sin \beta}$$

$$\text{for} \quad n_1 \longrightarrow 1.0$$

$$n_2 = \frac{\sin \alpha}{\sin \beta}$$

Conversely, if a light beam passes from a medium of high refractive index to a medium of low refractive index, the angle of reflection increases relative to the axis of incidence. If this reflection reaches a critical value **α^***, the reflection angle **β** becomes 90°, and the light will not exit the medium. If the reflection is greater, the reflection angle exceeds 90° and the light is totally reflected. The critical value **α^*** for total reflection depends on the refractive index (see equation for total reflection).

Equation 2.4: Total reflection

$$\sin \alpha^* = \frac{1}{n}$$

Polygonal objects with high refractive indices can cause total reflection in many directions (such as when diamonds are cut and polished to brilliants). When incident light passes into a medium of higher optical density, a specific fraction of the light is reflected at the surface of the medium. The intensity of this fraction of light I_R relative to that of the incident light I_0 also varies with the refractive index n of the medium. This relation is defined by Fresnel's law [12].

Equation 2.5: Fresnel's law

$$\frac{I_R}{I_0} = \frac{(n_2 - n_1)^2}{(n_2 + n_1)^2}$$

For example, the value of the surface reflection of a commercial glass is 0.16. Like the extinction coefficient, the refractive index has a specific value for each material; but it also depends on the wavelength of the light. Generally, light of higher energy, i.e. shorter wavelength, is refracted to a greater extent and, conversely, the refractive index decreases with increase in wavelength. Where objects exhibit absorption maxima at specific wavelengths, the refractive index also reaches a maximum.

Light scattering

Experience shows that the aforementioned laws governing the refraction of light do not hold for very small particles. The intensity I_R of light reflected by particles with diameters ranging from below 10 µm down to half the wavelength of visible light depends on the particle volume v, the wavelength of light l and a material-specific constant S (Rayleigh's law [13]).

Equation 2.6: Rayleigh's law

$$I_R = S \cdot I_0 \cdot \frac{v^2}{\lambda^4}$$

This Rayleigh scattering results from the fact that small particles oscillate with the incident light and become the centre of new light beams which are not directional but rather are scattered. For this reason, the specific constant S is called the scattering coefficient. Again, the scattering coefficient varies with the refractive index of the particles in relation to the surrounding medium. For example, barium sulphate (n = 1.64) is a white powder in air (n = 1.00) but it is transparent when dispersed in organic paint binders (n ~ 1.50). Another component of scattering is Mie scattering [14]. This is described mathematically as the scattering of light by spherical particles whose diameters are close to the wavelength of light. Unlike Rayleigh scattering, Mie scattering is independent of the wavelength of the incident light. Mie scattering is the reason for the Tyndall and interference effects.

Since Rayleigh's law applies to particles, the intensity of scattering increases with increase in the number of particles in a medium. However, if the particle size decreases beyond half the wavelength of the visible light, the particles are no longer able to oscillate. Such particles - although they might have high refractive indices - cannot reflect light, and so they are transparent. They are only visible if they absorb visible light or factions thereof. For example, titanium dioxide with a particle size in the nanometre range is transparent, but carbon powder, which has nearly the same particle size, is the most important black pigment. All the above-mentioned parameters work together to create colour. A layer of paint will be **white** if all incident light is reflected by scattering across all wavelengths. It will be **black** if all incident light is absorbed across all wavelengths. The layer will be **coloured** if specific fractions of the visible light are absorbed and other fractions are reflected by scattering or refraction from the surfaces below. The sum of all reflections is called remission.

Lightening power

Lightening (or reducing) power is the ability of a white pigment to increase the remission of a coloured or black system. The value of the lightening power depends mainly on the quantity and particle concentration of such a white pigment. Different white pigments have different lightening power, even though the particle concentration may be identical. Lightening power may also be influenced by the extent to which the pigment has been dispersed. It is quantified by comparing different pigments in different amounts [15] (PVC, pigment volume concentration).

Tinting strength

Like the lightening power of white pigments, tinting strength is the ability of coloured pigments to improve the depth of colour impression [16]. Of course, tinting strength, too, depends on the particle concentration (PVC) of a coloured pigment, but it is a specific property of different pigments. The depth of colour is made up of chroma and brightness. The relationship between these two parameters is not linear. The values therefore must be measured using comparison standards and different mixing ratios.

Hiding power

Hiding power is the ability of a coating layer to obliterate the colour of a substrate. Obliteration may proceed by both absorption and reflection of light. *P. Kubelka* and *E. Munk* [17] defined the total reflection of a pigmented coating layer as a function of the intensity of light (I) and layer thickness (x). Their definition consists of two coefficients that are specific to the pigment and film layer, namely the coefficient of absorption (K) and the coefficient of scattering (S). First, the basic equation for the change in the intensity of incident light (dI) is:

Equation 2.7:

$$dI = -(K + S) \cdot I \cdot dx$$

Similarly, the change in the intensity of the reflected light (dJ) is:

Equation 2.8:

$$dJ = (K + S) \cdot J \cdot dx$$

As both light beams are also scattered, the equations have to be expanded as follows:

Equation 2.9:

$$dI = -(K+S)\cdot I \cdot dx + S\cdot J\cdot dx$$

$$dJ = (K+S)\cdot J \cdot dx - S\cdot I\cdot dx$$

Combining the two equations by subtraction yields the following equation:

Equation 2.10:

$$\frac{dI}{I} = -(K+S)\cdot dx + S\cdot \frac{J}{I}\cdot dx$$

$$\frac{dJ}{J} = (K+S)\cdot dx - S\cdot \frac{I}{J}\cdot dx$$

$$\frac{dI}{I} - \frac{dJ}{J} = -2\cdot(K+S)\cdot dx + S\cdot\left(\frac{J}{I}+\frac{I}{J}\right)\cdot dx$$

If the ratio of the intensity of reflected light to the intensity of incident light is defined as reflection (R), the following equation is obtained:

Equation 2.11:

$$\frac{J}{I} = R$$

$$\frac{dR}{R} = -2\cdot(K+S)\cdot dx + S\cdot\left(\frac{1}{R}+R\right)\cdot dx$$

Total hiding by a coating layer occurs if the ratio of an infinitesimal change in reflection to a change in layer thickness becomes zero. In that case, the reflection value reaches a specific limit value (R_∞). The Kubelka-Munk equation can be rearranged to define the ratio of the coefficients as shown.

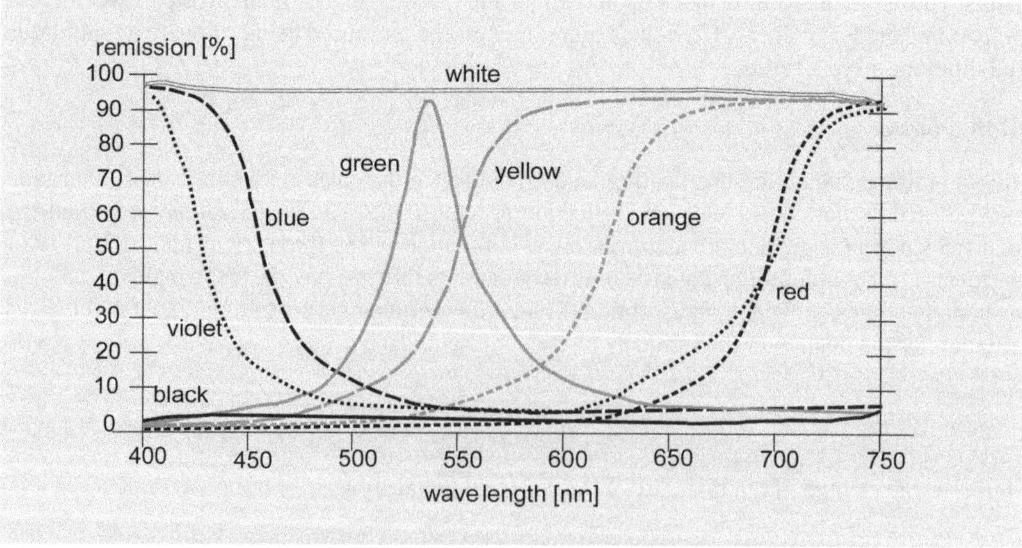

Figure 2.4: Remission curves for typical pigments as a function of the wavelength of visible light

Equation 2.12:

$$\frac{dR}{dx} = 0 \quad R \rightarrow R_\infty$$

$$R_\infty = 1 + \frac{K}{S} - \sqrt{\frac{K}{S} \cdot \left(\frac{K}{S} + 2\right)}$$

$$\frac{K}{S} = \frac{(1 - R_\infty)}{2 \cdot R_\infty}$$

The Kubelka-Munk equation contains some approximations. Nevertheless, it still provides the best description of the interaction between absorption and scattering that constitutes the hiding power of coloured coating layers. In practice, hiding power is measured by applying a wedge of the test material to a chequered surface of black and white squares resembling a chess-board. Optimum hiding power is obtained at that layer thickness which perfectly obliterates the contrast between the black and white squares. There are some hues which do not provide adequate hiding power at the film thicknesses encountered under standard application conditions. These are mainly yellow and orange, some red shades of paints consisting of organic pigments, as well as pure-white colours.

Classification of pigments

The various pigments are categorised by chemical class and colour. First, there are the groups of inorganic and organic pigments. Figure 2.4 shows remission curves for typical pigments as a function of the wavelength of visible light.

2.1.2.2 Inorganic pigments

White inorganic pigments consist of colourless solid substances of high refractive index. Coloured inorganic pigments consist of oxides and salts of heavy metals, the most important being mixed oxides. The substances absorb fractions of visible light by electron transfer to unfilled d-shells (e.g. chromiumIII oxide), charge transfer to different oxidation states (e.g. ironII,III oxide), absorption of free radicals (e.g. ultramarines), or transfer from metal ions to the basic crystal structure (e.g. cadmium sulphide, nickel titanium yellow). Basically, inorganic pigments cause more light scattering than organic pigments. With inorganic pigments, it is difficult to generate fine particles and narrow particle distributions. Typical inorganic pigments are described in Chapter 3.6.3.4.

2.1.2.3 Organic pigments

Organic pigments consist of fine particles of solid crystalline substances composed of relatively large organic molecules. The latter are made up of extended aromatic ring systems (conjugated double-bond systems). Additionally, they contain hetero atoms whose electron shells interact with the aromatic double bonds. Most of the molecules have some condensed aromatic rings which may be connected by diazo, azo, keto or imino groups. Substituents may be halogens, sulpho, hydroxyl, nitro and carboxyl groups. The colour effect stems from absorption of a fraction of visible light by mesomeric resonance of double-bond systems together with the electron shells of hetero atoms.

Choosing large molecules to achieve a greater mesomeric reaction confers an additional advantage. Such molecules are less soluble and relatively stable to migration. These are important properties for pigments for topcoats and effect coatings which have to be resistant to migration, solvents, chemicals, and weathering. Optimum organic pigments have particle sizes half the wavelength of visible light. The colour effect stems mainly from absorption and not light scattering. Some typical organic pigments are described in Chapter 3.6.3.5.

2.1.2.4 Effect substances

Effect substances for paints are pigment-like particles that, when incorporated in a film matrix, convey various brightness and colour impressions which change with the angle of view. This is also known as flip-flop effect. Initially, aluminium flakes were the most common pigments used in effect coatings. Added to basecoats and clearcoats, they are responsible for creating the metallic effect.

Then, pearlescent pigments were developed from mica particles. Pearlescence results from interference of visible light at very thin layers. If the mica particles are covered with layers of heavy metal oxides only a few nanometres thick, the pearlescent effect is extended to yield highly specific colour shifting (colour flip-flop).

Pearlescent pigments were followed by platelet-sized particles that were also covered with layers of heavy metal oxides. Very homogeneous thin layers of aluminium are created by depositing aluminium onto foil from the vapour phase. They are then stripped off and crushed to yield pigment-like particles. The result is a special metallic effect (so-called liquid metal). All the effect pigments mentioned so far are inorganic. But there are also organic effect pigments. These are produced with the aid of substances that form liquid crystals. The latter are immobilised by UV-crosslinking to yield lamellar molecular structures that create colour effects by interference. Effect substances are described in Chapter 3.7.3.5.

2.1.2.5 Functional pigments

Functional pigments are defined here as pigment-like substances whose purpose is not to generate colour or effects on coating layers, but rather to perform other functions.

Extenders

Some authors define extenders (fillers) as a special class of pigment-like substances in their own right. Extenders are particles of inorganic crystalline compounds which have low refractive indices. On account of the low refractive indices, they do not generate any colour in a film matrix. The main purpose of extenders or fillers is to impart fullness. This is the complex visual impression conveyed when a coating layer covers the substrate perfectly as regards smoothness, levelling, and uniformity. Gloss may also play a role in this impression. Fullness depends on the layer thickness and the surface structure of the coating. Extenders or fillers are mostly inexpensive paint raw materials. Extenders additionally have the task of improving the resistance of coating layers to mechanical impact, primarily stone-chipping.

Extenders are used in primers and primer surfacers. Typical extenders are described in Chapter 3.5.3.4.

Corrosion-protection pigments

The most important functional pigments in the narrower sense are the corrosion-protection pigments. Corrosion-protection pigments act in two principal ways, namely, by electrochemical interaction with metal surfaces (active corrosion protection) and by acting as a barrier (passive corrosion protection). For a long time, the most important active corrosion-protection pigments were lead and chromium compounds (red lead, lead chromate, zinc chromate). Due to their toxicity, they are no longer used in industrial paints. Until now, there have been only a few exceptions to this (e.g. wash primers). Initially, the toxic lead and chromate pigments were replaced by zinc phosphates. Zinc phosphates can also enter into an electrochemical reaction with metal surfaces, especially if they contain soluble fractions. The interaction is comparable to the reactions which take place in the pre-treatment of metals (see Chapter 3.4.3). But zinc phosphate pigments, too, have since been identified as posing an environmental risk. Paints which contain 2.5 % of zinc phosphates and more have to be labelled. Waste water may only contain a maximum of 2 mg/l zinc [18].

Since all anticorrosive pigments pose a potential environmental risk, the paint industry has turned its attention to pigments that offer passive anticorrosion, i.e. those which act as a barrier. Such pigments prevent penetration or diffusion by water and aqueous chemicals in the environment through the film layer, and so protect the metal surfaces against corrosion. Extenders with platelet-like particle structures, e.g. micas, wollastonite, and talcs, can form barriers [19]. Graphite and aluminium pigments have a barrier effect, too, but additionally generate coloured coating layers (black and metallic effects). A very special corrosion-protection pigment is zinc dust. Corrosion protection primers can contain very high amounts of zinc dust. The protective effect is similar to that of galvanising (see Chapter 3.4.1).

Pigment-like additives

Pigment-like additives contain particles that interact efficiently to create special rheological effects. Examples of such additives are silicas of very small particle size (pyrogenic silica or silica gel). Other pigment-like additives in this group are aluminium silicates with laminar crystal structures (bentonite, montmorillonite, hectorite) and talcum. The additives are used in both solventborne and water-borne systems. Also employed are organo-modified silicas (e.g. alkyl silanes) and bentonite modified with alcohols or amino alcohols to render them more compatible and more effective in solventborne systems. This group also includes nanoparticles of silica, titanium dioxide, barium sulphate and zinc oxide. Nanoscale pigment-like compounds bestow special properties on coating systems (see Chapter 3.8.3.11 for examples).

2.1.3 Solvents and dispersing agents

The form in which most paint materials are applied is attained with the aid of solvents or dispersing agents. All paints are liquids; the only exceptions are powder coatings. Most paints have very complex compositions and consist mainly of mixtures of colloidal solutions and dispersions.

Solvents

Solvents are low-molecular organic compounds. The molecules of solvent interact physically with the molecules of the polymer resins. This interaction yields an association of solvent molecules on the polymer molecules. Solvents form solvates comparable to the hydrates formed by water and inorganic salts. Solvation opens the coils of polymer molecules, but not completely. The solvate still contains several polymer coils, but it is penetrated and covered by solvent molecules and can move freely among the free solvent molecules. The free molecules and solvated molecules are in equilibrium. This type of solution is said to be colloidal. Such colloidal solutions give the impression of being homogeneous liquids, but physically there are some differences. The average number of molecules forming coiled particles depends on the interaction between the solvent and polymer, the molecular weight and molecular weight distribution of the polymer, the concentration of polymer in the solution and the temperature. The ability of solvents to form associations with polymers (solvating power) varies with the chemical structure of the solvent and the chemical structure of the polymer (solubility). There have been numerous attempts to quantify solvating power and solubility. Different solvent parameters have been defined [20] that take account of the polarity, ability to form hydrogen bonds and van der Waals forces.

Some solvent suppliers offer specific computer programs for calculating the optimum solvent composition for any resin type [21]. But paint formulators tend to ignore the solvent parameters in practice. There are a number of reasons for working without solvent parameters. Besides the parameters of the different solvents, it is necessary to take the corresponding solubility parameters of resins into consideration. These solubility parameters are difficult to define and each resin has its own individual solubility parameter value. Additionally, there is a limited number

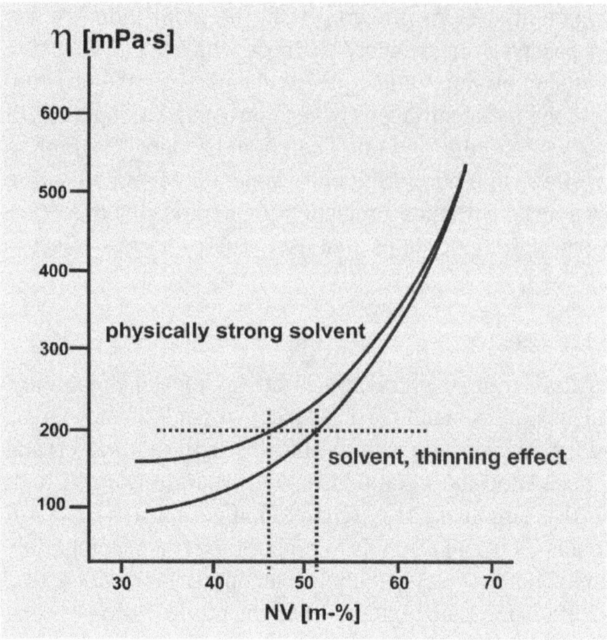

Figure 2.5: Solution viscosity as a function of solids content for a good solvent versus a solvent blend with a thinning effect

of suitable solvents and so paint formulators base their choices on experience and solvent availability. Besides solvating power, other parameters must be considered in relation to solvents. The most important are the application conditions and the behaviour of the paints after application.

The best solvents, in the physical sense, are those which interact most extensively with the polymer molecules of resins. Such solvent molecules penetrate the polymer coils effectively to form relatively large colloidal particles. The resultant solutions are very stable and are more viscous than other solutions (see Chapter 2.2.1 for a description of viscosity). Of course, high stability meets one of the basic requirements on paint properties, but high viscosity is a disadvantage when it comes to applying the paint. The optimum solutions for paints are those which have - for a given application viscosity - the highest solids content (polymer content, nonvolatiles). Low-viscosity solutions with the same solids content are made using less effective solvents in the physical sense. These may have a thinning effect. But the solution must remain stable. Paints therefore very often contain a combination of solvents in which so-called latent solvents serve to create either optimum solution stability and low viscosity or the highest solids content for a specific viscosity. Figure 2.5 shows the solution viscosity as a function of solids content for a good solvent in the physical sense and a solvent blend with a thinning effect.

Of particular importance for the application behaviour of paints is the solvent evaporation rate. For optimum film forming, solvents must evaporate from the film layer uniformly and completely. Thus, the choice of solvent must reflect the application conditions, e.g. application method, time, temperature, air circulation (see Chapter 2.3.1). Although solvent parameters are not widely used for calculations, there are some basic rules for choosing solvents. Polymers (resins) and solvents are classified according to the polarity of their molecules. Polarity is the efficiency with which solvents or polymers can acquire a charge on parts of their molecules. It is mainly exhibited by hetero atoms, e.g. the oxygen atom in hydroxyl, ester, carbonyl, and ether groups. Thus, it is possible to rank solvents by their polarity. The following list of solvent groups is ranked in descending order of polarity: alcohols, ethers, esters, ketones, aromatic hydrocarbons, terpene hydrocarbons, aliphatic hydrocarbons. Within these groups, molecules with long aliphatic chains are less polar than those with short alkyl chains.

Another important aspect is the environmental and physiological behaviour of solvents. Some solvents commonly used in the past have been defined as toxic or harmful to health. Their use has therefore been restricted or prohibited. There are also ecological reasons for limiting the use of certain solvents. For example, halogenated hydrocarbons and aromatic hydrocarbons are photolytic and may lead to smog. Table 2.1 presents a list of solvents which are no longer used for paints, along with the reason.

Table 2.1: Solvents restricted due to their physiological behaviour

Solvents	Reason
methanol	toxic
ethylene glycol methyl- and -ethyl ether	teratogenic
ethylene glycol methyl- and -ethyl ether acetates	teratogenic
dimethyl formamide	toxic
n-methyl pyrrolidone	hazardous
isophorone	hazardous
halogenated hydrocarbons	toxic, photolytic
benzene	toxic, cancerogen
toluene	hazardous
methyl ethyl ketone	unpleasant odour
methyl isobutyl ketone	unpleasant odour

Table 2.2: Suitable solvents for automotive paints

Solvents	Boiling temp. [°C]	Evaporation-rate (ether=1)	Vapour pressure [hPa] (20 °C)	Density (20 °C) [g/cm³]	Flashpoint [°C]
xylene	137–142	17	90	0.874	25
aromatic 100	155–182	43	3	0.877	41
aromatic 150	178–209	120	1	0.889	62
n-butanol	116–118	33	5.6	0.811	34
isobutanol	106–107	25	9.5	0.802	28
ethylhexanol	183–185	600	0.5	0.833	76
methoxy propanol	119–123	25	13.3	0.934	38
butyl glycol	167–173	163	0.9	0.901	67
butyl acetate	123–127	11	12.5	0.880	25
methoxy propyl acetate	143–149	33	3.4	0.965	45
butyl glycol acetate	190–198	250	0.4	0.945	75

Other solvents have to be labelled if their content in paints exceeds a specific level (xylene, turpentine oil, tetraline, and butylglycol). Additionally, there are restrictions and limitations as regards transportation regulations. For the purpose of fire protection, solvents are ranked by flash point. The flash point of a solvent is the temperature at which a mixture of solvent vapour and air is ignited by an approaching naked flame. An important temperature limit is 21 °C [22]. The auto-ignition temperature of a solvent must be borne in mind where solvents are used to prepare solutions at high temperatures (e.g. solutions of alkyd resins and polyesters after their manufacture). Table 2.2 lists the suitable solvents for automotive paints and their physical characteristics (boiling temperature, vapour pressure, evaporation rate, density, and flash point).

For several years, now, requirements have been imposed that are aimed at reducing emissions of volatile organic compounds (VOCs); in the paint industry, it is mainly solvents that are affected. The first response was the development of paints which can be applied at high solids content (high-solid paints). A parallel development came in the form of water-borne paints and solventless paints (e.g. powder coatings, coatings with reactive diluents).

Table 2.3: Physical constants of water in comparison to other solvents

	Water	n-Butanol	Butyl glycol	m-Xylene	Benzine 100/140	Aromatic 150
boiling temperatures (°C, 1013 hPa)	100	116–118	167–173	136–139	110–140	177–206
evaporation rate (diethylether=1)	80	33	160	13.5	7.6	115
evaporation enthalpy (kJ/kg on bp)	2.258	599	368	344	322	305
density (g/ml)	1.00	0.81	0.90	0.86	0.74	0.89
surface tension (mN/m, 20°C)	72.2	25.5	27.8	29.5	25.2	33.7
conductivity (kW/cm)	10	50	100	$>10^6$	$>10^6$	$>10^6$
amount of carbon (mg/g)	0	648	610	906	842	900

Water as solvent or dispersing agent

Water acting as a solvent or dispersing agent perfectly meets the requirements for cutting down on volatile organic compounds in coating systems. There are numerous ways to convert various polymers into aqueous phase with a view to creating the optimum application form (see Chapters 2.2.2 and 2.2.3). But the use of water creates some problems. Table 2.3 [23] shows the physical constants of water in comparison of those of typical organic solvents. Worth noting are the higher evaporation enthalpy, higher density, higher polarity, and higher conductivity of water. Water lends itself to the preparation of colloidal solutions and various polymer dispersions. When it comes to formulating water-borne paints, allowance must be made for its specific properties. The stability of colloidal solutions is often supported by adding organic solvents, also known as co-solvents. These also boost uniform evaporation of water during the film forming process. Special wetting agents are added to water-borne systems to ensure optimum wetting of various surfaces and pigments. Since water can exhibit anomalous viscosity behaviour and has a relatively low evaporation rate, rheological additives must be added if sagging and other film defects are to be avoided. The slow evaporation is also the reason why high air circulation rates and higher temperatures are used for the flash-off process.

2.1.4 Additives

Additives in coatings are used under the aspect "a little goes a long way". Additives sometimes constitute less than 1 % of the paint formulation. Some facilitate paint manufacture; others enhance the application properties of the liquid paint. Finally, there are additives which unfold their effects in the finished coating films. Additives are ranked mostly by their effects and then by their chemical composition. A widely known raw materials list for coating systems [24] lists 34 groups of additives. The following list here contains those additive groups which are important for automotive paints:

- wetting agents for substrates and pigments
- rheology agents
- catalysts for crosslinking reactions and initiators
- levelling agents
- light stabilisers

The composition and properties of the various additives are described in the chapters on the various paint systems.

2.2 Application form

Paints are made in different physical forms for application. These are

- organic solutions
- aqueous solutions
- aqueous dispersions
- non-aqueous dispersions (NADs)
- 100 % systems
- powder coatings

2.2.1 Organic solutions

The most important characteristic of organic solutions is their viscosity [25]. Viscosity (η) is the specific coefficient of shearing stress and shear gradient in liquid phases ($\eta = \tau/D$). Shearing stress has the same units as pressure; the vector is oriented not vertically towards the surface of the liquid, but horizontally ($\tau = F / A$). The shear gradient is the differential of the velocity and the thickness of the liquid ($D = d[v]/d[l]$) (see Figure 2.6). The unit of viscosity is Pa s (1 Pa s = N s/m^2).

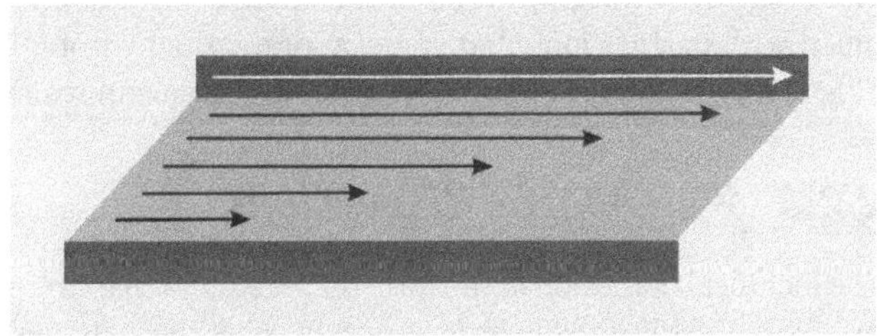

Figure 2.6: Shear gradient of a solution as a measure of viscosity

The viscosity of a colloidal solution varies with the type and viscosity of the solvent, the number of particles (concentration), the particle size, the interaction of polymer and solvent, and the temperature. Solutions of low molecular products have viscosities which are independent of shear gradients (viscosity behaviour is said to be Newtonian if it follows Newton's law of viscosity). On account of the specific particle interactions of polymers, the viscosities of most colloidal solutions vary with the shear gradient. With a few exceptions, the viscosity decreases with increase in shear rate. This behaviour is known as structural viscosity (pseudoplasticity) and is due to the stepwise decrease in particle interaction with increase in shear rate over time. If particle interaction of particles is very extensive, a minimum shear rate is often needed before a shear gradient can be created. The minimum shear rate is also called the yield point. Solutions with a viscosity below the yield point are gels. The viscosity of a colloidal solution may also

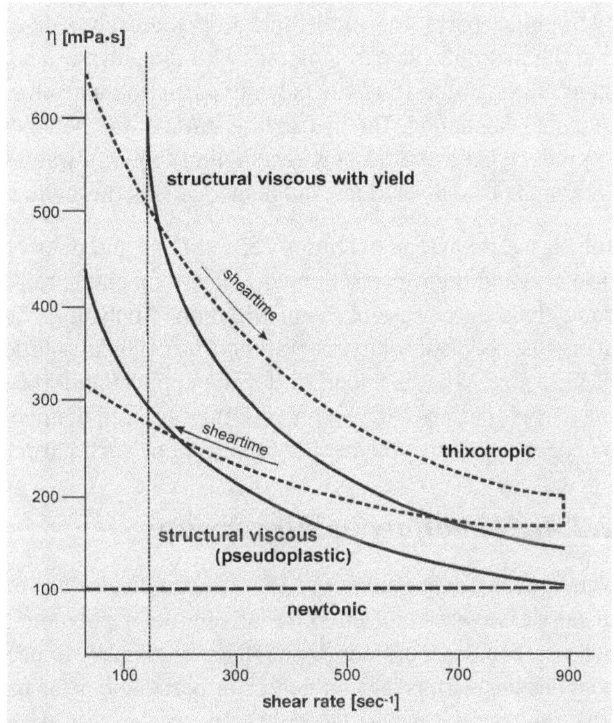

Figure 2.7: Viscosity of different types of solutions as a function of shear rate and time

be influenced by the shear time. The reason is that some time is needed for particle interactions to diminish as well as for them to increase again. Such viscosity behaviour is called thixotropy. Figure 2.7 shows various viscosity curves as a function of shear rate and time.

2.2.2 Water-borne solutions

Only a few polymers (binders) are water-soluble as such. However, it is possible to transfer resins bearing hydrophobic building blocks into the aqueous phase through modification with hydrophilic groups. There are two ways to do this. The first is to modify resins with ionic groups (anions or cations) which are hydrophilic and act as carrier groups in water-borne systems. The second is to modify resins with non-ionic but hydrophilic groups (e.g. with polyether side chains) to generate water tolerance. In both cases, the colloidal state is more pronounced than in the case of true water-borne solutions. The particles are often larger than in true colloidal solutions and virtually form dispersion particles. But the key difference is that such particles do not have well defined particle interfaces. Unlike dispersions, the particles can readily form homogeneous films. The first way of forming water-borne solutions is preferred. If the ions for modification are combined with partner ions that can evaporate, the hydrophilic character is lost after film formation. This is an advantage, of course, where films have to resist water and humidity.

2.2.3 Water-borne dispersions

The technical definition of the term dispersion is an extremely fine distribution of a polymer (resin) in a liquid medium which does not dissolve the polymer. The most important liquid medium is water. The distribution of the fine polymer particles must be stabilised. This is most commonly achieved with emulsifiers (surfactants), which consist of molecules that have hydrophilic and hydrophobic parts. The emulsifiers align themselves on the surface of the polymer particles such that the hydrophobic parts interact with the polymer and the hydrophilic part interacts with the outer phase, water. Thus, the polymer particles themselves can be prevented from interacting and causing coagulation. The hydrophilic parts of the polymer molecules may also act as a stabiliser to produce behaviour closely resembling that of a water-borne colloidal solution. The transition between colloidal solutions and dispersions is therefore a fluid one.

The classical physical definitions of solutions and dispersions characterise them in terms of particle size and their interaction with electromagnetic radiation (light). These classical definitions do not hold in the case of the distribution of polymers, however. There is evidence that particles of organic colloidal solutions, water-borne colloidal solutions, and dispersions can have the same particle size. What is important for the application form and film formation is the differences in the particle surfaces. In dispersions, they are well defined (sharp) and, for the purpose of creating homogeneous films, the barrier presented by such particle surfaces has to be overcome.

2.2.4 Non-aqueous dispersions

Non-aqueous dispersions (NADs) are stabilised distributions of polymer particles in organic dispersing agents (solvents). Relatively polar polymers and non-polar solvents (e.g. aliphatic hydrocarbons) are preferred. The dispersions have to be stabilised by surfactants. These consist of molecules with polar und non-polar parts. The polar part of the surfactant molecules interacts with the polymer molecules and the non-polar part, with the dispersing agent or non-polar solvent. There are different types of surfactants; polymers, too, are used to stabilise non-aqueous dispersions.

The advantages of non-aqueous dispersions over organic solutions include the fact that non-polar solvents (aliphatic hydrocarbons) are the most acceptable solvents from the aspect of environmental compatibility. They are not susceptible to photolysis through absorption of UV light. The viscosity curves of non-aqueous dispersions - like those of aqueous dispersions - are steeper than the viscosity curves of organic solutions. This opens up the possibility of achieving application viscosities at much higher solids content. The gain in solids is illustrated in Figure 2.8.

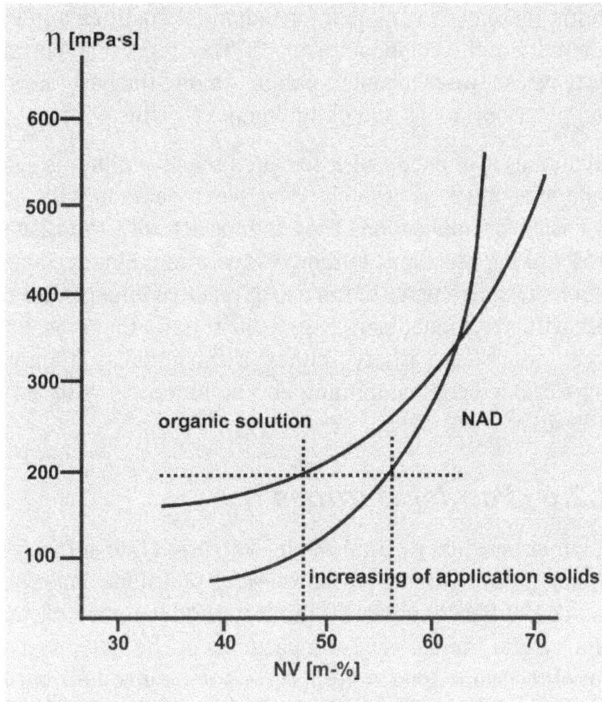

Figure 2.8: Viscosity curves of non-aqueous dispersions compared with organic solutions of polymers

Film forming by non-aqueous dispersions may resemble that of aqueous dispersions; the particles have to penetrate one another after the dispersing agent has evaporated. To optimise this film forming process, use is made of **organosols**. In organosols, the dispersing agent is mostly a non-polar solvent that has a specific amount of a polar solvent which can dissolve the polymer and which has a higher boiling point or lower evaporation rate than the non-polar solvent. After most of the non-polar solvent has evaporated, the polymer particles will dissolve in the remaining more polar solvent composition. The resulting solution forms films that feature optimum flow and levelling, as well as being glossy and dense. The forming of a solution during the second phase of film forming may be supported by elevated temperatures.

A special form such dispersions are **plastisols**. These consist of polymer particles dispersed in a plasticiser. The plasticiser does not act as a solvent for the polymer at ambient temperatures and so forms stable dispersions. However, it can dissolve the polymer at elevated temperatures. If the solution of the polymer in the plasticiser is cooled to ambient temperatures, a homogeneous film is obtained. Of course, the selected polymers have to be relatively hard in order that they may develop optimum film properties together with the quantity of plasticiser. The most important plastisols consist of polyvinylchloride mixed polymers (PVC-MP) in combination with phthalic ester plasticisers. The advantages of plastisols are: virtually no evaporation of solvents (zero VOC values). They can be applied in very thick coating layers without giving rise to film defects. The most important use of plastisols in automotive coatings is that of undersealing.

2.2.5 100 % systems

There are systems which have such low viscosities that they can be applied without any solvents - these are called 100 % systems. The components of 100 % systems are relatively low molecular compounds, but they have to be able to crosslink to large molecular networks very effectively. Only a few products can do that. In the past, drying oils were used. These are liquid, can be applied as such and crosslink with atmospheric oxygen. The crosslinking reaction is accelerated by metal salts (driers, siccatives). However, the film forming process takes a relatively long time and is therefore unsuitable for industrial coatings applications. For industrial application, high reactivity of the components of 100 % systems is important in addition to low viscosity. The best systems are two-component paints. Among the best known are combinations of polyether polyols and polyisocyanates and of liquid epoxy resins with polyamines as crosslinker.

Combination of resins with so-called reactive diluents also satisfies the goal of lowering the VOC content as much as possible. Reactive diluents initially act as solvents to confer an adequate low viscosity for application. They then react with the resins during film forming by entering into crosslinking reactions, without evaporation. Finally, they become an integral part of the coating film matrix. Examples of the use of reactive diluents are the combinations of unsaturated polyesters with vinyl monomers (e.g. with styrene) that crosslink under the influence of peroxides and accelerators (e.g. tertiary amines or cobalt salts) at ambient temperatures. Others are combinations of oligomers containing acrylic functions with acrylic monomers that are crosslinked by UV radiation [26].

2.2.6 Powder coatings

Polymer powders (resins) can be converted into aerosols with air. Such aerosols have properties which are comparable with properties of liquids. For example, they are as mobile as liquids and can coat different objects. Film forming by aerosols is a melting process. First, hot objects are dipped into the aerosol. The particles of the polymer powder impinging on the surface of the hot object melt and finally form a homogeneous film (fluid bed application). The second way to apply an aerosol is to charge the powder particles in a high voltage field. The particles then follow the electrical field lines and are applied to an object which is earthed. The object with the adhesive particles is transported into a stoving oven where the powder melts and forms a homogeneous film.

2.3 Film forming

2.3.1 Physical drying

After the polymer solution or dispersion has been applied, the solvent or dispersing agent must be removed so that a solid and resistant coating film may form. Removal of the solvent and dispersing agent occurs by evaporation. The evaporation process is called physical drying.

The reason that solvents or dispersing agents evaporate below their boiling temperatures is that they have a measurable vapour pressure at lower temperatures. The vapour pressure of solvents and dispersing agents results from the fact that the different molecules of liquids have different kinetic energy. The temperature of a liquid represents the average kinetic energy of all its molecules. But some of the molecules are moving so quickly that they can pass from the liquid to the vapour phase above, to become a part of this phase with a specific pressure. Since the fast ("hot") molecules leave the liquid, the liquid becomes cooler. If the space above the liquid contains a certain amount of vapour molecules, the molecules may re-enter the liquid. A dynamic equilibrium is then established in which the same numbers of molecules leave the liquid and re-enter it. The partial pressure in this equilibrium state is the so-called saturation vapour pressure. The saturation vapour pressure varies with the type of liquid and the temperature. Of course, the equilibrium state depends on the space above the liquid. If the liquid is allowed to take up heat energy from the surroundings and the space above the liquid is very large in relation to the mass of the liquid, equilibrium will never be achieved. It is then possible to evaporate the liquid completely in practice. That is the background to the physical drying process. For the efficiency of physical drying, it is important to replenish the evaporation energy (to maintain the given temperature) and to make sure that the volume of the vapour phase is large enough, for example by providing adequate air circulation. Such conditions are essential in all industrial application processes (e.g. automotive OEM paint application).

The quality of coating films also depends on uniform solvent evaporation; i.e. the films formed must be homogeneous, smooth, and glossy. Regular, uniform evaporation is achieved by blending solvents of different evaporation rates in one paint formulation. It must be taken into consideration that the vapour pressure and the evaporation rate of solvents are influenced by their interaction with the polymers. If polymer molecules form associations with effective solvents (solvates), the vapour pressure of such solvents decreases. However, the vapour pressure of less efficient solvents (thinners) will not decrease.

Water serving as solvent or dispersing agent will also evaporate as described above. But it should be remembered that water has very high evaporation energy (see Chapter 2.1.3). Thus, the drying conditions, and mainly temperature, have to be adapted accordingly when water is the solvent.

For optimum physical drying of dispersions, the polymer particles must fuse together minimally on the surface shell. Polymer particles are only able to do that if the ambient temperature is higher than the so-called minimum film forming temperature (MFT). The minimum film forming temperature of a polymer is slightly higher than the glass transition temperature (T_G). Basically, the very small particles of dispersions can fuse together easily if they have very large surfaces and if capillary forces are acting additionally to support the fusion process. It is also necessary to compensate for the barrier effect of the emulsifiers on the surface of the particles. All these conditions lead to problems if the dispersions have to form films at ambient temperatures (e.g. for house and wall paints). The resistance of polymer films above their minimum film forming temperature is not very good. A number of measures are available for ensuring that dispersions form films that are homogeneous and resistant:

- use of coalescing agents (solvents as additives to support film forming)
- use of co-resins in solution
- use of core-shell dispersions (comprising a hard core and soft shell)

2.3.2 Chemical film forming

The use of resins that form films by physical drying as a means of achieving optimum film properties (e.g. resistance to mechanical impact and chemicals) requires that the polymers chosen must have high molar masses and - if possible - high glass transition temperatures. But there are some disadvantages to using such polymers. In organic solutions, such polymers need a high amount of solvents to form solutions of sufficiently low viscosity for application. Since a major objective of the coating industry has been to reduce the emission of VOCs mainly during application, the use of solutions with low solids content is restricted. There is an additional objective to reduce the loss of materials with a view to saving on raw material costs. Finally, all polymers which form films by physical drying only have in adequate resistance to solvents and chemicals, as good solvents for the polymers are still able to re-dissolve the polymers.

If the application form of a coating material consists of a dispersion that forms films only by evaporation of the dispersing agent, the viscosity is virtually unaffected by the molar mass. But polymers in dispersions must have lower glass transition temperatures; otherwise film forming will not be optimum. Resistance to chemicals, humidity and solvents will therefore not be optimal either.

To overcome these disadvantages, most coating materials for industrial application contain resins that form very large molecules only after application. The molecules of such resins must therefore lend themselves to molecular extension. The molar masses of the resins in paints are relatively low. A better definition of such molecules is that they are not polymers, but oligomers. Such oligomers require a much lower quantity of solvent to achieve the viscosity for the application form. The process by which very large molecules are formed after application under defined conditions - mainly in a relatively short time - is that of crosslinking. Crosslinking is the reaction of different functional groups of the resins to form high molecular three-dimensional networks. The various functional groups which react with each other may be localised in one type of resin molecule. This type of reaction is called self-crosslinking. But in most cases the different functional groups belong to different resins and the reaction takes place between the molecules of different resins. This reaction is called co-crosslinking. The crosslinking reactions take place under different conditions. If the reactivity is high enough, crosslinking may take place at ambient temperature (room temperature). In that case, the reactive parts of the entire coating formulation have to be handled separately. In most cases, the coating material is supplied as two components (mainly a base component and the so-called hardener - the crosslinker which is normally the smaller amount of the two). Such paint systems are called two-component paints. Both components have to be properly mixed just before application. After mixing, the paint must be applied within a certain period of time if the coating is to form the optimum film. This time is called the pot-life. Most users of two-component paints ask for pot-lives of up to six hours.

In other cases, the crosslinking reaction takes place at ambient temperatures but the reaction has to be started by the addition of catalysts or initiators. Without those additions, the coating materials are stable in storage. Storage stability is defined as the period of time during which coating systems can be applied without undergoing a change in application form, mainly without changing the viscosity too much. And, of course, the resultant film properties must be the same as those of fresh materials. Users of various coating materials require a minimum storage time that has to be guaranteed by the supplier. The most common requirement is six months. Such storage stabilities are achieved by resin combinations which are less reactive at ambient temperature. After application, the resin combinations react at higher temperatures to form crosslinked film networks. These paint systems are called stoving lacquers (also stoving enamels or baking enamels). Although most of these paint systems consist of two or more different reactive resins, they are defined as one-component paints if the entire mixture can be supplied in one container.

Crosslinking efficiency

The functionality of resin molecules and the number of reactive functional groups affords the theoretical possibility that the crosslinked molecules are infinitely large. That would mean that the resin of a coating film on a substrate would consist of just one molecule. But, of course, there are several reasons why that is impossible. First, the reactivity of functional groups decreases rapidly with decrease in the number of functional groups. Furthermore, all chemical reactions need minimum impact energy between functional groups to form chemical linkages. The mobility of molecules in growing networks is controlled by diffusion processes. But that mobility will decrease as the molecular network extends and with increase in the crosslinking density. Here it is assumed the crosslinking reaction will ultimately be stopped due to thermodynamic reasons. The consequence of these conditions is that the coating film consists of very large, crosslinked molecules of different sizes and limited extent. But in comparison to the starting molecules of resins (oligomers), they strive to be infinite. Most of the film molecules are likely to be above 10^6 g/mol in size. But nearly all coating films also consist of some molecules which are not crosslinked because they can become dissolved by elution processes.

The efficiency of crosslinking significantly influences the film properties. Excellent resistance to solvents and chemicals is achieved by high crosslinking density. Until now in the literature, high crosslinking density stood for hardness and brittleness, but lower flexibility. However, there are examples of films with high crosslinking density that are also highly flexible: crosslinked polyesterimides for electrical insulation coatings (wire enamels) have a high crosslinking density but are exceptionally flexible (the films remain closed and show no cracks after elongation and bending over small diameters). UV coatings, too, have high a crosslinking density but may be flexible as well. The reason here is that molecules with high numbers of functional groups form narrow crosslinked molecules during chemical film forming, but due to the above mentioned thermodynamic reasons, the reaction stops after the formation of molecules which have no large extensions. They have a high crosslinking density, but cover only small areas of film layers. The films are hard and resistant to solvents but brittle and less flexible. Selecting resin molecules which contain fewer numbers of functional groups leads after crosslinking to networks with larger mesh links. For thermodynamic reasons, these networks may have larger extensions than the networks of high crosslinking density. Therefore, these network molecules cover larger parts of the film layers, and that results in greater flexibility and better adhesion properties. But molecular networks with larger mesh links are less resistant to solvents and chemicals, since, due to the possibility of diffusion of different agents, they may swell or allow the penetration of those agents. Of course, optimum coating films are expected to have both high flexibility and excellent resistance to solvents and chemicals. Besides crosslinking density and extension of network molecules, flexibility and adhesion properties are influenced by the mobility of the molecular chains of the building blocks of resins.

On account of the conditions described, in most cases it is necessary to arrive at a compromise on application properties, mainly through experiments. But, as mentioned above, there are coating systems which offer both outstanding flexibility and excellent resistance. The goal should therefore be to compensate for the paradigm of high crosslinking density combined with high flexibility. The reasons for high crosslinking density combined with high flexibility for wire enamels are the application conditions (extremely high crosslinking temperatures, low reactivity of transesterification process, and application of several thin layers). Crosslinked UV coatings have a high crosslinking density if the UV light is very reactive and crosslinking takes place even though the molecular mobility diminishes during the process. The resulting networks have relatively large molecular extensions and may be flexible as well. In the future, therefore, UV coatings will cover much more application fields than currently if it proves possible to overcome the existing application restrictions (see Chapter 3.8.8). The formation of interpenetrating networks (IPNs), too, is an excellent way to optimise film properties [27]. Interpenetrating networks are formed by a minimum of two different crosslinking reactions, and the idea is that the entire network consists of different types of linkages. But simply by mixing two different crosslinking methods (hybrid crosslinking), where in most cases the different reactions also have different reaction rates, leads to optimum film properties that are better than would be expected from a compromise [28] (see Chapter 3.8.3.5). Additionally, combinations of two crosslinking reactions are chosen if the application process offers restrictions on one of the methods. Such methods are called "dual-cure" processes [29]. For example, if it is difficult to crosslink complex three-dimensional objects and shadow zones with UV light, the reaction is combined with isocyanate crosslinking.

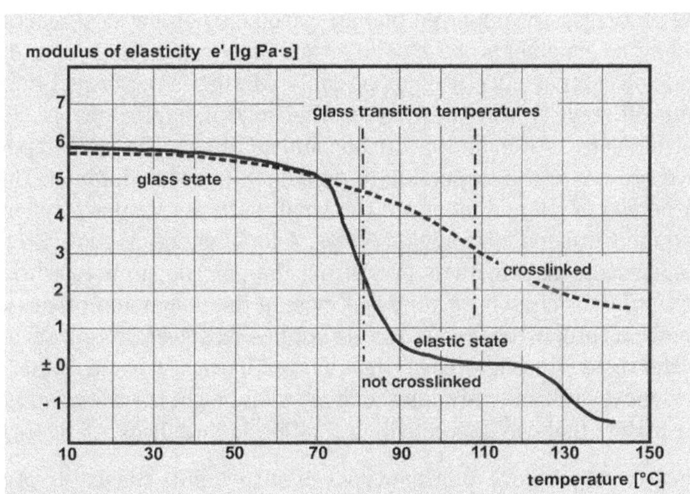

Figure 2.9: Elastic modulus of polymers as a function of temperature

The efficiency of crosslinking for use in coatings is determined by testing the film properties. An analysis of the elastic modulus (energy storage modulus) as a function of temperature [30] is considered to return a particularly good physical description. The curve of the modulus is an optimum way of describing the state of crosslinking. Figure 2.9 (page 35) shows the elastic modulus against temperature of a non-crosslinked polymer (thermoplastic) in comparison to a crosslinked polymer.

The unit of elastic modulus is the same as for viscosity. The values define the mobility of molecules or the resistance to deformation. Normally the resistance is relatively high at low temperatures. That is called the glass state, since polymers are comparable to glass. At a specific polymer temperature, the ability of polymer molecules to resist deformation decreases. The value may drop over several orders of magnitude. It is believed that the molecules of the polymer can uncoil themselves at this temperature. Over a certain temperature range, the polymer molecules can become uncoiled but still have the ability to return to the coiled state. That temperature phase is called the elastic state. The point of inflection on the modulus curve between the glass state and the flexible state is the glass transition temperature (T_G). Uncrosslinked polymers are changed thereafter by increasing temperature into a melt state that exhibits the typical viscosity behaviour of all liquids. Crosslinked polymers are unable to become a melt. The curve of elastic modulus against temperature is flatter than that for thermoplastic polymers and the glass transition temperature is higher. The values in the elastic state are much higher; no melting occurs, at best only decomposition of polymer. The glass transition temperature and the modulus values in the elastic state are measures of the efficiency of crosslinking - the higher they are, the more the polymer is crosslinked or the greater is the crosslinking density. The elastic modulus as a function of temperature is determined by making free films of polymers and subjecting them to dynamic forces. The modulus is interpreted from how the film strip responds to the elongation forces.

2.4 Production process

Resins (polymers) are made from their building blocks by chemical reactions. The reactions take place at elevated temperatures in reactors which are equipped specifically for the different reactions. On account of the number of different resin types, most resins are prepared in batch processes. But trials are also underway to introduce continuous production processes. It is standard practice to convert the resins into solutions or dispersions which are suitable for the production of paints.

Generally, the production of paints [31] does not involve chemical reactions. Instead, different mixing processes are employed that are based only on physical reactions. The goal is to disperse the different components of the paint formulation (resin solutions or dispersions, pigments, additives and solvents) as effectively as possible. Mixing is performed in containers which are equipped with different types of stirrers or other mixing aggregates. As already mentioned, some of the solvents are used to dissolve resins and additives. The most important and also the most expensive process is the dispersing of pigments in resin solutions. The pigment powders have to be dispersed in a liquid medium. The small particles have a very large surface area and adsorb air and moisture and form agglomerates. A lot of energy is required to negate the adsorption and to disperse the agglomerates by wetting the particle surfaces with liquid resin materials. Special equipment is chosen for this, and most of the dispersion processes contain two steps. The first, known as pre-dispersion, is carried out in a dissolver. Dissolvers are containers armed with large stirrer disks which may have edges shaped like a saw blade. For larger containers, the stirrer disks also move vertically through the dispersion batch. Dissolvers are equipped with high-power stirrer motors that can generate high rotation in liquids of relatively high viscosity.

The second employs different types of stirrer mills consisting of a horizontal or vertical milling container, with a stirrer and several stirrer disks as well as fixed disks for feeding the dispersing material intensively through the mill container. The mill contains grinding media that are special

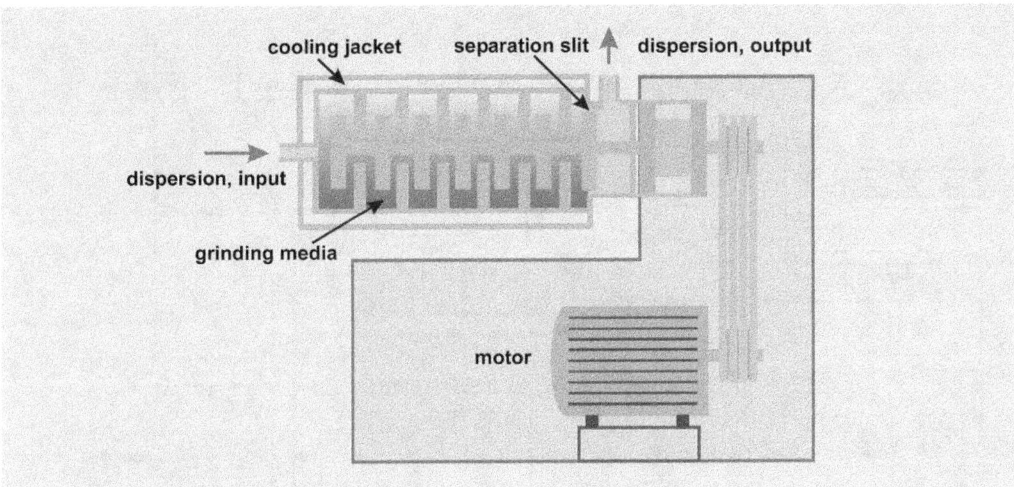

Figure 2.10: Principle of a horizontal stirrer mill

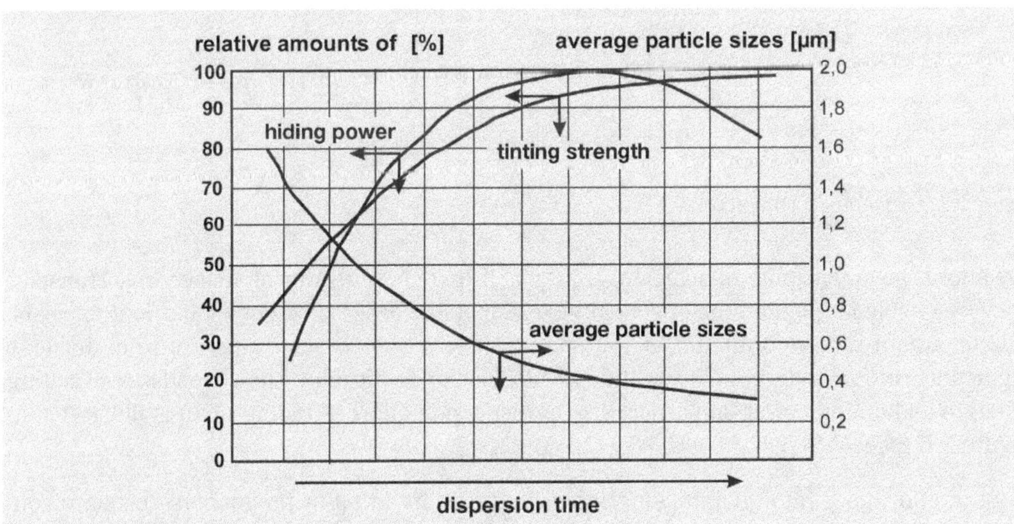

Figure 2.11: Dependence of tinting strength and hiding power on particle size or dispersion time

sand particles or beads that offer optimum resistance to abrasion (for example, they may consist of zirconium oxide). The dispersion batch is fed through the mill container continuously. The efficiency is controlled by stirrer speed and feeding velocity. After passing through the mill, the dispersion batch and the grinding media are separated by screens or by a separation slit. Figure 2.10 shows the principle of a horizontal stirrer mill.

The dispersion effect is created by the application of friction and shearing stress to the batch of pigment dispersion by the disks, blades and grinding media. Although the equipment is called a mill, there is no milling involved. Instead, the pigment agglomerates are dispersed. Most of the energy required goes to wetting the particle surfaces.

The efficiency of dispersing is controlled by measuring the tinting strength. The tinting strength increases continuously during dispersing, since it depends on the amount of particles. The hiding power of the pigment dispersion will pass through a maximum, since the optimum hiding power depends on the particle sizes which must be close to half the wavelength of visible light. Pigment dispersions containing small particles become transparent. Organic pigments prima-

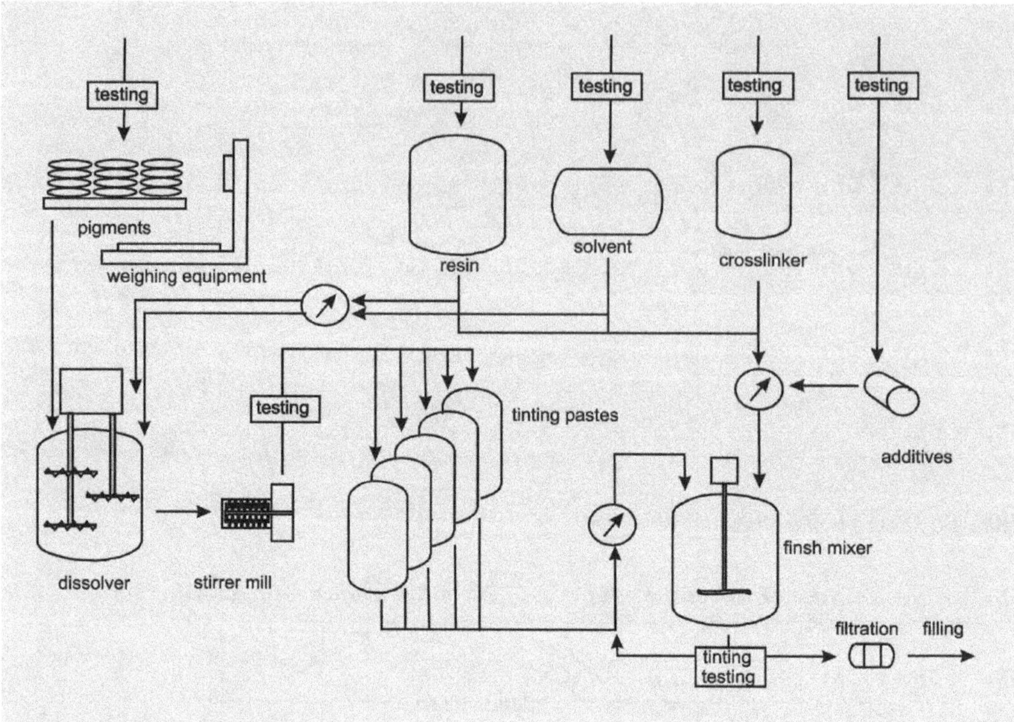

Figure 2.12: Paint production process

rily afford the opportunity of achieving very small particle sizes during dispersing. Transparency is suitable for pigmenting effect paints (e.g. metallic basecoats). For solid colour topcoats, it is important to have optimum hiding power. Thus, it sometimes makes sense to disperse pigments to different degrees to suit the intended use or application. The dependence of tinting strength and hiding power on particle size or dispersion effort (grinding time) is illustrated in Figure 2.11 (page 37).

A lot of effort goes into guaranteeing the reproducibility of paint production. Tests are conducted at several production stages. They start off with the raw materials, which are analysed for compliance with the specification given by the supplier. The storage stability of the raw materials is also checked. The various batches of resin solutions or dispersions must meet specific values that are checked before those intermediates are used. Most effort is expended on the reproducibility of colours. The pigment dispersions are defined and standardised by different measures. It must be possible to always mix pigment dispersions in the same way to prepare the same colour. Once the paint is made, the colour is tested intensively. It is adjusted by means of standardised pigment dispersions (pigment pastes) for tinting, which are controlled by computer programs. Often, it is also necessary to run an application test on each batch of paint. The final test concerns the cleanness of the paint, since even the tiniest contamination can seriously impair the paint finish. All the various tests and analyses take a lot of time, with occupation of mixing equipment being the most time-consuming. There are different ways to rationalise production and tests. For example, it is possible to use mobile mixing containers that are connected to the stirring unit only when a raw material or intermediate (tinting paste) is added. During the test period, the mobile container is temporarily stored in the background and the mixing equipment can be used for other production batches. Figure 2.12 is a schematic illustration of the full paint production process.

2.5 Application methods

As already mentioned, the viscosities of different paints are adjusted to suit the different application methods. The simplest application methods are manual **brushing**, and manual **roller coating**, which are suitable for house-paints, wall-paints and also for some corrosion protection paints. Fillers (putties) are also applied manually with a **filling knife**. These are direct application methods.

Indirect methods of application are *dipping process, flow coating, curtain coating,* and various *roller applications.*

In ***dipping,*** the objects are dipped into the coating material, which must adhere to the surface of the objects and form a film when the object is removed from the dipping bath. Originally, solventborne dip coatings were used that caused large emissions of volatile organic compounds. Solventborne dip coatings have now been widely replaced by water-borne dip coatings. A special dipping process is **electrodeposition coating**. This process uses water-thinnable resins that are stabilised by ions. The resin moves in an electrical field to the object which is connected to an electrode with a charge opposite to that of the resin. The resin carries the other ingredients of the paint, which are deposited on the object. This is followed by discharging, coagulation (additionally involving diffusion processes) and finally stoving to form a coating film. As deposition is independent of the wetting behaviour, this opens up the possibility of forming optimum films; the film thickness can be controlled very well. Another special dipping process is **fluid bed coating** In this case, a powder coating is transferred to an aerosol in a fluid bed (an open container with a membrane, through which compressed air is blown). Hot objects are dipped into the aerosol and powder particles melt on the surface of the objects to form coating films.

Wires for electrical equipment are also coated in a special dipping process. **Wire coating** entails feeding wires several times through a small dipping bath at relatively high speed. The layer thickness is controlled by removing excess material by nozzles or felt strips. The wires pass through a stoving oven with very high air temperatures (up to 400 °C).

Flow coating is the pouring of paint over objects and collecting the excess in a pan for removal. Coating here is also a wetting process.

Curtain coating is used for flat objects transported on a conveyor belt. The paint drops from a storage tank through a dosing slit in the form of a curtain onto the surface of objects moving below, and coating layers are formed.

In mechanical **roller application**, the paint is transferred from a storage container to a system of rollers and from there onto flat substrates in contact with the rollers. The substrates are metal panels and foils, but also plastic film. The metal panels are used for making containers (cans; **can-coating**). Continuous metal panels (from coils) are coated by **coil-coating**. These painted panels are used for architectural building elements and other articles.

The application of **printing inks** is a also special kind of roller coating process. Various **spray applications** are also indirect coating processes. If the coating materials are fed from a container under pressure and atomized in spray nozzles, the process is called **airless spraying**. If spray nozzles are used in which the paint material is transported by a supplementary jet of air, the method is called **pneumatic spray application**. It is possible to charge the spray paint particles with a high voltage field. The particles are transported to the earthed object, where they are deposited. This method is called **electrostatic spray application**. The goal of the process is to optimise the transfer efficiency of paint to objects (see Chapter 3.3). Powder particles, too, can be charged by a high voltage field in special spray guns, carried and fed by the field to earthed objects, where the particles adhere due to their charge before being melted in a stoving oven to form homogeneous films. That process is called **electrostatic powder application**.

The type of objects which have to be coated and the coating conditions influence the choice of application process. If numerous objects have to be coated serially, the application method must allow a high degree of automation, e.g. serial coating of cars (automotive OEM = original equipment manufacturing). The first layer of the automotive coating system (primer) is applied by electrodeposition. All other layers of the serial automotive coating process are applied by spray application methods. The preferred method is electrostatic spray application. Paints for plastic parts are also applied by spray coating. For automotive repair coating, pneumatic spray application is preferred as it offers the best flexibility as regards influencing the spray results (colour, effects, levelling, gloss) via the application parameters (amount of paint and air, and spray time; see Chapter 4).

2.6 Coating systems

The application and film forming conditions greatly influence the composition of paints. It therefore makes sense to divide the various coating systems into the respective application fields.

Decorative paints are paint systems used by craftsmen or DIY enthusiasts in houses (wall paints) or on parts of homes and buildings (e.g. window coatings).

Wood coatings are used mainly for furniture.

Corrosion protection coatings are all those paint systems which used to protect metal objects against the influence of environmental agents that may be corrosive. A great many objects require these coatings, ranging from a small metal grid to the metal superstructure of a large suspension bridge. Most **marine paints** are classified as corrosion protection paints.

Coatings for general industry cover all paint systems which are applied by industrial application methods, mostly by serial automotive processes. This group includes coatings for domestic appliances (refrigerators, washing machines, driers, mixers etc.), metal furniture, and plastic parts.

Special industrial coatings are those used for cans and other metal containers (**can coating**) or for continuous metal coils (**coil coating**) intended mainly for construction elements.

There are also **coatings for toys** and **leather articles.**

Automotive coatings are essentially industrial coatings. But there are so many different systems that paint systems for cars and other vehicles are arranged in a separate group. First, there are paint systems for serial production of new cars (**automotive OEM coatings**).

In the meanwhile, more and more car parts are being made from plastic. While there are numerous parts for car interiors, there are parts for car bodies (mirror housings, bumpers, headlamp reflectors, boot covers, and others). Most of these parts need special **plastic coatings**.

On account of the different application conditions, **automotive repair coatings** require types of paint that differ from coatings for new car production. Some of those paints are also used for inline repair coating, which can be necessary if the serial coating process has some flaws. But most repair coatings are used to remedy defects in coating layers arising from the use of a car.

Large vehicles, such as trucks, tractors and other agricultural implements, buses, train compartments and public transport vehicles use specific coating systems whose compositions often resemble those of repair coatings.

3 Automotive OEM coatings

3.1 History of automotive coatings

The first automobiles built at the **end of the 19th century** still looked like light horse-drawn carriages. The first motorcars from Benz and Daimler featured of a steel pipe construction which was coated with a black paint [32]. The paint material contained oxidative drying oils as binder (e.g. linseed oil). To improve the drying behaviour, the oils were combined or cooked with natural hard resins (e.g. copal). Nevertheless, the entire drying process often took several days if several paint layers were applied. Soon after that (around **1900**), the car bodies were made of wood panels and then finally steel panels. These were also coated with oil-based paints. Even at that time, car producers went to a lot of effort to make their cars appeal to potential customers. The coating process therefore became very important. It was very difficult to compensate for the surface structure of wood and hammered steel panels with a coating process. First, a primer had to be applied that took several days to dry. This was followed by up to three layers of filler (applied by brushing). All the layers had to dry for several days before they could be sanded. Sanding had to be very done carefully if optimum smooth surfaces were to be obtained. The real coating process started after that. Two layers of pigmented pre-coat were applied, then two layers of transparent topcoat, a clear flatting varnish, and finally the clear, glossy overcoat. Each of these coats took approximately two days to dry before the next one could be applied. In between, minor repairs and additional sanding would have been performed. While the transparent topcoats were drying that lasted up to eight days, the inside of the car was decorated with a wallcovering. The entire coating process required approximately 27 process steps and lasted from 200 to 300 hours. Although some white models were available (e.g. a white Mercedes car from Daimler in 1904), most of the cars were black. The colour was produced using tar and carbon black pigments (ivory black). It is believed that this colour was chosen to keep down the cost of the cars and because the weathering resistance was significantly better than that of coloured cars. The first cars were hand-made. Nevertheless, the goal soon became to reduce the very long application times for painting the car bodies. Meanwhile, car production companies were being founded in the USA, e.g. such as the later Ford Motor Company in Detroit in 1903. In **1908**, it started producing a less expensive two-seater cabriolet that became a very famous best-seller: the T-Model Ford. To boost productivity, Henry Ford introduced the first conveyor belt production line in **1913** [33]. As the long drying time of the paints was the most critical problem, drying ovens were installed to speed it up. These ovens were called light tunnels as they were fitted with carbon arc lamps. The freshly painted and pre-dried car bodies were stored in long rows of tents so that they could dry out. A significant improvement came with the development of the first synthetic resins (phenol resins) [34] for blending with the drying oils. Such combinations formed the basis for topcoats with bright colours, for example the famous, grass-green Opel *Laubfrosch* ("tree frog"). But the biggest step toward reducing the drying time was the introduction of paints based on cellulose nitrate in the **mid of 1920th**. These were not applied by brush, but rather by spray [35]. The primer and filler still consisted of a combination of phenol resins with drying oils, but the topcoat was based on a cellulose nitrate binder, along with plasticiser, and a quantity of hard resins. At the time, it was claimed that this slashed the time needed for the whole application process from 336 hours to 15 hours. But coatings based on cellulose nitrate also have disadvantages. First they have to be polished to produce an appealing surface effect. Second, they are not sufficiently weatherable. Even when paints based on cellulose nitrate were pigmented with carbon black, they still had to be polished from time to time. It was at this stage that Henry Ford made his famous comment: "Customers can have any colour they want as long as it's black" [36]. With the advent of alkyd resins [37], which were used instead of plasticiser for combination with cellulose nitrate (yielding a "combination enamel"), weathering resistance gradually improved. The first stoving enamels based on alkyd resins in combination with urea resins were introduced into automotive paint application in the **1930s** [38]. At that time, the coating systems for motorcar bodies (rolled steel) consisted of a primer, a primer surfacer, a pre-coat and a topcoat, which were all applied by spraying. The paints were dried in the oven for 30 to 60 minutes at 120 to 130 °C. The entire painting process took just 4 to 5 hours. Subsequent steps to improve the weathering resistance of automotive coatings included the replacement of urea resins by melamine resins and the introduction of alkyds modified with saturated fatty acids that no longer caused yellowing of topcoat films. In the **mid-1950s**, new organic pigments were launched onto the market which offered much greater lightfastness and resistance to chemicals and solvents. Such paint systems are still being used on mass-produced car bodies.

In the USA, the physically drying cellulose nitrate topcoats were partly replaced by topcoats consisting of physically drying acrylic resins which offer better weathering resistance. This led to the development of topcoats containing acrylic resins that could be crosslinked by melamine resins. These competed against the alkyd resin topcoats used in Europe. In the **early 1960s**, topcoats based on acrylic resins were also pigmented with aluminium pigments to yield metallic effects (one-coat metallics). These were followed in the late 1960s by so-called two-layer metallic topcoats, which offered optimised effects and resistance [39]. These systems consisted of a metallic basecoat containing the effect pigments and a clearcoat providing the gloss, and mechanical and chemical resistance. Finally, various suppliers developed solvent-borne metallic basecoats that consisted of cellulose acetobutyrate, polyester, and amino resins. Special-effect basecoat-clearcoat systems gained a high market share that is still increasing. Further progress in coating systems is influenced by demands for optimised technical properties and for new colours and effects. However, a major factor in such progress is the avoidance of both VOC emissions and the use of harmful or toxic ingredients. This explains the evolution of high-solids, water-borne paints, also called 100% systems, and powder coatings. On account of the different technical requirements of the various geographical regions and the different legislation in the various countries, there are still a lot of different paint systems in the market. All these paint systems are described in the following chapters (Chapters 3.4 to 3.8).

3.2 Automotive coatings as a multilayer system

As mentioned in the last section, automotive coatings have always consisted of several different layers. Despite all the trials aimed at streamlining the coating process, including reducing the number of paint layers, cars are still being coated with multilayer systems. The main reason is that, with a smaller number of coating layers, it is not possible for automotive paint systems to meet the imposed requirements, which are now much stricter than in the past. It is an ongoing requirement that the properties must be enhanced. The various layers have to perform different functions. In accordance with the aforementioned historical development, automotive coating systems now consist of three layers, namely the primer, the primer surfacer or filler, and the topcoat.

Primers, together with pre-treatment methods, have to generate adhesion on the substrate (car body) and to guarantee corrosion resistance. Fillers have to cover over any structures on the substrate and yield smooth surfaces. The second most important demand on filler layers is to absorb mechanical impact on the coating. Finally, the topcoat has to provide the colour, gloss and levelling, and resistance to weathering, solvents, and chemicals. This three-layer system is still in use today for solid colour coating systems (or straight shades, which are coatings without special effects). Figure 3.2.1 shows the structure of such an automotive three-layer system.

The late 1960s saw the introduction of effect coatings, which initially consisted of aluminium pigments. The effect stems from changes in brightness caused by variations in the amount of light reflected from the surface as the viewing angle changes. Since effect pigments consist of rather

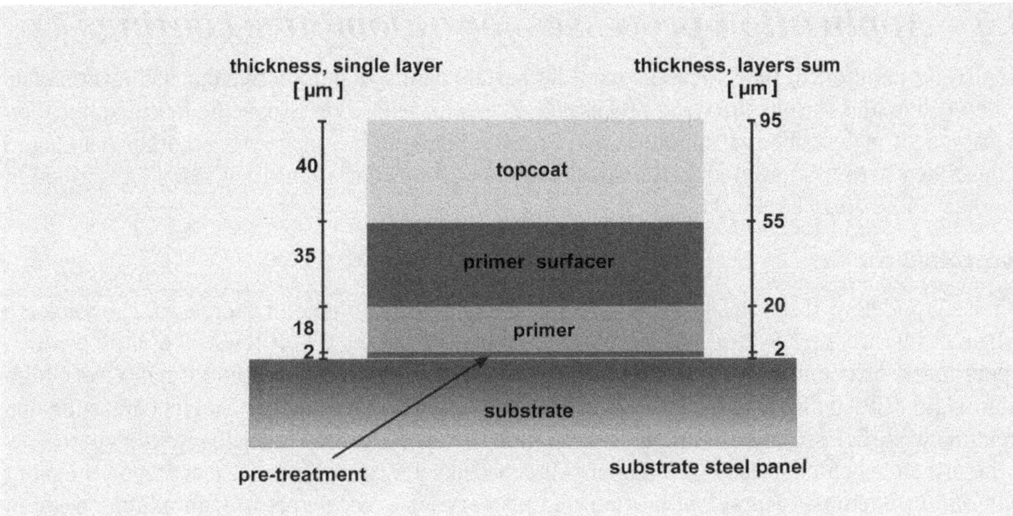

Figure 3.2.1: Structure of three-layer automotive coating

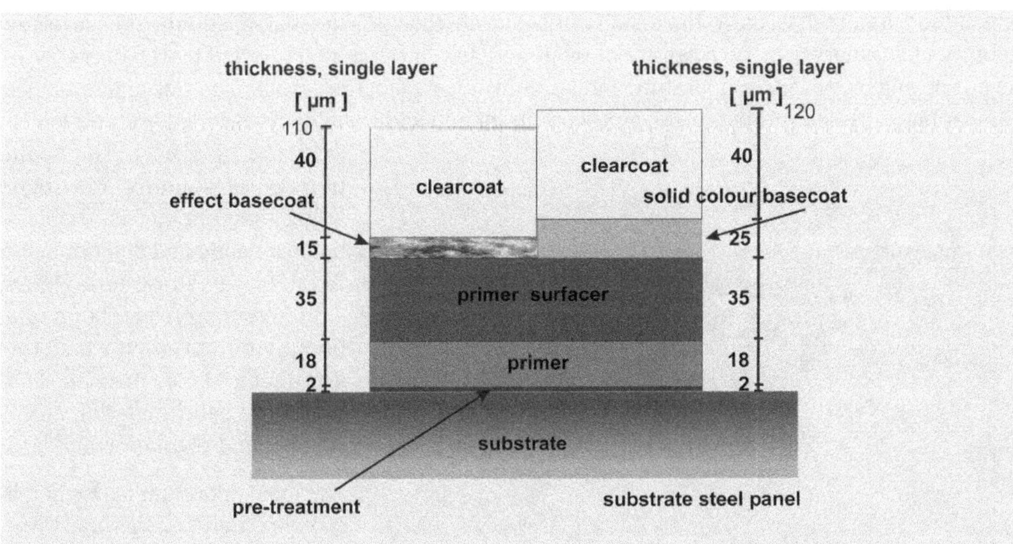

Figure 3.2.2: Four-layer coating with effect basecoat and solid colour basecoat

large particles which are sensitive to environmental influences, the coating layer containing those pigments is covered and protected by a clearcoat. In other words, the function of the topcoats is split across several layers. The effect basecoat provides the colour, effect, and attractive appearance. The clearcoat provides gloss, hardness, and resistance to weathering, solvents, and chemicals. On account of the good experience gained with such four-layer coating systems, not only effect paints, but also solid colour systems are applied as four-layer coatings. Since clearcoats offer the ability to protect even "normal" pigments against light and weathering, the three-layer coatings for solid colour systems will be replaced by their four-layer counterparts, despite the extra effort involved in the application process. One advantage of this is that all colours (solid and effect) can be applied in the same application line by the same method. Figure 3.2.2 shows a four-layer coating system consisting first of a effect basecoat and second of a solid colour basecoat.

Although four layers are applied that perform different functions, it must be remembered that the entire coating layer is just 90 to 120 µm thick, i.e. only slightly thicker than a human hair.

3.3 Application processes for automotive coatings

As already mentioned, the processes used for serial coating of automotive car bodies are: electrodeposition and various spraying processes. As electrodeposition is specific to the application of primers, it is described in Chapter 3.4.6. Spray application is suitable for all other paints and is described here in more detail. The special process of powder application is also described in Chapter 3.5.7.3 and Chapter 3.8.7.

Spray Application

When a liquid flows through a pipe, its velocity increases as the diameter decreases. The converse is true of the pressure - it decreases as the diameter increases [40]. At the end of a pipe which tapers into a nozzle, the velocity of the liquid is at a maximum and the liquid has a very high internal pressure on leaving the pipe. This causes the liquid to form an aerosol jet consisting of a great many particles. The built-up pressure behind the jet causes the particles to split apart. This is the principle behind airless spraying of paints based solely on the pressure acting on the paint material. In pneumatic spray application, the primary jet is accelerated by an additional jet of compressed air, which also generates more particles. This method allows optimum variation and dosing of the spray process. The results of pneumatic spray processes are governed by the pressure of the paint, the viscosity, the density or cohesion, the surface tension, and the pressure and quantity of auxiliary air. The quantity of paint and auxiliary air is controlled by the diameter of the nozzle and by valves. Temperature and humidity also affect the results. The size and structure of the spray jet is determined by the auxiliary air jet and additionally by so-called horn air which regulates the flow. The main purpose of shaping and regulating the spray jet is to obtain homogeneous and smooth paint layers on the substrates. Figure 3.3.1 shows a cross-section of a pneumatic spray gun and the arrangement of nozzles and valves for paint material, auxiliary air, and regulation air [41].

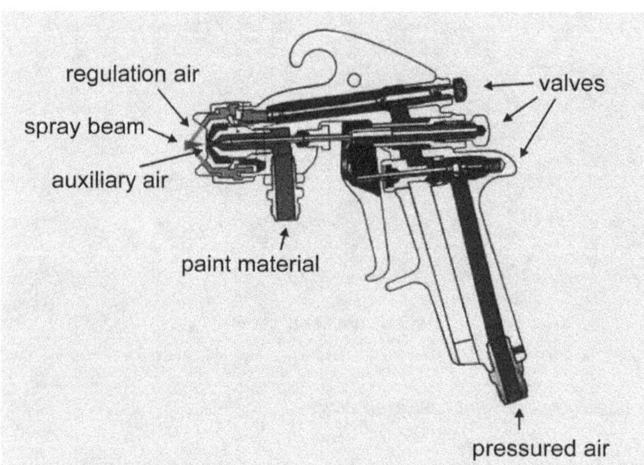

Figure 3.3.1: Cross-section of pneumatic spray gun

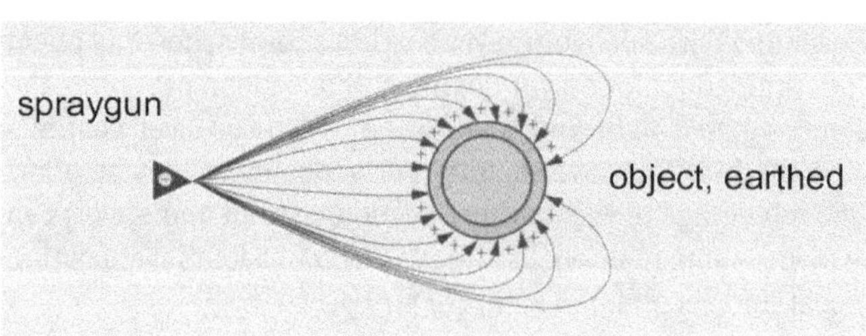

Figure 3.3.2: Principle behind electrostatic spray application

The most important task when it comes to optimising spray processes is to increase the transfer efficiency. This is the proportion of paint that forms a layer on the substrate, relative to the total amount sprayed. The residue is called over-spray. Theoretically, airless spraying has a better transfer efficiency than pneumatic spraying. However, the airless process does not allow regulation of the spray result, e.g. for obtaining special effects. Furthermore, reducing the air pressure increases the transfer efficiency of pneumatic application. However, that also reduces

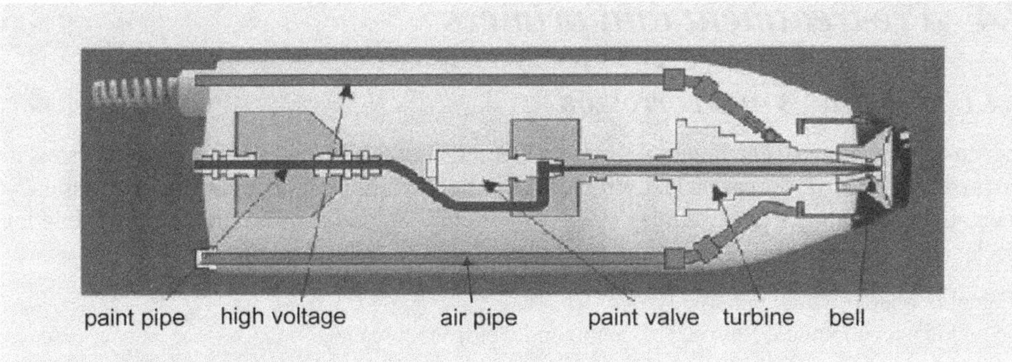

Figure 3.3.3: Cross-section of an electrostatic spray gun

material transfer per unit of time (lower layer thickness for the same amount of time) and limits productivity. This limitation has been overcome with optimised spray guns. Nowadays, automotive paint is applied by means of high-efficiency spray guns [42], which mainly use higher amounts of auxiliary air than the common pneumatic spray guns.

A significant improvement in transfer efficiency came with the introduction of electrostatic spray guns. In these spray guns, the paint is charged by a high-voltage electric field. The charged material is atomised on rotating disks or bells. The paint particles formed are transported along the electric field to the earthed object. The particles adhere on the surface of the object, where they form homogeneous coating films. The electric field reduces the level of overspray and increase transfer efficiency. The other advantage is that particles also follow the electric field to parts of the object facing away from the front of the object. This effect is called throwing power. However, it is not possible to coat the surface of hollow spaces if the electric field does not penetrate inside such objects (Faraday's cages). Figure 3.3.2 shows the principle behind electrostatic spray application with spray jets following the electric field lines.

Electrostatic spray guns contain a valve for the paint, a high-voltage electrode, an air feed, and a rotation bell driven by a motor. The rotation speed is very high in order that the finest particles possible may be produced. Figure 3.3.3 shows a cross-section of an electrostatic spray gun [43].

The transfer efficiencies of various spray application methods, independent of the shape of the substrates, are listed in Table 3.3.1 [44].

Optimisation of transfer efficiency is not the only criterion for resorting to spray application. For example, electrostatic spraying does not produce the optimum effects in the case of basecoats with very light and bright colours. These still have to be applied by normal pneumatic spray methods.

Table 3.3.1: Transfer efficiency of different spray application methods

Spray equipment	Transfer efficiency
pneumatic high pressure spray gun	50 %
pneumatic low pressure spray gun	60 %
airless spray gun	70 %
pneumatic gun with electrostatic support	80 %
electrostatic bell application	90 %
Reine ESTA-Applikation (pure electrostatic application (disc))	95 %

3.4 Pre-treatment and primers

3.4.1 Substrates and corrosion

The most important task of automotive primers is protecting against corrosion. Even today, steel is the preferred material for car bodies, although the quantity of plastic parts in cars is continually rising. Other metals are used besides steel, with aluminium and magnesium being suitable for special automotive parts. Positive properties of steel are [45]:

- Steel is relatively inexpensive
- Steel is easy to handle (to roll, cut, bend, and weld)
- Steel has excellent application properties (elasticity, tensile strength, hardness)
- Steel is relatively easy to recycle (to shred, scrap for blast furnaces)

In the past, steel was readily available, but now it is in short supply due to heavy demand from the Far East.

As already described in the chapter on the historical development of automotive coatings, the first cars were built along the lines of horse-drawn carriages. They had a chassis as a frame for the car body. The first car bodies consisted of wood, and then of steel panels. Eventually, the first so-called self-supporting car bodies were developed. The first car body of that type was constructed in 1922 [46]. The first mass-produced car to have a self-supporting steel body appeared in 1935 [47]. Now, nearly all passenger cars are constructed that way. For adequate stability, some parts of the body have folds or ridges and the outer shell is very compact. The various parts are welded together. But there are also alternative joining processes, which are mainly used for new construction materials. The most important goal of car body construction is to optimise structural stability (rigidity, crash safety) while reducing the weight (to save energy and fuel). To this end, special grades of steel have been developed that offer a combination of optimum shaping and optimum mechanical stability (e.g. IF-steel, DB-steel).

The most important disadvantage of using iron and steel is corrosion. Iron corrosion has been known since the Iron Age. There are steel grades that do not corrode, but they cannot be used for car-body steel, as they do not have the shaping properties and the required stability. Consequently, the new steel grades also have to be protected against corrosion.

All metals in contact with air acquire a layer of oxide on the surface after a while. The other components of air (moisture and carbon dioxide) can transform the oxides into hydroxides and carbonates. Some metals form dense, stable oxide layers, whereupon corrosion ceases. Unfortunately, iron and most grades of steel react with oxygen, moisture, and carbon dioxide to form oxide layers, which have totally different crystal structures (crystal lattice) than the metal. Thus, there is no dense oxide layer and so corrosion continues until nearly all the metal is transformed into oxidation products. The chemical process by which iron and steel corrode is clearly described by the electro chemical local element model [48]. The underlying reactions that trigger corrosion are illustrated in the local element model in Figure 3.4.1.

The key actor in the model is water, which supports all electrolytic processes. According to the model, metallic iron is oxidised to the iron(II) cation, which goes into solution. The electrons released by the oxidation reaction can reduce water and oxygen to hydroxyl ions (see Equation 3.4.1).

Equation 3.4.1

$$Fe \rightleftharpoons Fe^{2+} + 2e^-$$

$$H_2O + \tfrac{1}{2} O_2 + 2e^- \longrightarrow 2\,OH^-$$

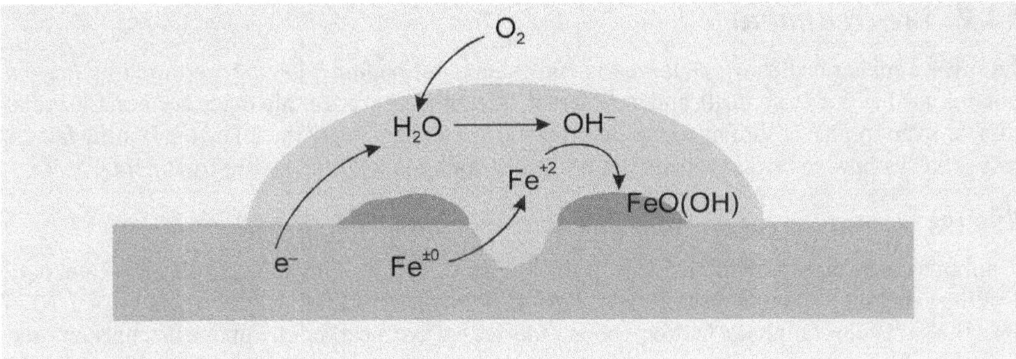

Figure 3.4.1: Local element model for describing the corrosion of iron

The iron(II) ions react with hydroxyl ions and further oxygen to yield a precipitate of iron(III) hydroxide. Iron(III) hydroxide can cleave water to form iron oxide hydrate or water-rich iron oxide and may also react with carbon dioxide to yield carbonates (see Equation 3.4.2). This mixture of the various products is commonly called rust.

Equation 3.4.2

$$Fe^{2+} + 4\,OH^- + \tfrac{1}{2}\,O_2 \longrightarrow 2\,Fe(OH)_3$$

$$2\,Fe(OH)_3 \longrightarrow 2\,FeO(OH) \cdot H_2O \longrightarrow Fe_2O_3 \cdot 2\,H_2O$$

The described process continues until most of the iron metal has been transformed into the reaction products. It is therefore of vital economic importance that corrosion is avoided and that iron and steel plant, equipment, vehicles and other objects are protected against corrosion.

Different measures can be taken to avoid corrosion of iron and steel [49]. One is to cover the iron or steel surface with a metal that oxidises more readily than iron. Such metals are more reactive, as defined in the electro chemical series of elements. In local electro chemical elements, that metal - instead of iron - acts as the anode and is oxidised to cations, which go into aqueous solution. If the metal takes the place of the iron, it is called a sacrificial anode, in which case the iron acts as the cathode. This process is called active cathodic corrosion protection. The most common sacrificial metal covering the iron is zinc. Zinc has the advantage that the oxide layers of zinc form a homogeneous crystal lattice with those of the metal, and also adhere very strongly. The corrosion process is stopped or significantly reduced if a thin layer of oxide is formed. Most car bodies nowadays are made from steel with a zinc coating. The zinc layer is applied by a thermal process (hot dipping) or, more commonly, by electro chemical reaction (galvanized steel) [50].

Another method is to cover the iron with a metal that is less reactive in the electro chemical series of elements. The most important metal used in the manufacture of objects for industrial use is tin. Metal sheeting coated with tin is called tinplate. It is used mainly for cans and containers, but also for some other objects. Many other iron or steel objects are covered with chromium or nickel. Tin, chromium and nickel react with atmospheric oxygen to form thin, homogeneous, dense oxide layers that escape further corrosion. Thus, as long as these metal layers are not damaged, the underlying steel is protected against corrosion. Basically, the metal coating on the iron presents a barrier to the corrosive agents. This method is called passive cathodic corrosion protection.

Similar methods are pre-treatment of the iron or steel and coating with a primer, or a combination of both. The outcome is layers which are intended to offer passive corrosion protection and additionally present a barrier to corrosive agents. Iron and steel are not the only surfaces to be protected by the aforementioned methods. Zinc, aluminium, and magnesium are also pre-treated and primed. Since all these metals are used for car bodies, primers need to be developed which offer optimum compatibility (adhesion, protection efficiency) with different metals.

3.4.2 Pre-treatment

The surfaces of car bodies which leave the body shop after bending, joining, and welding may be contaminated with grease, scale and maybe rust. Of course, there is a thin oxide layer on the metal surface. To guarantee optimum corrosion protection and adhesion of the following coating layers, the car bodies have to be cleaned very carefully. Cleaning is followed by pre-treatment [51].

Cleaning

Cleaning focuses on removing residues of the grease used for rolling and bending the steel panels. But scale and loose rust particles also have to be removed. In the past, solvents were used to clean the car bodies (aliphatic hydrocarbons, chlorinated hydrocarbons). But due to emissions and the harmful behaviour of chlorinated compounds, the solvents have been replaced by aqueous surfactant solutions. Preferred surfactants are alkali salts of alkyl or aryl polyether sulphonates. Cleaning takes place at a pH values of 8 to 9 and a temperature of 60 to 80 °C. The ester greases are hydrolysed under these conditions. Those greases which cannot be hydrolysed are removed by adding colloidal emulsifiers or wetting agents. Cleaning is done either by dipping or by spraying. The cleaning solutions for dipping contain 50 g/l surfactant while the spraying solutions contain about 1 g/l. Both these processes have to be followed by rinsing with pure water.

Pre-treatment

Cleaning is followed by the real pretreatment. In the past, the steel panels were treated with chromate solution, sometimes in combination with chromium(III) salts. On account of the toxicity of chromium VI compounds (they are carcinogenic), their use is restricted. The current preferred method for treating car bodies utilises zinc phosphates [52]. Treatment is effected either by dipping or by spraying. In the low-zinc process, the concentration of zinc is 0.7 g/l for spraying and 1.5 g/l for dipping. The aqueous solutions contain an excess of phosphoric acid (6 to 15 g/l, calculated as phosphorus pentoxide). Additionally, the solutions contain oxidation agents, such as nitrites, peroxides, hydroxylamine, and nitroguanidine. Other cations, e.g. nickel, manganese, and calcium, may be added to improve the corrosion resistance. Treatment with zinc phosphate is suitable not only for iron and steel, but also for zinc surfaces and aluminium parts. In the case of aluminium, some fluoride salts must be added, as otherwise the aluminium oxide on the surface of the metal is unable to dissolve in the treatment solution. The pH values of the solutions are 2.0 to 3.6 and the temperatures are 40 to 70 °C. Dipping lasts 3 to 5 minutes. Spraying takes just 1 to 2 minutes.

The first chemical reaction which occurs in pre-treatment is pickling. The metal ions and oxides react with the acidic agents to form a solution. Oxidation agents then transform iron(II) ions into iron(III) ions. Since the pH value is relatively low close to the surface of the metal, sparingly soluble phosphates are formed and precipitated on the surface. The products of the reactions are zinc phosphate ($Zn_3[PO_4]_2 \cdot 4\ H_2O$, known as hopeite) and zinc iron phosphate ($Zn_3Fe[PO_4]_2 \cdot 4\ H_2O$, phosphophyllite). As phosphophyllite forms smaller crystals which are oriented parallel to the metal surface, the process is carried out such that the yield of phosphophyllite is a maximum. The important influences here are the concentration of zinc ions and the temperature. The smooth surface of the pre-treatment layer is of advantage to the following coating layers. The application quantities are between 1.2 and 6.0 g/cm², corresponding to a layer thickness of 0.5 to 2.0 µm. A typical pre-treatment solution for car steel bodies has the following composition: 1.2 g/l Zn^{2+}, 0.1 g/l Ni^{2+}, 15 g/l H_3PO_4 or $H_2PO_4^-$, and 0.1 g/l NO_2^-.

The precipitate of the pre-treatment solutions consists of iron(III) phosphate. For the pre-treatment of aluminium, the precipitate consists of sodium hexafluoroaluminate (Na_3AlF_6). The pre-treatment solution can be regenerated by means of ultra-filtration. The pre-treatment layer thus formed protects the steel surface against corrosion and promotes adhesion of the following coating layer (primer). The corrosion protection effect stems from the fact that the phosphate layer is less reactive (in the sense of the electro chemical series) than the metal. The layer acts as a barrier to corrosive agents (electrolytes, water, products of air pollution). A precondition for optimum adhesion is the formation of homogeneous crystal structures between compatible crystal lattices of phosphates and steel. Some inorganic compounds are surprisingly compatible with organic structures; they include chromates, phosphates, and phosphites, as well as some silanes and siloxanes that have been modified with special organic compounds.

3.4.3 Advent of electro deposition primers

The first car bodies were assembled from separate parts during production. Such parts were easy to paint manually. As the number of cars increased and spray coating was introduced, the process had to be optimised. By the time self-supporting car bodies were being made, spraying was no longer efficient. Apart from the time needed to spray the car, the spray was not reaching hollow spaces, which were therefore at risk of corrosion. This led to the introduction of dip priming for applying solventborne primers [53]. Dipping the entire car body required large dipping tanks and large, storage-stable batches of paint material. A compromise solution that used less primer was "slipper dipping" in which the lower car body (including the doors) was dipped into a shallow tank, with the upper section primed by flow-coating or spraying. But although slipper dipping economised on material, there were additional problems. Coating of interior body parts and in hollow spaces was not optimal since evaporating solvent reduced the thicknesses of the layers there.

The solventborne primers consisted of a combination of alkyd resins modified with unsaturated fatty acids and amino resins (mainly urea resins) for crosslinking at elevated temperatures. The level of corrosion protection improved when these resins were replaced by a combination of epoxy esters and phenol resins. The coloured pigments employed in the primers were red iron oxide or titanium dioxide, which were tinted mostly grey by adding carbon black and yellow iron oxide. Extenders were barium sulphate and calcium carbonate. Functional pigments added to boost the corrosion protection were white lead (basic lead carbonate) and zinc chromates. The solvents were mainly high-boiling aromatic hydrocarbons, combined with solvents of greater polarity, such as butanols and glycol ethers. The additives for such primers were wetting agents (for pigments and surfaces) and rheological products to prevent pigment settling and paint sagging during dipping and film forming. Modified bentonites and colloidal silicas proved to be very efficient rheological additives. The thickness of the film layers and the coating's performance were influenced by the application conditions and the primer formulation. Critical formulation parameters were the solids content, the viscosity and the solvent evaporation rate. The most important dipping parameters were lifting velocity, air circulation, and air temperature. These primers were mostly stoved at relatively high temperatures (up to 180 °C).

The biggest problem of the dipping tanks described was the extent of solvent evaporation. Besides pollution of the environment, there was the risk of explosion, which meant that fire insurance costs were high. Research then got under way to find alternatives to solventborne systems for automotive dip priming - this was one of the first industrial coating processes to introduce water-dilutable paint systems. Water-dilutable stoving enamels were developed in the early 1950s [54]. They were based on polyester, alkyd, and phenol resins doped with quantities of free carboxyl groups. The carboxyl groups were neutralised with amines to yield anions that could carry the polymer molecules in aqueous phase. Neutralisation and addition of water yielded colloidal solutions. The colloidal solutions may also have contained some solvents, called co-solvents that stabilised the colloidal solution and helped optimise film formation. The carboxylic groups were incorporated into resin molecules in various ways: addition of anhydrides to hydroxyl groups, reaction with hydroxy carboxylic acids, e.g. etherification with the methylol groups of phenol resins, and diene-addition of maleic anhydride across conjugated double-bond systems. The addition of maleic anhydride had the advantage that the carboxyl groups were bonded via carbon-carbon linkages, which were much more stable to hydrolysis than ester linkages. Polyolefin emulsions also served as alternative resin systems [55].

Water-borne dipping primers had the advantage of reducing emissions of organic solvents and eliminating the risk of explosion and thus led to savings on the costs of fire insurance. But they also had some disadvantages. On account of the specific physical properties of water (high surface tension, high polarity, and high evaporation enthalpy), it was very difficult to make homogeneous coating films without defects, such as sagging and craters. Water-borne primers enjoyed only a short reign in the coating industry - from 1960 to just 1965. Further intensive research to find better solutions led to the advent of **electro deposition**. This is based on electrophoresis, which is the transport of electrically charged particles in a medium to an oppositely charged electrode in an electric field. Relatively large particles can be transported, and these may also have nonionic moieties, e.g. pigments, which are wetted and covered by resin material. At the electrode, the film is formed by discharge and deposition. Discharge at the anode yields acids and is an oxidation reaction. Discharge at the cathode yields bases and is a reduction process. The discharged resin particles coagulate to form homogeneous films of high dielectric constant. These films permit migration and osmotic processes to occur, however. Water and soluble moieties in the film are driven out by electroosmosis, for example. Film properties are determined by a number of different influences: the resin type and quantity of ionic groups, the degree of neutralisation (charge density), the concentration in the tank (non-volatiles), temperature, current density, coating time, and also the circulation system, which is important for the stability of the primer. Careful control over all the parameters yields highly uniform, reproducible coating layers. The films are homogeneous, smooth and cover all parts of the object to virtually the same thickness, even at the edges and to a large extent in hollow spaces, too. Developed in the mid-1960s, electro deposition has gained acceptance throughout the automotive coating industry. It can be extensively automated.

Anodic electro deposition

The first resins to be used in electro deposition primers contained carboxylic groups. These are anionically stabilised. The object (car body) therefore acts as the anode.

Composition of anodic electro deposition primers

The preferred resin system for anodic electro deposition primers comprises adducts of maleic anhydride to unsaturated triglycerides (natural oils), more particularly linseed oil, isomerised linseed oil, and linseed stand oils (maleinised oils). The anhydride adducts are made to react with monoalcohols or polyols to form half-esters and free carboxylic groups. These carboxylic groups are at least partially neutralised with amines. The anions thus formed act as the carrier groups for the colloidally dissolved resin particles in aqueous phase. Figure 3.4.2 shows the chemical reactions that occur during the manufacture of such resins.

In the second development phase of anodic electro deposition primers, the linseed oil used for forming maleinised oils was replaced by butadiene oils. Unlike linseed oil, these were totally stable to hydrolysis (saponification-resistant). Other resins used for anodic electro deposition primers were epoxy esters with maleinised fatty acids, which offered better adhesion and superior corrosion protection.

Neutralisation agents for anodic electro deposition primer resins are amines, e.g. triethylamine, N,N-dimethylethanolamine, diisopropanolamine, 2-amino-2-methyl-1-propanol (AMP). Since maleinised oils contain double bonds, they can undergo self-crosslinking, mainly at elevated temperatures in the presence of catalysts (siccatives). Anionically stabilised resins containing free hydroxyl groups - these include maleinised oils and epoxy esters where the anhydride group is opened by adding polyols - can be co-crosslinked. The **crosslinkers** of choice are phenol resins (resols), melamine resins and benzoguanamine resins.

Anodic electro deposition primers contain **coloured pigments** (titanium dioxide, mostly tinted with small amounts of carbon black and yellow iron oxide to form grey shades, but red iron oxides are also employed), **extenders** (mainly barium sulphates) and **corrosion protection pigments** (lead or zinc chromates, lead silicates). Maleinised oils have excellent wetting properties that are conducive to the manufacture of pigment dispersions. The pigment volume concentration for electro deposition primers, calculated on solids contents, is a relatively low 5 to 10 % by volume. **Additives** for anodic electro deposition primers are levelling agents and anti-settling additives. The primers contained up to 10 % **co-solvents**, with glycol ethers preferred (e.g. monobutyl ether of ethylene glycol = butyl cellosolve, or the monobutyl ether of diethylene glycol = butyl carbitol). The solids content of the anodic tank was about 10 to 15 % by weight.

Application process

Anodic electro deposition tanks [56, 57] are operated at a temperature of 25 to 30 °C, a pH value of 7.5 to 8.5 and a specific conductivity of 5,000 mS/cm. The car body is totally immersed for 2 to 3 minutes; the deposition current potential is 150 to 250 Volts. The deposition equivalent is 50 to

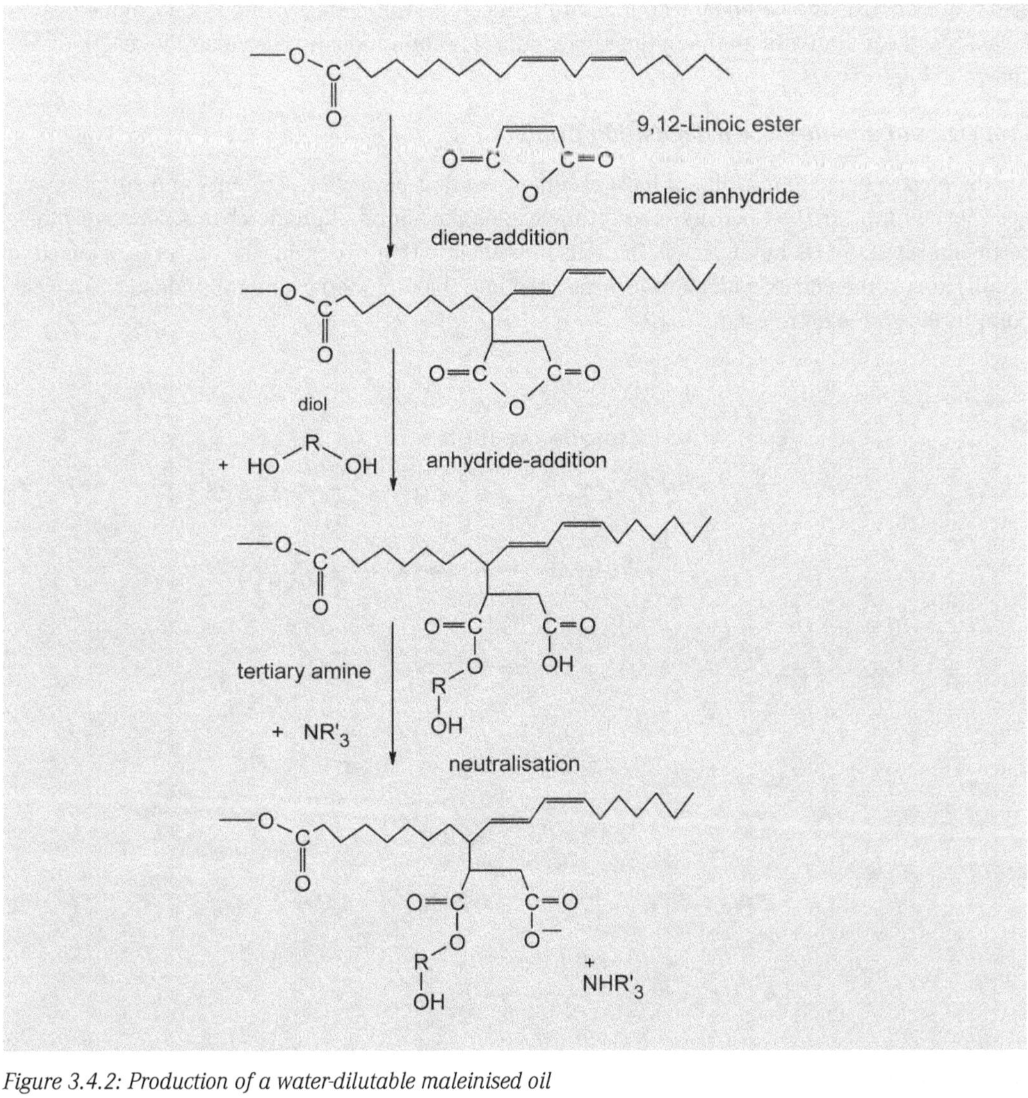

Figure 3.4.2: Production of a water-dilutable maleinised oil

70 mg/C and the resultant films are 25 to 35 µm thick. After removal from the tank, the coated car body is carefully rinsed to remove all soluble materials from the film. The primer is then stoved for 20 to 30 minutes at approx. 170 °C. The quantity of coating material removed from the tank during deposition is replenished with concentrated coating material.

Electrode reactions

As already mentioned, the charged particles move in the direct current towards the anode. At the anode, electrolysis splits the water into protons and into molecular oxygen, which escapes from the tank. The protons discharge the carboxylate anions from the neutralised resin into carboxyl groups. As the carboxyl groups are not polar enough to serve as carrier in aqueous phase, the resin coagulates and is deposited on the surface of the object to form a film. There is the additional possibility that iron from the object will be oxidised to iron(II) ions, which may react with the ions of resins to form insoluble iron salts which become incorporated into the film matrix.

At the cathode, electrolysis yields molecular hydrogen and hydroxyl ions by reduction of water. The hydroxyl ions form partner ions for the ammonium ions of the neutralisation agent that also move towards the cathode. Iron ions are reduced to iron metal. Without any compensation measures, the quantity of amine in the tank would rise continuously. The electrode reactions are shown in Figure 3.4.3.

Properties of anodic electro deposition primers

Anodic electro deposition primers have excellent wetting power for pigments and surfaces and good levelling properties because the oil molecules have long aliphatic chains. However, their corrosion resistance is not optimal. The reason is mainly that iron from the object is oxidised to iron(II) ions at the anode, which are incorporated into the film matrix and thus constitute a weak point in the corrosion resistance.

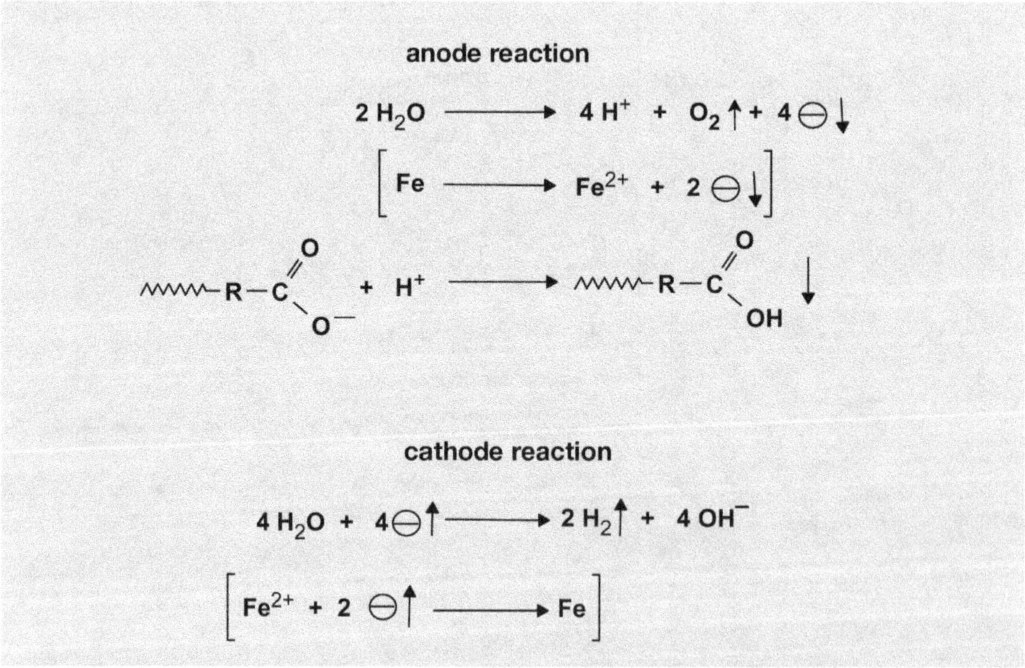

Figure 3.4.3: Electrode reactions in anodic electro deposition

Cathodic electro deposition primers

Cathodic electro deposition primers were introduced in the mid-1970s in an attempt to greatly improve the corrosion resistance. These primers offered much better corrosion resistance, improved throwing power, and better crosslinking efficiency. The ingredients of cathodic electro deposition primers were tailored systematically to the application requirements and conditions (see Chapter 3.4.5) [58]:

- Aromatic epoxy resins, the main component, offered optimum corrosion resistance and adhesion
- Plasticising compounds guaranteed optimum levelling and adequate film flexibility
- Addition of amines to epoxy groups solubilise and functionalise the colloidal dispersion
- Evaporable organic acids to neutralise the amines
- Blocked polyisocyanates, whose reaction is catalysed by the basic behaviour of the amines in epoxy adducts, to serve as crosslinkers
- Molecular network containing urethane groups after crosslinking to support primer adhesion

Alternative developments aimed at introducing cationic carrier groups into epoxy resins [59], but they failed to gain acceptance in automotive electro deposition primers, although they enjoyed some use in general industrial coatings. There were also further developments in the aforementioned cathodic electro deposition primers. Hardly any other field of coating chemistry has witnessed so many patent applications as that of electro deposition primers. It is all the more remarkable, then, that current cathodic electro deposition primers are still based on the same chemistry as the first commercial systems dating from the 1970s. The following advances were made:

- optimisation of application behaviour
  - variation of film layer thickness (e.g. thick coating systems)
  - improvement in throughput speed
  - optimisation of throwing power
  - improvement in edge covering
- reduction in emissions of volatile organic compounds (VOCs, co-solvents)
- abandonment of harmful pigments (e.g. lead pigments)

Current developments are focusing on:

- reducing the stoving temperatures (low-stove systems)
- replacing the tin catalysts (by more harmless compounds)
- improving the weatherability (particularly for combinations involving effect topcoats but not primer surfacers)

In the past, changes to the application process sought to improve the properties of the entire car-body coating system. The goals were:

- an increase in the layer thickness to eliminate the primer surfacer
- optimum edge covering
- improved throwing power

The methods used to achieve these goals were the:

- reverse process
- electro deposition of a powder slurry (EPC or electro powder coating)
- use of two dipping steps
- use of nonionic electrophoresis
- studies of autophoresis

Reverse process

The reverse (or inverted) process was developed when electro deposition primers still contained anionically stabilised resin systems. Pre-treatment is mainly followed by spray priming (primer-filler) the exterior to form layers 30 to 50 µm thick. The primer is then stoved at relatively low temperatures, e.g. at 120 °C for 10 to 20 minutes in a step called pre-gelation. This is followed by electro deposition. Since the primer-filler layer on the exterior prevents deposition of the electro deposition primer, it is mostly the interior parts and hollow spaces of the car body which are coated, albeit very effectively. The car body is then stoved in the oven for 20 to 30 minutes at 170 °C. The reverse process was still used for a relatively long time after cathodic electro deposition was introduced [60]. The process itself is followed by conventional topcoat application, but without additional primer surfacer. It offers the advantages of optimised throwing power and savings on paint material. However, the appearance of the complete coating (including topcoats) is inferior to that of the conventional process. Additionally, it is costly to switch the application line over to the reverse process.

Electro powder coating

The coating material in electro powder coating (EPC) is a dispersion of a typical powder comprising epoxy resins, crosslinker and pigments stabilised in an electro deposition primer. The process was developed in the mid-1970s for anionically stabilised resins. In electro powder coating, it is possible to deposit the primer in a layer thickness ranging from 40 to 80 µm (dry film). The deposition conditions are nearly the same as in conventional electro deposition applications: 15 to 60 seconds at a potential of 100 to 200 Volt. To avoid popping, the thicker layers are pre-dried for approx. 10 minutes at 80 to 90 °C and then stoved for 20 minutes at 180 °C. On account of the high electrical resistance of the deposited film, the throwing power is relatively low. The process had been developed to eliminate the need to apply a primer surfacer. However, it suffered from the disadvantage that the surface structure, which was typical of a powder coat, impaired the smoothness of the subsequent topcoat. Further development work to enhance the surface structure led to electro slurry coating (ESC) [62]. The slurry consisted of a dispersion containing the main resin and the crosslinker, as well as additives for boosting flow and levelling.

Electro deposition using two dipping steps

Since edge covering and throwing power of electro deposition primers were initially very poor and since electro powder coating and electro slurry coating failed to improve the throwing power, it was postulated that better results might be achieved by electro depositing two layers. After the first dipping step in a conventional electro deposition tank, the resulting primer layer was pre-gelled by stoving. The car body was then fed into a second dipping tank, where a second layer was applied. Those parts of the object which had been coated with just a thin layer in the first step exhibited a higher specific conductivity than the others. Those parts were preferentially coated in the second dipping step. The car body was then rinsed and both coating layers were stoved under the usual stoving conditions. The composition of the first tank met the demand for optimum corrosion resistance, while the second tank conferred excellent flow and levelling to provide an optimum surface for the following topcoats [63]. In the meantime, conventional electro deposition primers had been optimised to yield better edge covering, throwing power and levelling. It was therefore possible to forgo the two-step process, not least because it was very expensive to run two dipping tanks.

Nonionic electro deposition

Nonionically stabilised dispersions can be deposited at the cathode due to the formation of protons there from water. The resultant acids make the dispersion coagulate. If such systems do not contain volatile acids as neutralisation agents, it is possible to formulate with corrosion protection pigments, which would otherwise be diluted in aqueous phase under acidic conditions, and to use metal pigments, e.g. zinc or aluminium, which are also sensitive to acids. This renders any special pre-treatment unnecessary. Alternatively, such a process can be combined with a conventional electro deposition method [64].

Autophoresis

Colloidal solutions of polymer which develop acidity in aqueous phase (pH values of 2 to 3) can coagulate on the surface of metals. No electric field is needed for this. Coagulation occurs when the acid groups react with the metal atoms to form ions (salts) that are insoluble in water. The applied film layers are immobilized at relatively low stoving temperatures. Although this has not gained acceptance as a way of applying primers to entire car bodies, it is suitable for pre-coating metal parts for industrial uses and car accessories.

Other optimisation measures

Research work was not restricted to developing new or different materials for electro deposition primers - ways of improving the process itself were also explored. Some advantages in the homogeneity of primer layers were gained by the roto dipping process in which the car bodies, mounted transversely on a spit, were continually rotated as they were carried through the various steps of pre-treatment, rinsing, dipping and - most importantly - stoving [65]. It not only improved the homogeneity of primer films, it also boosted the throwing power by effectively covering surfaces in hollow spaces. While used by some car makers, it was very expensive to rotate the car bodies throughout the various steps. Most car makers therefore focused on the products as a way of improving the process.

As the problems persist to this day, though, the idea has arisen that electro deposition primers might be dispensed with totally, the reason being the increasing use of other materials in car body production, e.g. aluminium, magnesium and plastic. Plans and trials are in hand aimed at making entire car bodies from pre-coated panels (coil coating) or exclusively from plastics (see Chapter 7).

3.4.4 Requirements and properties

The requirements imposed on an electro deposition primer relate to the primer material, the application behaviour and the film properties. The primer material must be stable in the dipping tank. There must be no settling, no aging, and no sensitivity to microorganisms. During application, the primer must provide optimum wetting, levelling, edge covering, and throwing power. After film formation, the primer has to offer adhesion, flexibility, resistance to corrosion and other influences. The stability of the dipping tank has been covered under application conditions (see Chapter 3.4.6).

To avoid **settling**, functional pigments are added to the primer to control the rheology. Nevertheless, the tank material still needs to be continually circulated. In the laboratory, settling behaviour is tested by storing the thinned tank material in a glass cylinder under defined conditions (temperature, time) and observing the results. Electro deposition primer materials nowadays contain very little or no solvents, and no hazardous components, such as active corrosion protection pigments: they contain only water as solvent, and relatively inert pigments. The raw materials and tanks are therefore at risk of contamination by microorganisms. The raw materials and products need to be protected. Of course, the additives selected for protection must neither be hazardous nor cause new environmental pollution. The types and quantities of **fungicides** and **bactericides** selected must not adversely affect humans or the environment. Nor must they have a deleterious influence on the application behaviour or primer film properties (levelling, yellowing, corrosion resistance).

Wetting properties

The resins of anionically stabilised electro deposition primers, the maleinised oil and butadiene oils, have excellent wetting properties (for pigments and surfaces). Aromatic epoxy resins have inferior wetting properties due to their polarity and relatively high glass transition temperatures.

The epoxy resins are therefore modified. To wet the pigments, a special binder, called a grinding resin, is prepared. Wetting of surfaces can also be effected with additives. Wetting agents influence the surface tension of the primer material. Yet foaming must be avoided. Most additives of this kind are surfactants with ionic structures. However, they are not effective in electro deposition processes due to the difficult deposition behaviour of soluble ions. The focus of attention therefore has to be on nonionic compounds. Of course, the wetting properties are influenced by the surface itself. The aforementioned pre-treatment materials (see Chapter 3.4.3) have also been optimised as regards wetting behaviour.

Adhesion

Good wetting of the surface is a pre-condition for good adhesion. Modified aromatic epoxy resins afford excellent adhesion if the surface is prepared well. Various explanations are given for this. One is the positive interaction between the aromatic ether structure of the epoxy resins and the metals or metal ions (akin to complex bonding). Aliphatic epoxy resins do not have this property. The amine groups might also support adhesion. In addition, crosslinking of modified aromatic epoxy resins is associated with only a small variation in specific density, compared with other crosslinking reactions. That is why the initial physical molecular contacts between resin and surface remain stable throughout the film forming process, which is important for adhesion. The type of pigments, too, may be beneficial to the adhesion properties. The physical background to adhesion is described in more detail in the chapter on primer surfacers (see Chapter 3.5.2).

Flexibility

Flexibility is closely related to adhesion properties. It is measured by deforming the coated substrate and inspecting the coating. For this, a spherical segment is pressed (mainly from the rear) into a test panel at different velocities. In the Erichsen test [66], the hemispherical indenter is driven relatively slowly into the back of a panel. The limit value for indentation is approx. 10 mm, as that is the value at which most panels are damaged. The quoted value is the depth in mm at which small cracks in the coating layers first become apparent under a microscope.

Other impact tests use test pieces with an attached hemispherical segment on one end and various weights that fall on the panel from different heights (onto the front or the reverse face of the coated panel). The impact resistance is then quoted as the weight of the test piece or the dropping height at which damage to the coating just fails to occur [67].

Levelling

Although the primer is the first of three or four layers, its purpose is to create smooth surfaces and to cover the surface structures of the car body panels and those resulting from pre-treatment. Electro deposition does not lend itself to meeting this goal since current density is higher on edges and structures than on flat surface parts, and the amount of deposition may be higher. The efficiency of levelling also depends significantly on wetting of the surface. During the stoving phase, the dispersion particles of the deposited primer material must fuse together efficiently to form homogeneous, smooth films. The viscosity and rheology of the primer play an important role in that process. Aromatic epoxy resins have relatively high viscosities that decrease as the temperature rises. When the crosslinking reaction starts, the viscosity increases again. Periods of low viscosity can support flow and levelling. On the other hand, the coating layer must not sag on vertical parts or escape from edges. The optimum viscosity for achieving smooth surfaces, no sagging and optimum edge covering depends on a number of influences: the molar mass and the type and extent of modification of the aromatic epoxy resin (see Chapter 3.4.5), the type and quantity of blocked polyisocyanate crosslinker, the rheological effects of the type and quantity of pigments, and the effects of catalysts and additives. Additives for improving levelling work by floating on the surface of the coating layer during film forming. Such additives are suitable for primers only if they do not interfere with wetting or adhesion of subsequent film layers. The smoothness of film layers is determined by so-called pertometer measurements (mechanical or optical methods). The measurement is mainly performed on the complete coating, not just on the primer alone.

Layer thickness

The layer thickness is influenced by the electrical resistance of the layer, which increases with the layer's thickness and reduces the deposition yield. The specific electrical resistance of a primer material depends on the value of the glass transition temperature, the content of polar molecular groups, and the quantity of solvent (if used). Naturally, the physical application parameters of the deposition process influence the electrical resistance, the current density, the current yield, and the temperature during deposition.

Throwing power

Throwing power is the ability of paints applied electrophoretically to be deposited in hollow spaces in adequately thick layers. This is naturally determined by the layer thickness that is generally attainable in the first place. The layer thickness generally depends on the specific electrical film resistance, which - as already mentioned - is influenced by the properties of the primer material (glass transition temperature, amount of polar groups, residues of solvents). The higher the electrical resistance, the lower is the attainable layer thickness, and, accordingly, the lower is the film thickness in hollow spaces (low throwing power). Throwing power is also affected by the conductivity of the material in the dipping tank. This, in turn, is influenced by the solids content, temperature, degree of neutralisation, and the type of ions. But, as mentioned above, electro deposition is generally suitable for coating the hollow spaces on surfaces. That is essential for guaranteeing corrosion resistance over several years. This aspect is growing increasingly important as car bodies become more complex. This not only has been driven by design considerations, but is the result of lighter car bodies, which has necessitated the incorporation of felts and tubes into the design to provide stiffness and stability.

Throwing power is measured with the aid of hollow test pieces. These may be cubes with one open face or else pipes of specific inner diameter. The test pieces are coated under defined conditions. The throwing power is determined by measuring the film thickness as a function of the distance to the opening of the test piece. Throwing power can be optimised under the same application conditions by varying the primer composition or the resin composition.

Edge covering

As the electric field is relatively dense (higher current density) on edges, good deposition and film thickness (immediately after the dipping stage) can be expected. However, if film formation entails fusion of the dispersion particles, which is a kind of melting process at elevated temperatures, there is the possibility that the film material will contract around the edges. It is important to avoid such effects because thin layers on edges or edges without film material are at a major risk of corrosion, which may spread outwards over the object from the edges. Again, edge covering is governed by the glass transition temperature of the resin composition, the viscosity of the film material during film formation, the polarity of the resins, and the velocity of the crosslinking reaction. Stoving conditions therefore also play a role. Edge covering is determined by measuring the breakdown voltage on well-defined test specimens.

Corrosion resistance

Surface wetting and adhesion are important pre-conditions for corrosion resistance. The most important is to avoid subsurface infiltration of moisture. The interphase between metal surface and primer layer, including the thin layer of pre-treatment material, must therefore be as homogeneous as possible. The other requirement is for the film layer to act as a barrier to diffusion processes.

The diffusion density of the film depends on the glass transition temperature, the hydrophobic behaviour, and the crosslinking density. Disadvantageous to diffusion density are low-molecular resin fractions, residues of ionic compounds, residues of solvents and any contamination. Furthermore, it must be ensured that any additives which remain in the film do not adversely affect the corrosion resistance. Finally, pigmentation can also support corrosion resistance. The active pigments interact with the metal surface and the pre-treatment layer to form crystalline associations. However, the amount of active pigments in electro deposition primers is restricted because of their irregular behaviour in a direct current. Optimum pigment packing can also act as a barrier to diffusion. Pigment agglomerates in the film matrix generate additional interfaces that would support diffusion. It is therefore important to check the degree to which the pigments are dispersed during grinding and to avoid re-agglomeration during film forming and crosslinking.

The corrosion resistance of primers is tested on primed test panels that are exposed in cabins to high moisture levels at various temperatures (climate cycling at -20 °C to 60 °C). Other methods are immersion in or spraying with water or salt water. To achieve meaningful results as soon as possible, the primer layers are cut with a knife (in an X-shape) down as far as the metal, and the panel is then exposed to the humidity cabin or salt-spray equipment. The degree of corrosion, as measured from the X outwards, is measured and expressed as the degree of subsurface infiltration in mm for a defined exposure time (240 h, 720 h or more) [68]. The results of the test indicate how well the primer layer and metal surface interact. This test model simulates total damage to the coating, for example, stonechipping that occurs when a car is driven at high speed. However, other influences, such as diffusion, require that the entire coating afford corrosion resistance. The other layers must make a significant contribution to preventing diffusion and mechanical damage to the coating layers. Although such tests are performed, it takes too long to obtain meaningful results by testing the entire coating structure. Such tests would take as long as the guarantee that the manufacturers themselves offer on their cars. For this reason, further development work aimed at continuous improvements is based on the results of more powerful, shorter tests.

Other film properties

Surprisingly, there is some debate about the weatherability of electro deposition primers. Normally, weatherability is the task of topcoats. The reason for the debate is that weathering tests have revealed instances of adhesion loss by primer layers. An analysis of the weathering conditions employed showed that UV light can damage the primer layer if there is no primer surfacer anywhere in the coating structure; some effect basecoats have a relatively low pigment volume concentration (e.g. contain a large aluminium pigment platelets), and a clearcoat is also known which transmits some UV light, even though it contains a UV absorber package (see Chapter 3.4.8). Aromatic epoxy resins, the principal component of all electro deposition primers, are renowned for their sensitivity to UV light. Even low-energy UV light, in combination with moisture, can damage them. To circumvent this problem, it has been proposed that resins with improved UV stability be used. Plans are afoot to replace some of the aromatic epoxy resins by aliphatic or cycloaliphatic epoxy resins [69]. In that event, it must be remembered that aliphatic and cycloaliphatic epoxy resins do not offer the level of corrosion resistance afforded by their aromatic counterparts.

Another possibility would be to add UV absorber to all the layers in the coating structure, including the primer. But experience has shown that it is best to spread the various functions imposed by the different requirements across the layers of the entire structure. In other words, the structure should consist of a primer surfacer with optimum pigmentation, which can absorb the last vestiges of UV light that have managed to pass through the layers of topcoat. At first sight, it would seem logical to eliminate the layer of primer surfacer as that would save on material and application costs. However, it makes no sense to omit one layer in order to save on costs, and then to have to go to a great deal of effort to rectify the problems resulting from that very saving (see Chapter 3.5.1).

3.4.5 *Composition of cathodic electro deposition primers*

Cathodic electro deposition primers for automotive OEM coatings consist of modified aromatic epoxy resins. In most cases, two resin modifications are employed, namely the resin for the main dispersion and the grinding resin for the pigment paste. The first modification consists of amine adducts that are neutralised to generate cations which can be hydrolysed to groups capable of creating colloidal water-borne dispersions. The second modification contains building blocks that confer flexibility. The crosslinkers for the modified epoxy resins are blocked aromatic polyisocyanates. There are coloured pigments, extenders and some corrosion protection pigments. The additives employed are wetting agents, levelling agents, bactericides and fungicides, and crosslinking catalysts [70].

3.4.5.1 *Epoxy resins*

Aromatic epoxy resins, the primary and critical resin component of cathodic electro deposition primers are notable for their excellent corrosion protection. Films formed by crosslinked epoxy resins have high hardness and excellent resistance to solvents and many chemicals. They are relatively resistant to high temperatures. But aromatic epoxy resins are neither weatherable nor resistant to light and yellowing.

The most important starting components for the production of aromatic epoxy resins are bisphenol A (4,4'-dihydroxy-2,2-diphenylpropane), a condensation product of two molecules of phenol and one molecule of acetone, and epichlorohydrin (1-chloro-2,3-epoxypropane). The epichlorohydrin adds across the acidic OH-groups of the bisphenol A. Treatment with concentrated sodium hydroxide solution generates hydrochloric acid and ultimately sodium chloride. An epoxy group is also formed. The simplest compound contains two molecules of epichlorohydrin and one molecule of bisphenol A. It is the bis-glycidyl ether of bisphenol A (Figure 3.4.4 shows the reaction). This low-molecular product ranges from being liquid to crystalline in accordance with its purity.

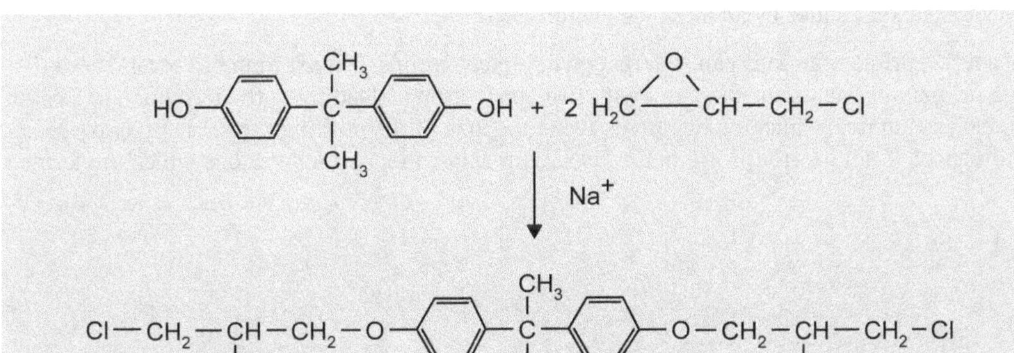

Figure 3.4.4: Preparation of the bis-glycidyl ether of bisphenol A

High-molecular epoxy resins can be made by making epichlorohydrin react with bisphenol A in a ratio of less than 2:1. The preparation of high-molecular epoxy resins in this way is shown in Figure 3.4.5

High-molecular epoxy resins can also be prepared by an alternative process. This involves making the bis-glycidyl ether of bisphenol A react with more bisphenol A, and is shown in Figure 3.4.6.

The second method is advantageous because it does not require handling of the highly toxic epichlorohydrin, which is only used for the preparation of the bis-glycidyl ether. Another advantage is that high-molecular resins prepared by this method are much more bi-functional than those resins prepared by the first. With the first method, there is the risk that - at high molecular weights - the epoxy groups will not be formed due to back reaction with concentrated sodium hydroxide solution; in that event, hydrolysis leads only to bishydroxyl compounds. It is very important for the chlorine content in the epoxy resins to be as low as possible, as it would be risky to use the resins for corrosion protection. The target is less than 50 ppm saponifiable chlorine.

Epoxy resins are capable of crosslinking. Crosslinking may take place at the epoxy groups or at the secondary hydroxyl groups. The epoxy groups react with amines or carboxylic acids in an addition reaction. The reaction with amines occurs at ambient temperatures, while that with carboxyl groups needs an elevated temperature. The addition reaction is beneficial to paints that are applied in thick layers. There is a risk, however, that reactions involving cleavage of products will lead to film defects. The secondary hydroxyl groups of epoxy resins are less reactive. They react at elevated temperatures with phenol resins or amino resins. These are condensation reactions in which water or alcohols are cleaved. Aromatic epoxy resins are widely used, not only for coatings, but also for plastic parts. Before they can be used in cathodic electro deposition primers, they have to be modified with amines.

3.4.5.2 Amine modification of epoxy resins

Terminal epoxy groups of aromatic epoxy resins react readily with secondary amines to form β-hydroxyalkyl amines by addition (see Figure 3.4.7)

Of course, primary amines can also react with epoxy groups. It must then be remembered that one primary amine group can react with two epoxy groups, leading to the formation of a chain of tertiary amines, which on average is twice the size of the starting molecule of epoxy resin. If the tertiary amine groups are neutralised with acids, the ammonium ions which are formed

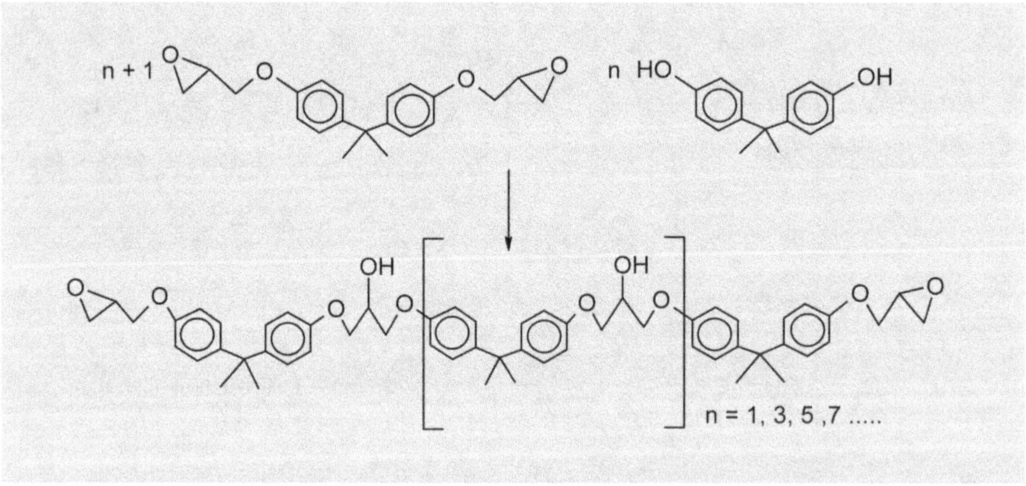

Figure 3.4.5: Preparation of high-molecular epoxy resins with a deficiency of epichlorohydrin

can render the polymer containing them soluble in water. This means that the amine-modified epoxy resins, after neutralisation with acids, form aqueous solutions containing colloidal particles. As neutralisation is only partial in most cases, the particles are relatively large; this type of dispersion is commonly called a water-borne dispersion, although there are some differences from, e.g. primary acrylic dispersions. The neutralisation agents employed are acetic acid, formic acid, and lactic acid.

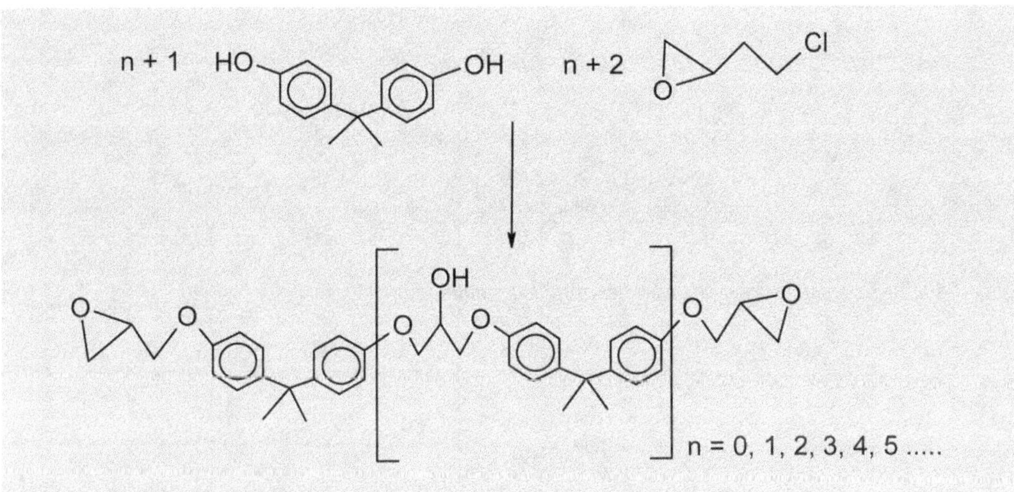

Figure 3.4.6: Preparation of high-molecular epoxy resin from bis-glycidyl ether and bisphenol A

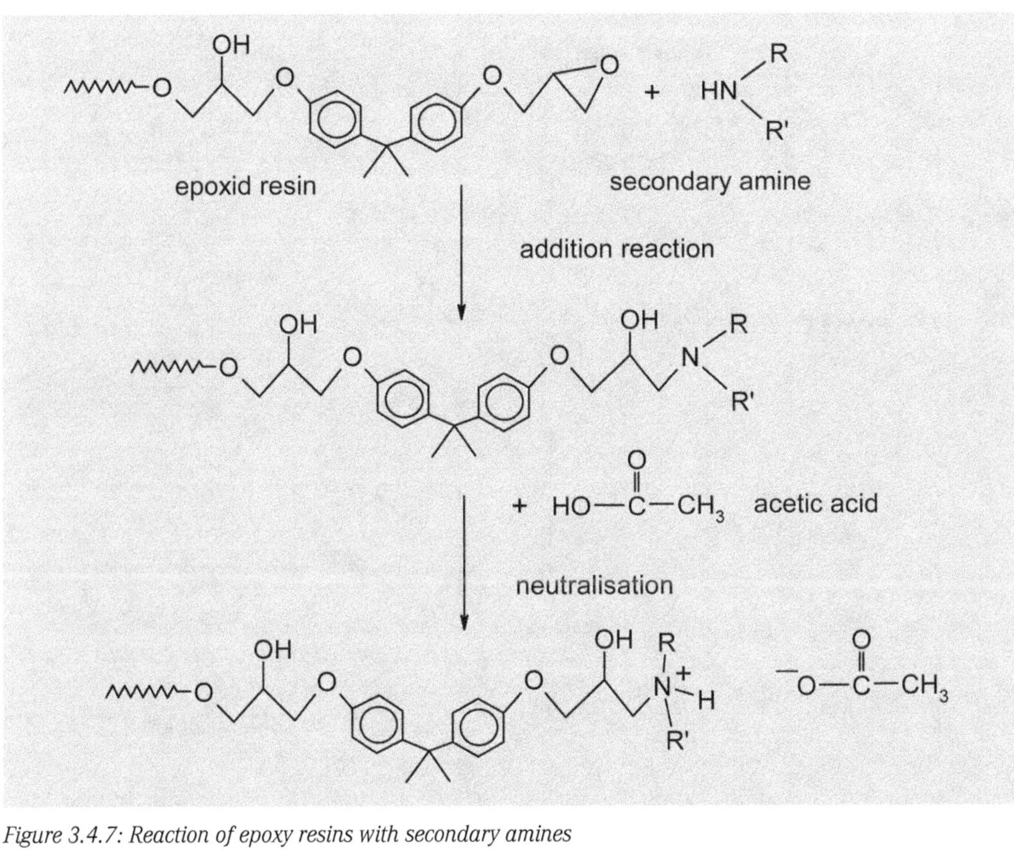

Figure 3.4.7: Reaction of epoxy resins with secondary amines

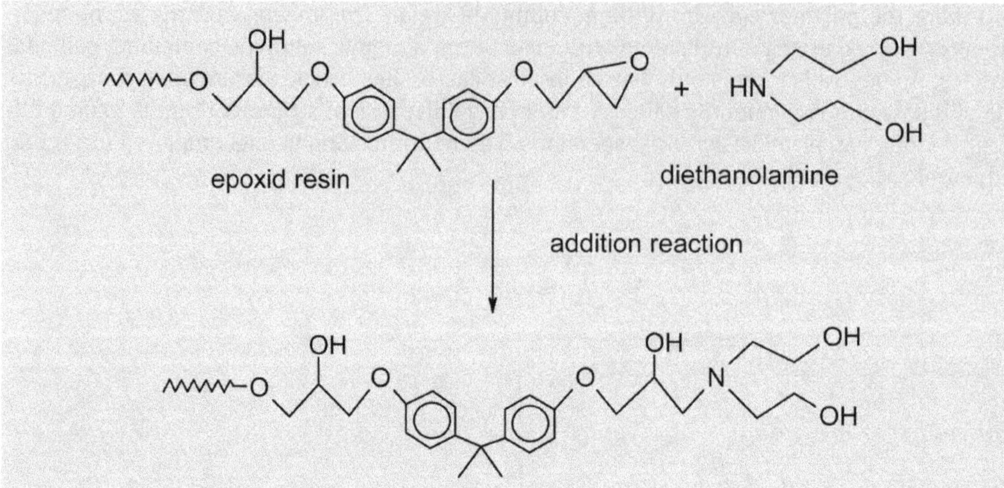

Figure 3.4.8: Reaction of epoxy resins with hydroxyalkyl amines

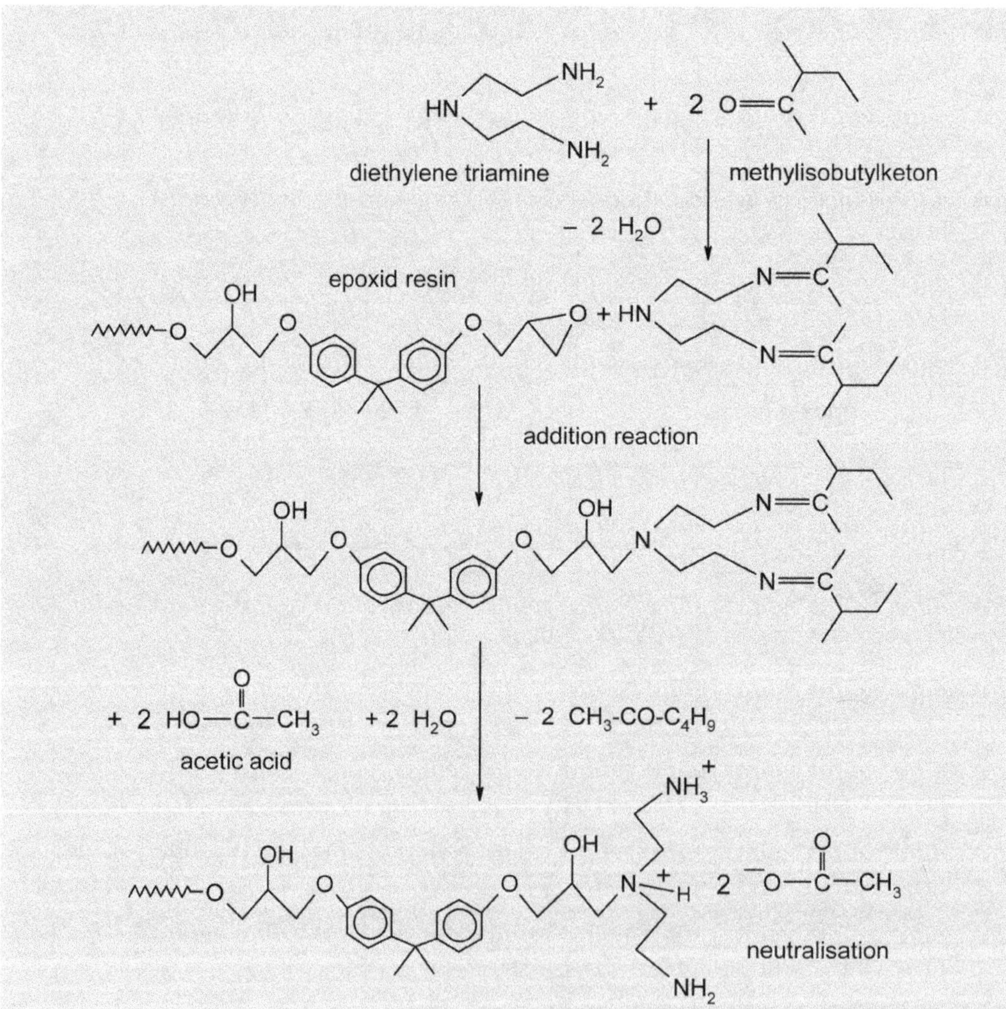

Figure 3.4.9: Addition reaction for ketimines and hydrolysis during neutralisation

As already mentioned the amine-modified epoxy resins contain secondary hydroxyl groups, which are capable of crosslinking reactions. It is also possible, though, to introduce further functional groups. For example, the addition of hydroxyalkyl amines (e.g. N-methylethanolamine, diethanolamine) yields amine-modified epoxy resins with terminal hydroxyl groups (see Figure 3.4.8).

Products containing terminal primary amine groups can also be made. This is done by making primary-secondary amines (e.g. N-methylaminopropylamine, diethylene triamine) react with ketones to form ketimines. The remaining secondary amine group, which is unable react with ketones, then adds across epoxy groups, as described above. The ketimine is subsequently hydrolysed, and the primary amine groups are re-formed. This happens when the resins are neutralised during transfer into the aqueous phase. Primary amines are highly reactive partners for the crosslinking reaction with blocked polyisocyanates. Of course, the amine groups are also neutralised - at least partly - by the acids. The addition reaction for ketimines and hydrolysis during neutralisation are shown in Figure 3.4.9.

The content of amines or ammonium ions determines the particle sizes of cationically stabilised aqueous colloidal solutions or dispersions. Optimum deposition depends heavily on the particle size of the colloidal solution or dispersion. It is therefore important to define and control the amount of ions via the type and quantity of amines and the degree of neutralisation.

3.4.5.3 Further modifications

The specific properties of aromatic epoxy resins in electro deposition primers are influenced by further modifications. The main goals are of such modifications are to optimise flexibility, as well as flow and levelling during film formation. This can be accomplished in various ways. Nearly all of them are based on the introduction of building blocks or additive molecules that act as plasticiser. The compounds more or less contain long aliphatic chains, with some additionally containing ether or ester groups. These are aliphatic polyethers, polyesters, or phenols with aliphatic side chains.

Extending the molecular weight of aromatic epoxy resins proceeds from bis-glycidyl ether and bisphenol A. Plasticisation is achieved by replacing a certain amount of the bisphenol A with other components which react with the epoxy groups of the bis-glycidyl ether. These are polyester polyols (e.g. polycaprolactones) and polyether polyols (e.g. polypropylene glycol). It is also possible to use low-molecular polyols for the extension reaction. The reaction consists in the addition of hydroxyl groups across the epoxy groups. It is necessary to use catalysts for the reaction of aliphatic hydroxyl groups with epoxy groups, as the hydroxyl groups are less acidic than the phenol hydroxyl groups. The reaction conditions also enable the secondary hydroxyl groups of the epoxy resin to react with epoxy groups. The increase in molecular weight from that side reaction must be taken into consideration.

Similar effects are achieved through the use of alkyl phenols having long alkyl side chains (e.g. dodecylphenol). These alkyl phenols not only form terminal groups since the secondary hydroxyl groups also enter into a side reaction under the chosen conditions and since strong catalysts are used. The entire reaction sequence is fairly complex. Since the amine addition reaction is much faster than hydroxyl addition, the latter must be allowed to go to completion before amine modification is performed. Bisphenol A-bis-glycidyl ether, more bisphenol A, alkyl phenols, and amines ultimately lead to high-molecular epoxy resins. The molecules may be not only linear, but also branched due to the side reactions of the secondary hydroxyl groups. The tertiary amine groups are neutralised by volatile organic acids and transferred into aqueous dispersions, whose state is close to that of aqueous colloidal solutions. The average number molecular weight of the dispersion resins lies between 3,000 and 5,000 g/mol. The basic chemical structure of these resins is shown in Figure 3.4.10.

The process for making the resins is carried out at temperatures of 110 to 180 °C. The various reaction catalysts named in the corresponding patents are tertiary amines (e.g. N,N-dimethylbenzyl amine) and other Lewis bases (e.g. triphenylphosphine). Another variant of the extension process consists in replacing parts of the aromatic epoxy resin (the bisphenol A-bis-glycidyl ether) by aliphatic bis-glycidyl ether with long side chains (e.g. polypropylene glycol bis-glycidyl ether or polytetrahydrofuran bis-glycidyl ether).

Other patents describe combining the amine-modified aromatic epoxy resins with aliphatic diamines, e.g. a polypropylene oxide bearing terminal amine groups (polyether amine). Both these methods introduce molecules or parts of molecules which act as plasticiser. It is also possible to add plasticising components, such as polypropylene glycol or ethoxylated bisphenol A, during manufacture of the dispersion, without their reacting with the epoxy resins or undergo-

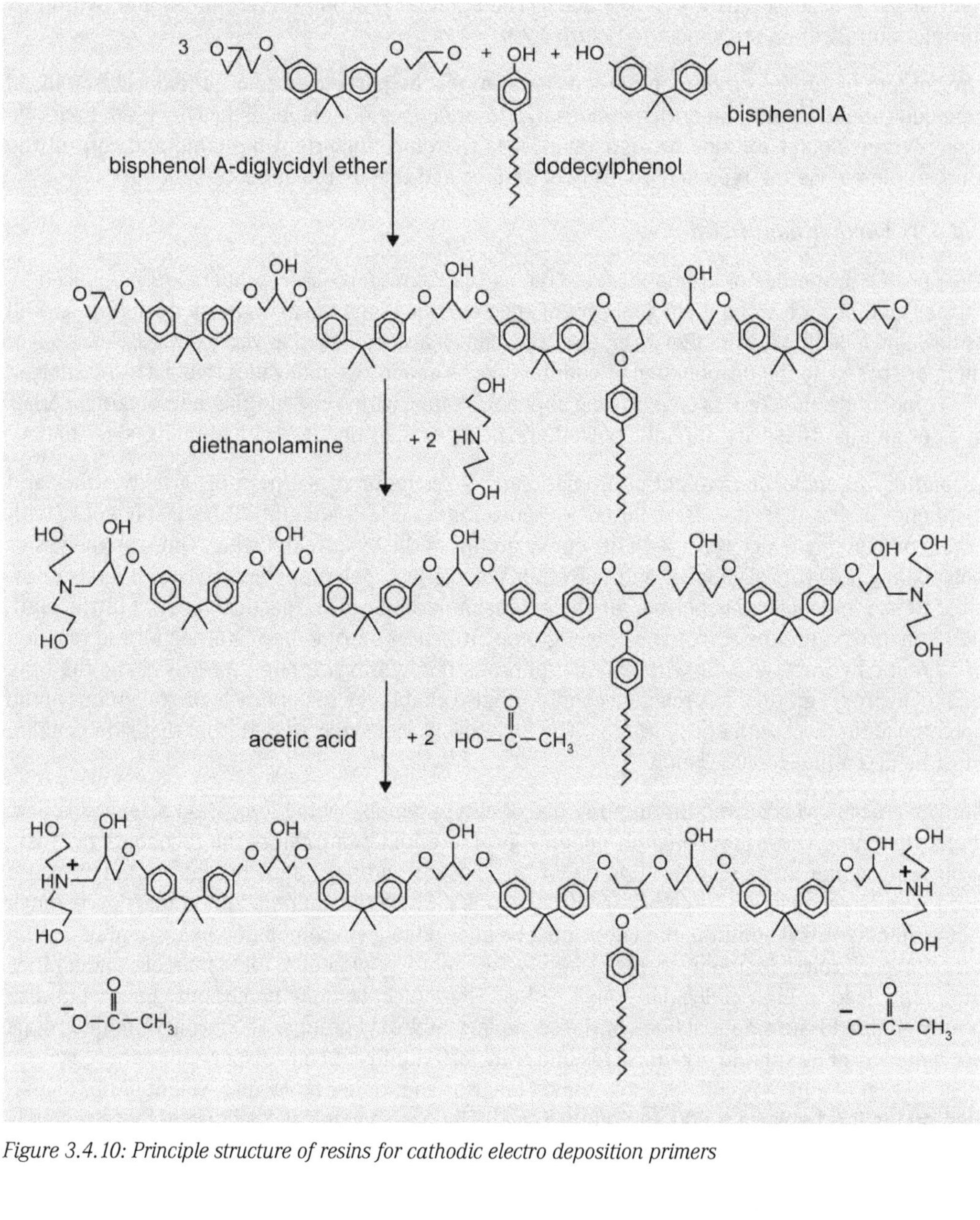

Figure 3.4.10: Principle structure of resins for cathodic electro deposition primers

ing neutralisation. These plasticising components form part of the dispersion particles and are deposited with them on the object, although they do not form cations. It may be assumed that the free hydroxyl groups of those components react with the blocked polyisocyanates under stoving conditions and so become part of the molecular network of the primer film.

3.4.5.4 *Grinding resins*

The relatively large particles of the dispersions of modified epoxy resins, described above, do not provide optimum pigment wetting. Separate resins are therefore produced which are optimised for pigment wetting. These are called grinding resins and are amine-modified epoxy compounds; they therefore belong to the same group as the main resins. The principal differences between grinding resins and the main resins are that they contain molecular groups that can interact with the surfaces of the pigments. Although these resins do not have the structure of true surfactants, it is helpful to think of them acting as such. The wetting properties of these resins can be boosted in particular by the use of quaternary ammonium ions. Quaternary ammonium salts are prepared by making an ammonium salt of one of the acids used for neutralisation react with epoxy groups. The modification of epoxy resins with quaternary ammonium salts is illustrated in Figure 3.4.11.

An example of a grinding resin containing polypropylene glycol bis-glycidyl ether is given in Table 3.4.2 (see Chapter 3.4.5.8).

3.4.5.5 *Crosslinkers*

The crosslinkers used in the modification of epoxy resins are blocked polyisocyanates. Isocyanates are capable of reacting with resins bearing free hydroxyl groups or secondary or primary amine groups at ambient temperature. That is why isocyanates are crosslinkers for two-pack paints, in which the crosslinker, called the hardener, has to be provided separately from the base component (main component). The two components are mixed just prior to application, whereupon the reaction between the functional groups of the main resin and the polyisocyanate starts immediately.

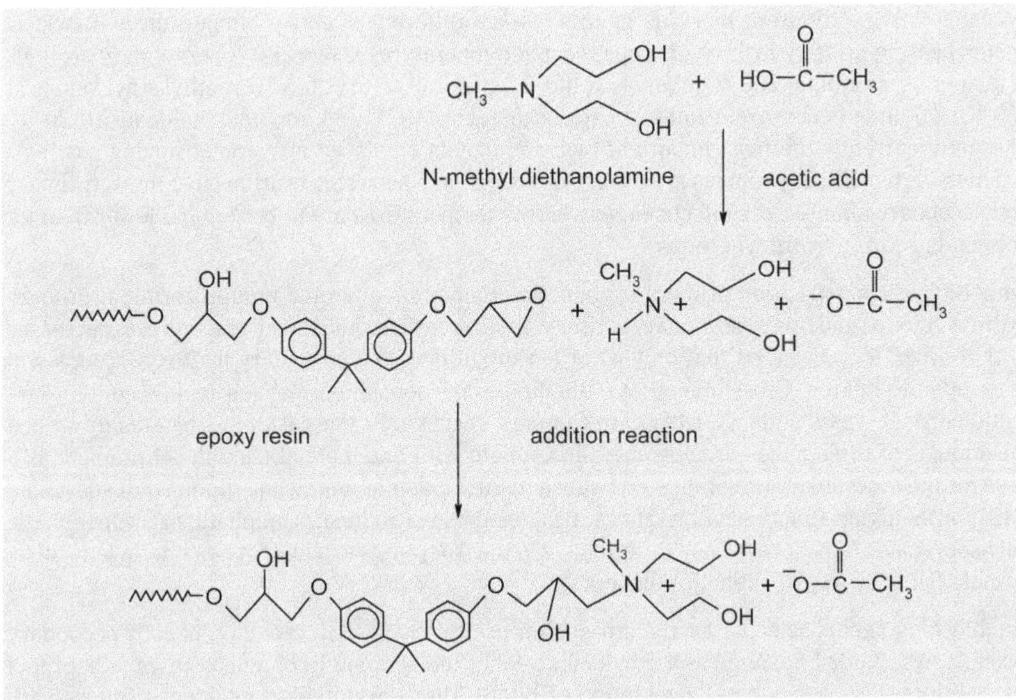

Figure 3.4.11: Modifying an epoxy resin with a quaternary ammonium salt to prepare a grinding resin

The mixtures can therefore be handled and applied only for a certain amount of time, known as the pot-life. Two-pack paints are used for automotive clearcoats (see Chapter 3.8.3.4) and automotive refinishes (see Chapter 4.4). Of course, such crosslinkers cannot be used in dipping paints as the latter have to be stable for a relatively long time. But systems crosslinked by polyisocyanates boast properties that are attractive to primers, too. There is a way of modifying polyisocyanates to render them stable in mixtures with reactive partner resins. The isocyanates are made to react with a compound that can be cleaved at elevated temperatures. Such "pre-reacted" isocyanates are no longer reactive at ambient or room temperature and mixtures of them with the partner resins which they are to crosslink are relatively stable in storage. The pre-reacted isocyanates are known as blocked or capped polyisocyanates. Blocked polyisocyanates are only reactive at elevated temperatures; they crosslink the partner resins after the blocking agent has been cleaved. The effective de-blocking or cleavage temperature depends on the type of polyisocyanate, the type of blocking agent, and the reactivity of the functional groups on the partner resin. In the past, it was assumed that the reaction of blocked polyisocyanates involved two steps, namely cleavage of blocking agent followed by reaction of the freed-up isocyanate group with the partner functional groups of the resin to be crosslinked. But our investigations show that the reaction of blocked polyisocyanate is a substitution reaction, and as such is comparable to transesterification and other reactions [71]. One reason why the reaction takes place in this way is that the same blocked polyisocyanate reacts at different temperatures to an extent which depends on the reactivity of the functional groups on the partner resins. For example, resins bearing primary hydroxyl groups react effectively at lower temperatures than resins bearing secondary hydroxyl groups. In addition, such differences can be revealed by comparing primary hydroxyl groups on different extended aliphatic chains (spacers). The greatest difference is observed when the reactivity of NH-groups is compared with that of hydroxyl groups. On the other hand, isocyanates differ in their reactivity and effective crosslinking temperature. Aromatic isocyanate groups react much faster than aliphatic isocyanate groups. These, in turn, react faster than cycloaliphatic isocyanate groups. The type of blocking agent mainly influences the effective crosslinking temperature. Comparison of the crosslinking of polyesters containing hydroxyl groups with aliphatic polyisocyanates bearing different blocking agents reveals different effective temperatures. Blocking agents bearing primary hydroxyl groups react at temperatures above 200 °C, secondary alcohols at temperatures around 200 °C, phenols at 180 °C, ε-caprolactam at 165 °C, methyl ethyl ketone at 150 °C, 3,5-dimethyl pyrazole and ethyl acetoacetate at 140 °C and dimethyl malonate at 130 °C. These are currently the most important blocking agents employed for various stoving coatings. When these blocked isocyanates are used to crosslink partner resins bearing NH-groups (primary and secondary amines), the effective crosslinking temperatures are at least 50 °C lower than for polyesters bearing hydroxyl groups.

Cathodic electro deposition primers contain small amounts of amine groups, terminal primary hydroxyl groups and finally secondary hydroxyl groups on side chains. The reaction rate decreases in that order. It is assumed that only a certain quantity of the secondary hydroxyl groups will react by crosslinking. Crosslinkers for cathodic electro deposition primers mainly contain aromatic polyisocyanate adducts, which are the most reactive. In the past, most patents described the adducts of three moles of toluylene diisocyanate with one mole of trimethylol propane as a base for the crosslinker. Preference was subsequently given to methylene diphenyl diisocyanate (MDI) or its higher functional derivatives. Besides these crosslinkers, aliphatic or cycloaliphatic polyisocyanate adducts have been described (e.g. the isocyanurate trimer of hexamethylene diisocyanate [HDI] by way of aliphatic compound).

The blocking agents selected for the crosslinker are relatively slow-reacting, usually secondary alcohols (e.g. 2-ethyl hexanol) and ethylene glycol or diethylene glycol mono ethers (e.g. propyl glycol [propyl cellosolve], butyl digol [butyl carbitol]). They are suitable for current stoving conditions involving temperatures of 160 to 175 °C. The catalysts for the isocyanate reactions, as well as for the reactions of blocked polyisocyanates, are heavy metal cations or tertiary amines. The preferred catalyst for cathodic electro deposition primers is dibutyltin oxide, which is described in most of the patents. Since tin compounds are currently classified as physiologically harmful, alternatives are currently being researched [72]. As already mentioned, the main resin of the primer, the modified epoxy resin, contains tertiary amines. These amines also have a catalytic effect on the crosslinking reaction of the primers. The blocked polyisocyanates are themselves insoluble in water. But when they are mixed with the cationically stabilised epoxy resins in organic phase, they are transferred along with them into the aqueous phase, where they form stable aqueous dispersions. After neutralisation with volatile organic acids, the amine-modified epoxy resins are able to carry the blocked isocyanates in aqueous phase. Other blocked polyisocyanates act as crosslinkers for OEM clearcoats and are described in Chapter 3.8.3.5.

3.4.5.6 Pigments

Cathodic electro deposition primers contain coloured pigments, extenders, limited quantities of active corrosion protection pigments and perhaps some pigment-like additives. In most cases, the coloured pigment is a rutile titanium dioxide. There are numerous commercial titanium dioxides. Few types are suitable for electro deposition primers since they have to suit the specific conditions of electrophoresis (water as dispersing agent, stability in dipping tanks, low pH value, electro deposition process, high stoving temperature, crosslinking reaction). Most of the primers are grey. The titanium dioxide is therefore combined with carbon black. However, there are also black electro deposition primers. In this connection, it should be noted that there were attempts to formulate electro deposition one-layer coatings of different colours.

The extenders or fillers for the primers are mainly barium sulphates and aluminium silicates. A combination of titanium dioxide of very small particle size and extenders of somewhat larger particle size bestows optimum mechanical resistance on the primer layer. This effect is described in more detail in Chapter 3.5.3.4.

As active corrosion protection pigments contain quantities of water-soluble compounds, they are not so suitable for electro deposition paints because they may interfere with the deposition process. Accordingly, only a few pigments are chosen for use in primers and the quantities employed are restricted. Most older patents mention lead silicates as corrosion protection pigments. Although lead pigments are widely known to support corrosion protection, they are no longer used due to ecological reasons. New formulations are labelled "lead free", and this is used as a selling point. The cationically stabilised epoxy resins in the tank dispersions have to carry pigments to the cathode and deposit them there. They therefore have to be incorporated into the dispersion particles very well. That might explain why electro deposition primers have only a relatively low pigment volume concentration (PVC) of just 5 to 10 %.

3.4.5.7 Additives

Cathodic electro deposition primers may contain wetting agents, defoamers, and levelling agents. It is important to select nonionic materials, as they do not impair tank stability or the deposition process. The levelling agents employed are polyacrylates or polyvinyl ethers. The products have to be organophilic because they must diffuse into the dispersion particles and be deposited along with them so as to be active during film formation. Hydrophilic compounds may be released in the dipping tank. Some formulations contain modified siloxanes. Since levelling agents act on the surface of the paint layers by floating to the surface during film formation, it is essential to test their influence on intercoat adhesion. Naturally, wetting and adhesion of the following paint layer (primer surfacers) must not be impaired.

Several patents describe special alkyne diols as defoamers. Co-solvents can also support levelling and may act as defoamer. For this reason, some primers in the past contained small amounts of glycol ethers. But since emissions of volatile compounds are to be reduced, current formulations dispense with co-solvents altogether. As already mentioned, the crosslinking reaction of blocked polyisocyanates and modified epoxy resins is catalysed. The preferred catalyst is still dibutyltin oxide, which is incorporated into the entire formulation after having been dispersed in resin along with pigments. Alternatives are also currently being studied, e.g. bismuth compounds.

3.4.5.8 Overall primer formulation

A fully formulated cathodic electro deposition primer consists of the main resin dispersion, comprising the modified epoxy resin, the crosslinker and the volatile organic acid which acts as neutralisation agent, along with the pigment paste containing the special grinding resin, pigments, and the crosslinking catalyst.

Main resin dispersion

For optimum film properties, it is important that all the dispersion particles contain the ingredients necessary for film formation by means of crosslinking. The first step in the preparation of the main resin is the reaction of epoxy resin with plasticising compounds. The second step is modification with amines. Both reactions are carried out in a melt or in organic solution. The preferred solvents are polar, water-tolerant, but inert (not functional), e.g. low ketones. These reactions may be followed by the addition of protic solvents, e.g. glycol ethers. The crosslinker - the blocked polyisocyanate - is then admixed in the organic phase. Other ingredients may then be added, such as polypropylene glycol, which acts as a solvent - but is no longer volatile - and additionally as a plasticiser. The amines of the modified epoxy resins are then neutralised by adding the volatile organic acids, such as acetic acid, formic acid, or lactic acid. After that, deionised water is added to prepare the composition for transfer to the aqueous phase. The degree of neutralisation is between 0.4 and 0.7 acid equivalents, expressed in terms of one amine equivalent. The resulting pH values are 5.5 to 6.5. After that the process solvents are distilled off, mainly by vacuum distillation. The particle sizes of the resulting water dispersion range from 80 to 150 nm. Film forming additives are added to the dispersion or to the organic phase.

Pigment paste

Aqueous pigment pastes (pigment dispersions) are prepared from the grinding resin, selected pigments and the solid catalyst. Some neutralisation acid may be added to adjust the pH value. The mixture is pre-dispersed with the aid of a "dissolver". This is followed by grinding in a stirrer mill (sand mill). The grinding efficiency is checked by measuring the particle size to avoid residual agglomerates.

Dipping tank

The content of dipping tank is prepared by mixing the main dispersion, the pigment dispersion and water. The solids content is adjusted to 20 to 25% by weight. If the process is operated in-line, this adjustment only occurs at the start of the line. Thereafter, material removed by the coating process is replenished with concentrated main resin and pigment paste. The replenishment materials are added when the concentration of the tank ingredients falls below specific values. It should be mentioned in this regard that the dipping tanks always contain both old and new material. That is why tank stability is so important. Parameters affecting the tank stability are the degree of neutralisation, the viscosity and polarity of the resins, the solvent content (if solvents are used), and the size of the dispersion particles. Contamination by electrolytes can adversely affect tank stability. Since electro deposition tanks are in use continuously, specific tank parameters are measured regularly in order that replenishment may be initiated as soon as possible. If the electro deposition process is the first element in a coating process comprising several application stages, and if the quantity of tank material is very large and the application equipment large and expensive, any problems and failures that occur during electro deposition lead to major economic problems and increase the costs of the overall process. Thus, a breakdown in electro deposition is followed by shutdown of the entire application line, loss of productivity and incurs high failure costs. Since the commercial risks are so critical, paint and car producers agree that the paint producer is responsible for failures related to paint materials and must build up reserves of paint materials or guarantee immediate replacement material. The composition of the various parts of an electro deposition primer is described in the following examples [73].

Examples: composition of an electro deposition primer

Main resin dispersion

In a laboratory reactor, 3 moles of bis-glycidyl ether bisphenol A (technical grade, epoxy equivalent mass 188 g/mol) are made to react at 130 to 160 °C with 1 mole bisphenol A and 1 mole dodecyl-phenol in the presence of a triphenyl phosphine catalyst to yield a modified epoxy resin with the epoxy equivalent mass of 532 g/mol. The resulting product is made to react at 110 to 130 °C with 1 mole diethanolamine and 0.5 mole N,N-dimethylaminopropylamine. The reaction is carried out with the aid of polypropylene glycol (average molar mass 900 g/mol) as process solvent. Since the epoxy equivalent mass of the reaction mixture tends to become infinite, small quantities of butyl glycol are added. The blocked polyisocyanate is then added by way of crosslinker. It contains 1 equivalent of MDI derivatives and 1 mole diethylene glycol monobutyl ether (butyl digol). The blocking reaction is carried out at temperatures below 70 °C in the presence of dibutyltin dilaurate as catalyst. The quantity of added crosslinker is 888 g, which represents 3 potential NCO groups intended for reaction with the functional groups of the already prepared modified epoxy resin. The mixture is transformed into an aqueous dispersion with deionised water containing 0.8 moles of concentrated acetic acid. More deionised water is added to adjust the solids content of the dispersion to 35 % by weight. The composition [74] of this dispersion is given in Table 3.4.1.

Table 3.4.1: Composition of the main resin dispersion for ED-primer

	Moles	Components	m	wt-‰
01	3.0	bisphenol-A-diglycidyl ether (EW 188 g/mol)	1128.0	141.6
02	1.0	dodecyl phenol	262.0	32.9
03		xylene	31.4	3.9
04	1.0	bisphenol A	228.0	28.6
05		triphenyl phosphine	1.6	0.2
06		polypropylene glycol 900	297.5	37.4
07	1.0	diethanol amine	105.0	13.2
08	0.5	N.N-dimethyl amino propyl amine	51.0	6.4
09		butyl glycol	58.5	7.3
10	3.0 equivalents	blocked polyisocyanate 3 equiv MDI-oligomer + 3 mol butyl diglycol	887.8	111.5
11	0.8	acetic acid	48.1	6.0
12		deionised water	2825.6	354.8
13		deionised water	2039.7	256.2
		sum	7964.2	1000.0

Characteristic values

solids (1h 130 °C)	35.7 %	neutralisation degree:	43 %	particle size (average):	125 nm
base (FK)	0.657 meq/g	pH-value:	5.4	viscosity (DIN 3511. 4/23 °C)	15"
acid (FK)	0.283 meq/g			(equivalent ISO cup)	58"

Grinding resin

The grinding resin [75] consists first of the product yielded by the reaction between 2 moles of the bis-glycidyl ether of bisphenol A (technical grade, epoxy equivalent mass 188 g/mol), 1 mole bisphenol A and 1 mole dodecylphenol, which are made to react at 130 to 150 °C in the presence of triphenyl phosphine as catalyst. The process solvent is butyl glycol. The second step is the addition of 1 mole polypropylene glycol diglycidyl ether (epoxy equivalent mass 333 g/mol). After addition of some more butyl glycol by way of solvent, the mixture is made to react with 0.5 moles N,N-dimethylaminopropylamine and 1 mole aminoethoxyethanol for two hours at 90 °C. Since

Table 3.4.2: Grinding resin formulation

	Moles	Components	m	wt-‰
01	2.0	bisphenol A-diglycidyl ether (EW 188 g/mol)	752.0	250.7
02	1.0	dodecyl phenol	262.0	87.3
03	1.0	bisphenol A	228.0	76.0
04		butyl glycol	64.0	21.3
05		xylene	50.0	16.7
06		triphenyl phosphine	1.6	0.5
07		butyl glycol	300.0	100.0
08	1.0	polypropylene glycol diglycidyl ether	666.0	222.0
09	1.0	diethanol amine	105.0	35.0
10	0.5	N,N-dimethyl amino propyl amine	51.0	17.0
11		butyl glycol	520.0	173.5
		sum	2999.6	1000.0

Characteristic values

solids (1^h 130 °C)	695.7 %
viscosity (40 % in PM)	2.2 dPa·s

Table 3.4.3: Pigment paste formulation for electro deposition primer

	Components	wt-‰
01	grind resin, neutralised (60 % in butyl glycol/water)	2770.7
02	lead silicate	14.2
03	aluminium silicate	77.3
04	carbon black	5.1
05	titanium dioxide	339.3
06	dibutyl tin oxide	16.7
07	deionised water	269.6
	sum	1000.0

Table 3.4.4: Composition of the dipping tank for an electro deposition primer

	Components	wt-‰
01	resin dispersion (see table 3.4.2)	401.2
02	pigment dispersion (see table 3.4.3)	127.4
03	deionised water	471.4
	sum	1000.0

the epoxy equivalent mass of the reaction mixture tends to become infinite, butyl glycol is added to adjust the solids content of the mixture to 70 % by weight. The formulation of the grinding resin is shown in Table 3.4.2.

Pigment paste

A grinding resin [76] comparable to that described above is neutralised with acetic acid and the solids content of the delivery form is adjusted to 60 % by weight with deionised water. To this resin solution is added aluminium silicate, titanium dioxide, carbon black, dibutyltin dioxide, and more deionised water. The mixture is dispersed on a stirrer mill (sand mill) to form a pigment paste. The formulation of this pigment paste [77] is given in Table 3.4.3.

Dipping tank

The aforementioned main dispersion, pigment paste, and additional deionised water are mixed to form a dipping tank for the deposition process. The composition [78] of the tank is described in Table 3.4.4.

The dipping tank has a solids content of 22 % by weight, a pH value of 5.67, a temperature of 29 °C, and a conductance of 1.34 mS/cm. Steel panels pre-treated with zinc phosphate are coated for 2 minutes at a deposition voltage of 220 Volt. After dipping, the panels are rinsed carefully with deionised water and stoved for 15 minutes at 175 °C. The resulting primer layer has a thickness of 22 µm. Levelling of the primer surface is rated as being perfect, and the Erichsen indentation test yields a figure of 6.5 mm. The reverse impact test returns 40 kp/cm², and the subsurface rusting result following 100 hours in the salt spray test is less than 1 mm.

3.4.6 Application of electro deposition primers

3.4.6.1 Deposition

The pre-treated car bodies are transported into the dipping tank on a conveyor belt. They form the cathode of a direct current circuit during dipping and are coated with the electro deposition primer. The coating process lasts 2 to 3 minutes and the voltage is 250 to 350 Volt. It makes sense to switch on the current only once the bodies have been dipped. The tank material has a pH value of 5.5 to 6.5 and a temperature of 25 to 30 °C. The attainable film thickness (dry film) ranges from 18 to 25 µm. Specialty primers can achieve up to 35 µm (thick layer primer). On account of the effects of electrophoretic deposition and osmosis, the freshly deposited film is relatively mechanically stable. Consequently, the film can be rinsed thoroughly. After rinsing, the film is stoved for 20 to 30 minutes at 160 to 175 °C [57].

Electrode reactions

During the deposition process, water is reduced (electrolysed) at the cathode to molecular hydrogen and hydroxyl ions. The hydroxyl ions neutralise the cations of the resin compounds which also approach the cathode, transforming the ammonium ions to amines and forming water molecules. The resins, which are no longer cationically stabilised, lose their water solubility and coagulate on the cathode (the car body) to form a film. In addition, the iron ions at the cathode are reduced to iron metal, which boosts corrosion resistance. Osmosis drives the water out of the film layer, as a result of which the film has adequate mechanical stability. During film formation, the layer thickness increases continuously and so, too, does the electrical resistance of the film. Once the electrical resistance reaches a certain level, deposition stops.

At the anode of the electro deposition tank, water is oxidised (electrolysed) to protons or hydronium ions and molecular oxygen, which escapes. Free acids are therefore formed at the anode, along with the anions of organic acids. The equations for the electrode reactions are described in Figure 3.4.12 (page 72).

3.4.6.2 Process sequence and equipment

All materials which are removed from the tank by deposition have to be replenished periodically. Replenishment consists in adding concentrates of the main resin dispersion and pigment pastes. The replenishment materials for anodic electro deposition tanks are either not neutralised or only partly neutralised. In contrast, due to stability requirements, the dispersion for cathodic electro deposition tanks must be neutralised with acids to the pH level at which the deposition process is to be performed. That is why, during deposition, which removes the basic substances from the tank, the concentration of acids formed at the anode increases continuously. Since the pH value decreases accordingly, application problems occur. The excess acid must therefore be removed from the tank as well to prevent a drop in pH value. Installed around the anodes are so-called anode boxes, which consist of semipermeable membranes and prevent the acid from entering the tank. The contents of the anode boxes are pumped out and passed through an ion exchanger, where the acid is collected; the water is returned to the tank. This part of the process is called the anolyte circuit.

After the dipping process, the coated car bodies are rinsed carefully to remove all soluble film material, which would otherwise pose a corrosion risk once film formation was complete. The rinsing water is recycled to avoid generating large amounts of waste water. The recycling process for the rinsing water, which contains some low-molecular compounds, consists in first feeding the rinsing water into the dipping tank. The additional volume in the dipping tank is removed continuously and fed to ultra-filtration equipment, where contamination in the form of low-molecular products is removed. At the same time, the material is also fed through a heat exchanger. As a result of the energy released during the electrolysis, the tank temperature would increase continuously and lead to variable results. The heat exchanger ensures that the deposition conditions are constant. Residue from the ultra-filtration process is recycled to the dipping tank, mostly along with the replenishment material. The filtrate (water) is stored in a tank and used for the first rinsing step for the car bodies (ultra-filtration circuit). However, fresh, deionised water is always used for final rinsing. The process and equipment are shown schematically in Figure 3.4.13 [79].

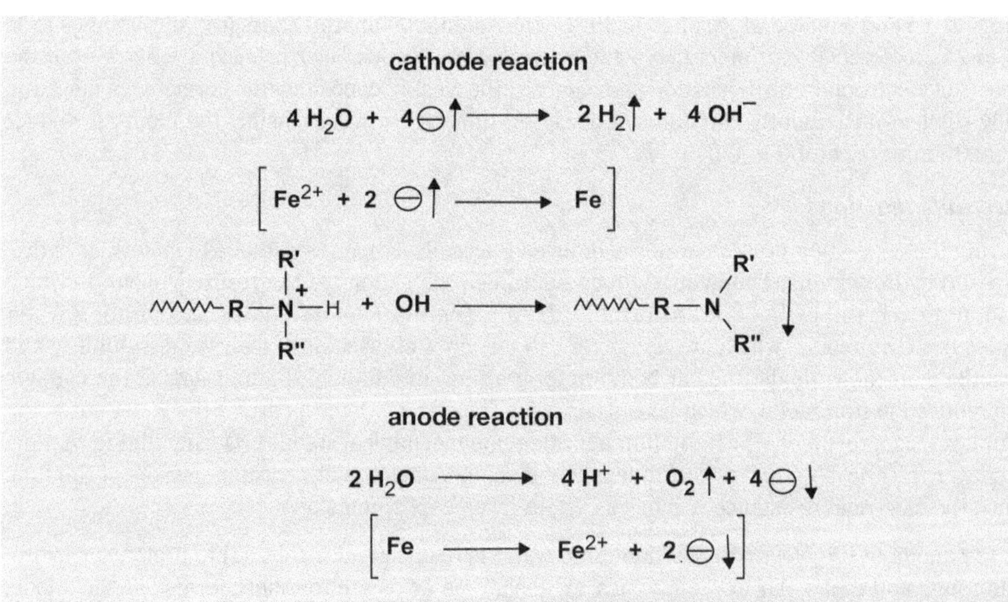

Figure 3.4.12: Equations for the electrode reactions that occur in cathodic electro priming

The size of the dipping tank depends on the number of car bodies to be coated. There are tanks which hold "only" 60 t of material, while others have a capacity of 500 t. These are sufficient for priming 15 to 60 car bodies per hour. Large tanks, in which deposition is performed on a conveyor basis, may be up to 50 m long. For small tanks, up to 10 m in length, the car bodies are lowered in and then lifted from the conveyor belt. After dipping and rinsing, the car bodies are transferred from the overhead line to a floor conveyor. The entire process, comprising cleaning of the car bodies, pre-treating, dipping, rinsing, and stoving, followed by underbody protection and joint sealing, is shown schematically in Figure 3.4.14.

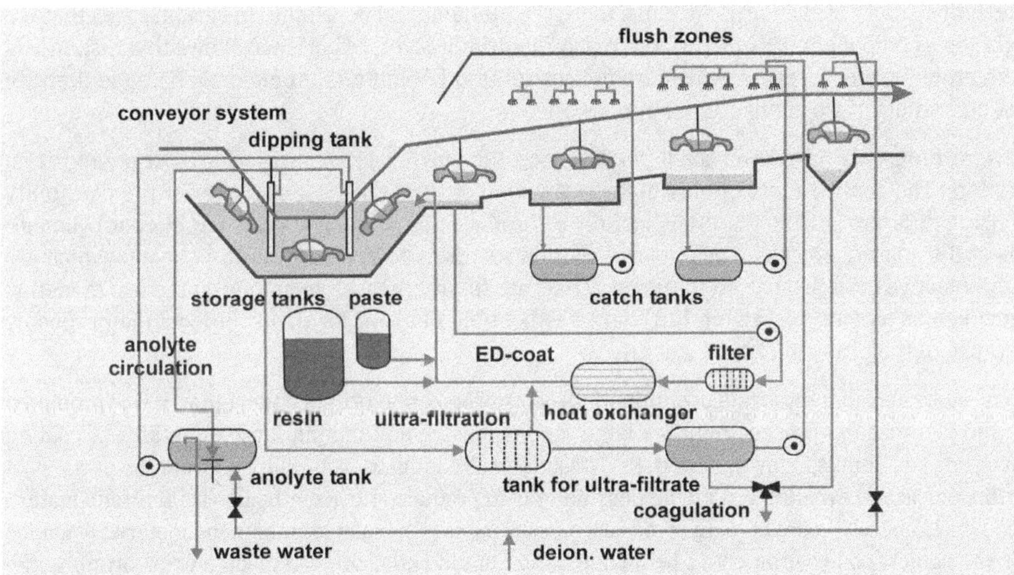

Figure 3.4.13: Process and equipment involved in electro priming

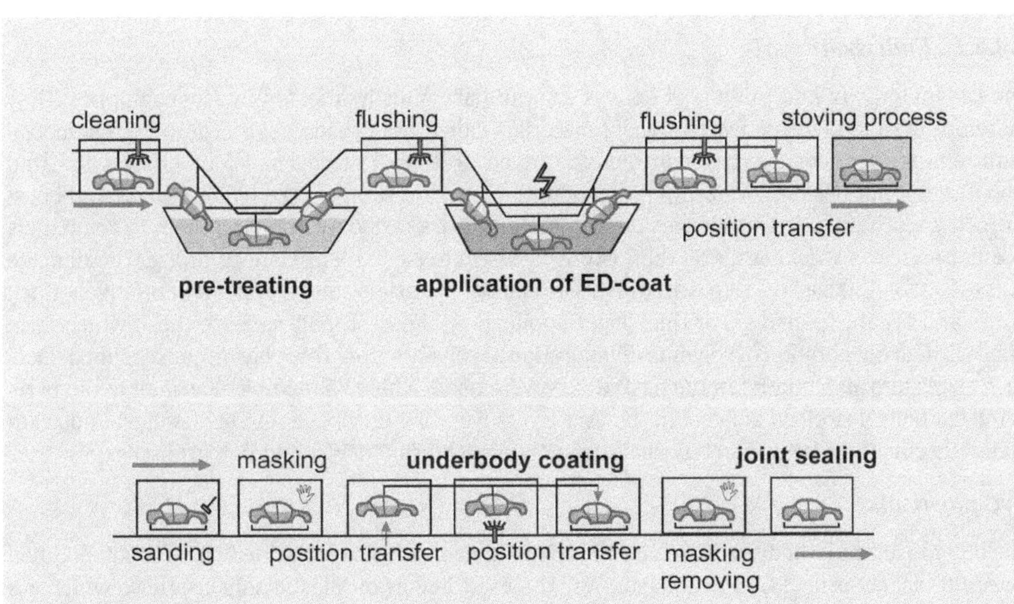

Figure 3.4.14: Production line for pre-treating, electro priming, and underbody protection

3.4.7 Prospects for electro deposition primers

Studies of electro deposition primers represent the bulk of the development activities in this section of the paint industry. Numerous patents have been issued for electro deposition primers, many more than in most other fields of coating technology.

Currently, systems are being introduced onto the market which are lead-free and contain less than 1% by weight volatile organic compounds (VOCs). There are also primers which avoid the use of tin catalysts. The issue of thick-layer electro deposition primers is only of interest where the car producer dispenses with primer surfacers. In those cases, it is additionally important to improve the light stability of such primers since there is a fundamental possibility in certain cases that UV light (as a component of sunlight) will reach the primer layer. Experience shows that dispensing with aromatic epoxy resins, which are not resistant to UV light, is problematic because they are the guarantors of excellent corrosion resistance.

Several proposals have been made for lowering the stoving temperature for electro deposition primers. The goal here is not only to lower energy consumption, but also to offer the possibility of using different sealing materials, mainly prior to electro priming, as this might greatly benefit the entire coating process. The obvious solution seems to be to select blocking agents which are more reactive than those used until now. Trials are in progress, for example, on the use of methyl ethyl ketone as blocking agent. But there has been no demand for this solution so far, probably because of the problem of tank stability.

Some years ago, there was talk of eliminating electro deposition primers altogether. It was prompted by the fact that the process requires large and expensive equipment, large quantities of coating materials and entails many quality risks. The automotive industry is mulling alternatives as a way of cutting costs and avoiding risks. One possibility is to produce entire car bodies from plastic materials. This is already being done in some cases (see Chapter 7). If steel remains the material of choice for car bodies, coil-coating could be a candidate. Should coil-coating replace electro priming, the basic materials for the production of car bodies would be pre-primed metals (see also Chapter 7).

3.4.8 Underbody seals and seam sealants

3.4.8.1 Underbody seals

The car underbody and the wheel houses are the parts mostly affected by stonechipping. They therefore have to be coated with special materials called underbody seals. There are no special requirements here concerning an attractive surface structure in terms of colour, effect, gloss, and smoothness. But besides stonechip protection, the seals must offer excellent adhesion and boost corrosion resistance. The properties of such materials are tested by applying them to test panels in a defined layer thickness, and then exposing them to permanent stonechipping. The damage is assessed on the basis of exposure time. The chosen materials can be applied in relatively thick layers and are inexpensive. For thick-layer application, the best materials are those which form films without evaporation of solvents. Evaporation of solvents from thick layers causes film defects such as blisters and craters. In the past, waxes were used, which were applied as melt to the parts of the car body described above. But the wax layers were not durable enough. The most important materials currently for underbody seals are polyvinylchloride (PVC) and polyurethanes.

PVC plastisols

The binders for underbody seals consist of a dispersion of polyvinylchloride (which is not modified by additional chlorine, PCU) in a plasticiser. The most important plasticisers are those which are not saponifiable, and, because of their lower cost, esters of phthalic acid and adipic acid. The PVC is produced by emulsion polymerisation in water. The molar masses of the polymer are 30,000 to

100,000 g/mol. The primary particles of the dispersions are 0.1 to 2.0 µm. The PVC dispersion is spray-dried to yield a particle size of 10 to 20 µm. Those particles are dispersed in the plasticiser to form a plastisol (an extremely fine distribution of polymer particles in a plasticiser).

The underbody coating materials contain plastisols and coloured pigments and extenders. The colours are usually grey. The pigments and extenders are dispersed in the PVC dispersion. Film forming takes place at elevated temperatures. In most cases, stoving is carried out along with that of the primers or primer surfacers. During stoving, the PVC particles soften and dissolve in the plasticiser to form a homogeneous film matrix. After cooling, the resultant coating layers are resistant to stonechipping, moisture, and chemicals. No volatile organic compounds are omitted during film formation. Film thicknesses of approx. 600 µm (in some cases up to 1 mm) are attainable.

Some criticism surrounds the use of PVC. At high temperatures - e.g. in a fire - PVC can decompose into hazardous chlorine compounds.

Polyurethanes

So-called 100 % materials can be formulated from liquid binders that contain free hydroxyl groups, and liquid polyisocyanates. The liquid binders are mainly polyether polyols and castor oil. The liquid polyisocyanates, which are the crosslinkers, are polyfunctional derivatives of MDI (methylene diphenyl isocyanate or 4,4'-diisocyanato diphenylmethane). The binders containing hydroxyl groups serve to disperse the pigment. Since polyisocyanates react at room temperatures, both parts are supplied separately and mixed just prior to application. Application is therefore performed with a two-component spray gun. The products yield highly flexible films. As the films consist of crosslinked molecular networks, their resistance to solvents and chemicals are much better than those of the PVC systems. The required stonechip resistance can be achieved at a much lower film thickness. This offers the possibility of weight savings. However, polyurethane systems are more expensive than PVC systems.

Alternatives

Some car bodies do not need underbody seals. The underbody has a plastic cover integrated into the subassembly.

3.4.8.2 Seam sealing

The sealants used for car bodies are thermosetting adhesives and sealants [80]. They are used for door folds and joints, roof tops and other parts of the body. They also find use in acoustic insulation layers. In addition, they serve to boost the rigidity of body parts.

The materials used during car body construction (before the coating process) are reactive adhesives which mainly consist of epoxy resins and amine hardeners. Materials used alongside coatings are PVC plastisols, polyacrylate plastisols, and rubber compounds. Sometimes the sealant systems may be applied manually, but automatic application equipment are also used. Gelation and solidification of the sealants occurs in the stoving ovens for electro deposition primers or primer surfacers. The rubber systems vulcanise under these conditions.

A special type of sealing is that used in windscreen frame sealers. When car windows are incorporated into the car body, they are fixed in place with a special sealant material. These sealers contain polyurethane compounds that solidify by reaction with atmospheric moisture. It is essential that these materials adhere well to the topcoat or clearcoat layer and, of course, to the window glass (see Chapter 3.8.2.6).

3.5 Primer surfacers

3.5.1 Development of primer surfacers

The second layer usually applied after the primer is the primer surfacer or filler. These two names reflect the fact that the second layer has to cover the surface structure of the primer and to provide a degree of "filling". The concept of filling is very difficult to define and the influencing factors are described in more detail in Chapter 3.5.2. First, filling power is a subjective visual impression of a coating system that forms a layer of adequate thickness and attractiveness on the surface of a substrate. Some of the requirements are fulfilled when the layer covers the structure of the substrate or the underlying layer effectively. To achieve that, the primer surfacer should form films with excellent levelling and smoothness. Smooth surfaces are achieved very simply by sanding a coating layer. In the past, primer surfacers were sanded extensively to create optimum application conditions for topcoats. The main goal of primer surfacers at that time was to provide a combination of optimum levelling and sandability after film forming. The primer surfacers contained resins which achieved the goals of optimum pigment wetting and high hardness by physical drying and crosslinking. Those primer surfacers had relatively high pigment volume concentrations (PVC) and the most common colour for primer surfacers is still grey. The high quantities of extenders and pigments are the reason that, as mentioned above, the resins have to achieve optimum pigment wetting. Products used in the past were a combination of alkyd resins, modified by smaller quantities of long-chain, unsaturated fatty acids (e.g. technical grade 9,12-linoleic acid obtained from cottonseed oil, soy oil, sunflower oil, and 9,11-linoleic acid from dehydrogenated castor oil). The crosslinkers were urea resins, which are notable for their high hardness and adhesion. The stoving temperatures for such primer surfacers were 160 to 180 °C. Film thicknesses were 35 to 40 µm. Urea resins were preferred to melamine resins because of their greater stability to overstoving. Besides the aforementioned properties, primer surfacers, both past and present, have to bestow excellent adhesion on the primer and to produce layers that offer adequate intercoat adhesion for topcoats. Since the main demands on primers surfacers were initially "merely" to provide filling and sandability, they were additionally expected to be inexpensive. The ingredients of those primer surfacers, namely alkyd resins, urea resins, and extenders, were relatively cheap. Today, such venerable primer surfacer systems, with their high levels of extenders and the need for extensive sanding, would be classified as spray fillers. When primers had to be sanded (due to coating failure), there was the possibility that the metal surface would be exposed. In that event, the primer surfacer, applied after the primer, had to provide some corrosion protection. Primer surfacers modified so as to support corrosion resistance are defined as **primer fillers**. In most cases, modification consisted in adding epoxy resins and defined quantities of pigments that actively supported corrosion resistance.

For top-of-the-range cars, the primer surfacer was followed by a **pre-coat**. The entire coating layer then consisted of four parts. The goal was to offer a surface with excellent filling power, topcoat holdout, and bright colour effect. For the latter, the hiding power of the topcoat had to be as high as possible. The undercoat was therefore formulated with coloured pigments that would generate shades close to that of the topcoat; it contributed to hiding power and boosted colour brightness. However, for the purpose of providing filling, the resin composition of an undercoat resembled that of primer surfacers, and so too did the stoving conditions. The additional layer required its own application line with spray booth, air blowing and stoving oven. Accordingly, this additional application step was reserved for relatively expensive, prestige cars. With the further development of topcoats and, primarily, the transition to basecoat-clearcoat combinations (see Chapter 3.6.1) the reasons for using undercoats disappeared.

One of the biggest problems to do with the stability of automotive coatings has always been their stonechip resistance. Stonechipping was initially observed in the car underbody. To this day, therefore, car underbodies are protected by an underbody seal (see Chapter 3.4.8.1). The second area of a car body that is often treated with an anti-chip coating is the sill area. A **stonechip protection primer** was applied there in a separate working step (in Germany: stonechip intermediate primer). It contained resins with flexibility akin to that of elastomers. The stonechip protection primer was covered with the same topcoat used for the entire car body. However, the smoothness and appearance in that area were unsatisfactory.

Since the 1980s, two new requirements have been imposed on primer surfacers. First, the need for greater unit output per unit time necessitated optimisation of the coating process. As most of the time and effort were spent on sanding, the sanding process was reduced step by step. Finally, to cut down on personnel, primer surfacers were only sanded to rectify flaws in the coating layer (specks, inclusions).

With the advent of streamlined car bodies (the latest developments being wedge-shaped) and increases in the average speed of cars, stonechipping can now occur anywhere on the car body, and not just on the underbody and sill parts. As the appearance of the stonechip protection primer was not good enough for use on the entire car, there arose a need for primer surfacers that would protect the whole car against stonechipping, while still providing filling and smoothness for topcoat systems. Finally, variations in the composition of the old primer surfacers led to so-called stonechip primer surfacers, which met the two requirements of protecting the entire coating system against mechanical impact and contributing to optimum appearance. Apart from the specific pigment composition, it was mainly the resin system which was modified to this end. The urea resins which had served as crosslinkers in the earlier primer formulations were replaced first by melamine resins and later by a combination of melamine resins and blocked polyisocyanates. The short oil alkyds were replaced by saturated polyesters containing building blocks that could provide flexibility. Some products additionally contained polyurethanes that provided excellent elastomeric properties, too. To highlight the new properties of these products, they were dubbed stonechip primer surfacers. Since the ingredients of these stonechip primer surfacers (saturated polyester, blocked polyisocyanate, melamine resin, and polyurethane) were all more expensive than those of the earlier primer surfacers, they too were more expensive. Nonetheless, the extra requirements imposed and the goal of guaranteeing longer coating service life on cars meant that the new generation of primer surfacers soon penetrated the whole market. The result was that there were fewer complaints. Today, nearly all available primer surfacers for automotive coatings meet the demands of stonechip resistance and the other properties described. It is therefore no longer necessary to preface the term automotive OEM primer surfacer with "stonechip".

Where car makers needed very bright colours, which require pigments with effective hiding power, coloured primer surfacers were developed. The requirements were comparable to those of undercoats, but now the primer surfacers themselves contained some coloured pigment in addition to the typical pigments for surfacers. For some automotive coating lines, **coloured primer surfacers** are available in 5 to 10 different shades. They are chosen for their ability to support the colour brightness of the different topcoats.

Some car producers called for primer surfacers that could form films and crosslink effectively at **lower temperatures** (130 to 140 °C) than the common primer surfacers (160 to 175 °C). The idea was not only to save energy costs, but also to facilitate car designs made of heat-sensitive materials.

Conventional solvent-borne primer surfacers have solids contents of 55 to 60 % by weight. In line with general calls on the coatings industry to reduce the quantities of volatile organic compounds (VOCs) employed during coating, solids content were raised to 65 to 70 % by weight. These **high-solid primer surfacers** contained special resin systems (see Chapter 3.5.5).

The next step in the reduction of solvent emissions was the development of **water-borne primer surfacers** (see Chapter 3.5.6).

Totally solvent-free primer surfacers are available in the form of **powder primer surfacers** (see Chapter 3.5.7), but powder systems are not very widespread in the automotive coatings market.

In contrast to the aforementioned principle of spreading the requisite properties of the automotive coating system across different layers, there have been several attempts to avoid the use of primer surfacers, mainly in the USA. Of course, that would yield not only cost savings for coating materials, but also reduce the installation and operation costs of application equipment, including man power and energy. It would also mean a reduction in VOC emissions. However, some of these savings would be lost again if the elimination of primer surfacers necessitated greater outlay on the application of the other coating layers, e.g. the application of a thick-layer electro deposition primer, perhaps in two dipping tanks (see Chapter 3.4.3). Where the requirements are for filling power, topcoat holdout, stonechip resistance, and resistance to weathering and chemicals, it makes sense to use primer surfacers, especially for medium-range and top-of-the-range cars, and mainly in Europe.

Another attempt to save material and application costs and additionally to reduce the weight of the entire car body was the introduction of thin-layer primer surfacers. The film thicknesses (dry film) of these primer surfacers were 18 to 22 μm, compared with 35 to 45 μm of the conventional primer surfacers. The requisite stonechip resistance was obtained by means of resins of defined elastomeric character. However, when primer surfacers are applied at such low film thicknesses, it is more difficult to provide adequate filling power and topcoat holdout. To highlight the difference between such materials and primer surfacers, they were called **functional layers**. From the outset, functional layer materials were designed to be water-borne resins.

It made sense to call them functional layers as they were expected to meet additional requirements. Water-borne solid-colour basecoats suffer from a problem (see Chapter 3.6): they need be applied in a minimum layer thickness in order to provide adequate hiding power and colour brightness. However, it is very difficult, especially in the case of water-borne solid colour basecoats, to apply high film thicknesses without flaws such as sagging and popping. Layer thickness was limited by these application problems. Consequently, in comparison to the use of coloured primer surfacers, the functional layers were pigmented to match the following basecoat colour as closely as possible to improve the hiding power and colour brightness of the entire coating system. Not only that, but they still had to provide excellent stonechip resistance.

From functional layers, it was only a small step to the **integrated coating concept.** In this, the entire coating structure consists of electro deposition primer, two basecoat layers (made from water-borne products), and the clearcoat. The first basecoat layer acts as a primer surfacer (stonechip resistance, intercoat adhesion), but also contains pigments to support the formation of an aesthetic colour. The second basecoat is the main source of the effect and the colour. Two basecoats are a precondition for achieving optimum brightness and hiding power. Since most developments must also have a rationalization step before they can be become introduced into the market, the two basecoats are applied without separate drying processes. They are applied by wet-on-wet application of three layers. Application of the first basecoat is followed by a brief flash-off and then the second basecoat is applied. After that has been flashed-off, the clearcoat is applied and then film forming for all three layers is effected in the oven [4]. Some producers are seeking to take the development a step further. They conceive of producing two basecoats from one primer formulation [81].

3.5.2 Requirements and properties

The initial demands on primer surfacers, **covering of structure** on the surface of the primer and providing **filling** and **topcoat holdout** are still in force and are being observed more closely. The requirements are closely connected to the **application behaviour**. Primer surfacers must form flawless films with optimum flow and levelling under the given application conditions (line speed, electrostatic spraying equipment, flash-off time and temperature, air flow velocity, oven length and temperature). As primer surfacers are intermediate layers, they must provide excellent **wetting** and **adhesion**, much more so than other layer materials. Although both these properties are important for all other layers, they are described here. Last but not least, primer surfacers are layers that have to compensate or absorb the mechanical impact, primarily **stonechipping**, on the entire coating system.

Structure covering

Although the electro deposition primer has to provide flow and levelling, the deposition process chiefly generates a specific structure. The reason is that, on the edges of surface structures (metal panels), the electrical field density is a maximum, which leads to a high deposition yield. The first goal of the primer surfacer is to cover such structures perfectly. The second goal is to produce smooth surfaces during film forming. This is essential if the following topcoats are also to form smooth layers. The third goal is to generate a layer that is influenced not by the application of the topcoat (holdout), but instead only by wetting. Whereas smoothness is related to flow, low viscosity, and softness, holdout is related to hardness and resistance. Primer surfacer formulations have to achieve a balance of both.

Filling power

Paints of the same colour and layer thickness but different surface structure can convey totally different impressions. They can create the impression that the various layers have different thicknesses. Therefore filling power depends not only on the thickness, but also on the surface structure. Surprisingly, a totally smooth surface does not create the impression of high filling power. Nor does a layer that reproduces the structure of the object below the surface (the structure is said to be telegraphed). Furthermore, a surface structure with very short waves does not permit high filling. The impression of optimum filling is conveyed by small quantities of a long wave structure. However, too much waviness conveys a negative impression. Such structures give rise to orange peel. This description shows that it is very difficult to define the value of filling power accurately in terms of physical parameters.

The way for primer surfacers to provide optimum filling power is for them to cover the structure of the electro deposition primer very well, and to form their own structure with at most a small quantity of long waves and virtually no short waves. Primer surfacers contain ingredients not only designed for excellent flow and levelling but also for generating structural viscosity during the film forming process with a view to forming their own structure. After the stoving process, the surface of the primer surfacer has to be wetted by the following topcoat, but must be so dense and resistant that there is no *strike-in* by the topcoat. Conversely, a positive definition would be that the primer surfacer has to offer excellent *hold-out* of the topcoat.

In the past, since the primer surfacers were sanded extensively, the surface had to be relatively hard process and matt. Nowadays, the surfaces of primer surfacers are relatively glossy; the gloss values are 60 to 90 % (surface light reflection at a 60° angle). First and foremost, gloss depends on the pigment volume concentration (PVC). However, extenders with very small particles and those with platelet-like particles can reduce the gloss in low quantities. Matt surfaces have a very fine structure that reflects the light diffusely. With matt surfaces, it is impossible to recognize the effect of levelling, but, with glossy surfaces, the human eye is sensitive enough to perceive different surface structures.

Wetting and adhesion

Surface tension is responsible for interactions at interfaces. Unlike molecules of gas, the molecules of liquids and solids occupy a defined volume. The reason is that these molecules associate more by intermolecular forces. The effect of these intermolecular forces is called cohesion. However, on the surface of liquids and solid substances, cohesive forces are not used for interaction with the molecules of air. Instead, they increase the level of interaction with neighbouring molecules. Although the molecules in liquids can move past each other, it takes additional energy to create a new surface. The energy (work) needed for this compensates for the additional energy between surface molecules. That energy is surface tension. The underlying principle is described schematically in Figure 3.5.1.

Surface tension is the reason that liquids form droplets. Water has a relatively high surface tension. That is why some insects can run on its surface. In principle, the surface tension is a specific interfacial energy between liquids and air. Interactions also exist between liquids and solids. Those interactions are responsible for wetting and capillary forces.

Surface wetting

The surface tension of liquids is the reason that droplets of different shapes form on surfaces. The contact angle (wetting angle) of the various droplets is a measure of the surface tension. Liquids of low surface tension create flat droplets with small wetting angles. Increasing surface tension generates droplets with increasing wetting angles until eventually a hemisphere is formed at a wetting angle of 90°. Liquids of very high surface tension form droplets with wetting angles greater than 90°. The area in contact with the substrate is then relatively small (see Figure 3.5.2 for schematic description of droplets of different wetting angles).

The first precondition for a liquid to wet a surface is a low surface tension at the interface with air. In addition, wetting depends on the interaction of the surface of the liquid with the surface of the object (interfacial tension). The principle of interfacial tension is similar to that of solubility.

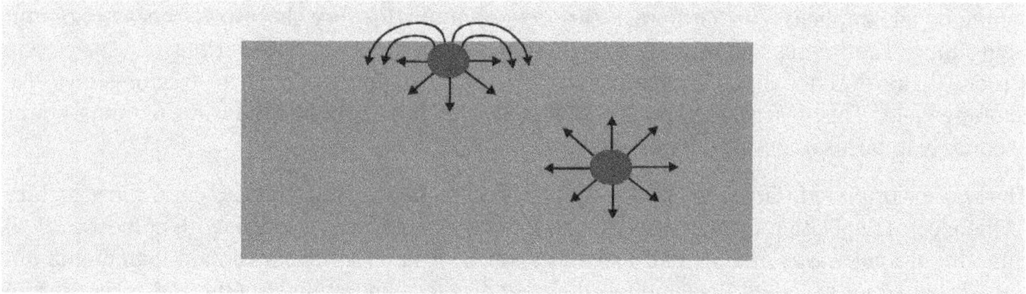

Figure 3.5.1: Principles behind cohesion and surface tension

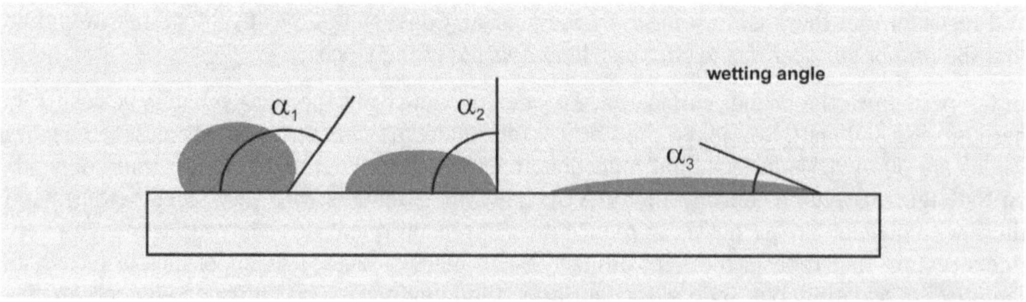

Figure 3.5.2: Droplets of different wetting angle on a substrate

Materials of comparable molecular structure - at any rate, of comparable polarity - improve the interactions of their surfaces. Therefore, wetting efficiency depends on the degree of surface interaction (affinity). Surface structure, too, exerts a significant influence on wetting properties. A surface that has a high affinity for a liquid and high roughness is wetted much more readily than one with a smooth surface. However, a rough substrate with less affinity for a liquid is more repulsive than the same object with a smooth surface (see Chapter 3.8.3.10).

Adhesion

Adhesion is closely related to wetting. The extent of adhesion is defined by measuring the force per unit area that is needed to remove a layer of film from the substrate. After application of paint material and wetting of the substrate, film forming takes place through evaporation of solvents or dispersing agents and, in most cases, by crosslinking reactions. Both of these processes can change the volume and density of the paint material. The impression is that wetting is the generation of physical bonds between paint material and a substrate. For optimum adhesion, it is important that these physical interactions be rendered permanent during film forming (drying by evaporation, crosslinking, e.g. in a stoving process), even if the density changes. It is also conceivable that the physical interactions of wetting are replaced by other types of physical bonding.

A number of debates seek to define the specific nature of the physical interactions needed for adhesion. Very appealing is the explanation that polar groups from the film matrix, e.g. carbonyl groups, interact physically with electrophilic compounds of the substrate surface, e.g. metal cations. Such physical interactions could approximate those of complex chemical bonds. In keeping with this, it should then be possible for other polar groups, such as hydroxyl, ether, carboxyl, amide, and urethane groups to form such physical bonds with suitable partner groups on the substrate. The totality of all those physical bonds would be what we term adhesion. Although this seems to be a plausible explanation, it fails to adequately describe all adhesion effects. Special materials are needed for providing good adhesion to very non-polar surfaces. A prime example here is polypropylene plastic parts. With one exception, no paint materials (polymers) will adhere very well to unmodified and untreated polypropylene. The exception is highly chlorinated polyolefines, which adhere excellently to unmodified polypropylene. This cannot be explained away in terms of an interaction between non-polar molecules as there are polymers which are even less polar than chlorinated polyolefines. It has been proposed that the main reason for adhesion is that the molecular moieties are able to get close to each other. It may be that van der Waals forces of attraction play the most important role in adhesion, and that all the effects of polar groups and others are only part of the overall phenomenon. In this regard, the influence of crosslinking on adhesion was studied. If a layer consists of a film matrix with a high crosslinking density, adhesion of the following layer is poor. The model for describing this is that, during application, the paint material has to diffuse partly into the previous layer in order to effect adhesion. Some swelling of the first layer may take place, but the layer of high crosslinking density is unable to swell and is incapable of molecular interactions. In such cases, it is possible to improve adhesion by incorporating very strong solvents into the second paint material that are capable of swelling the first layer. It has also been established that excellent adhesion to the first layer is obtained if the second coating layer is optimally crosslinked. A model will now be proposed to explain this effect. As the source of adhesion is the physical interaction between specific molecular groups on the second layer with the surface of the substrate, the type of network behind those groups is important. If the crosslinking is relatively dense, but the molecular network is not large, the number of adhesion groups per molecular area is low. Such networks do not adhere very well and, in addition, they are relatively brittle. The reason that the crosslinking density is high in the case of the smaller molecular flakes described in Chapter 2.3.2. However, where molecular networks extend over large areas, those

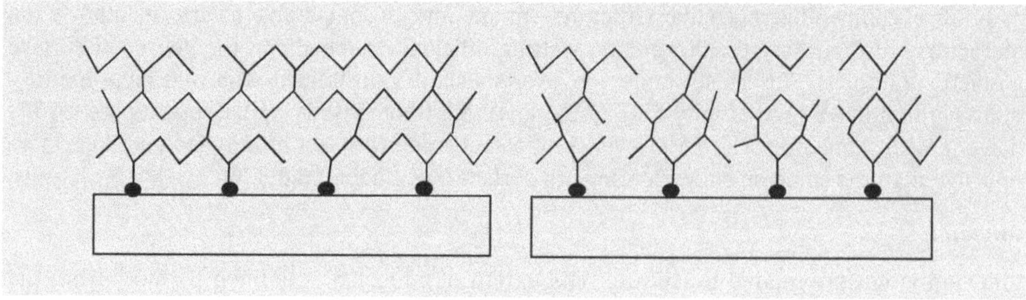

Figure 3.5.3: Model of adhesion as a reason of crosslinking

parts of the network are connected by more adhesive groups, and so adhesion is optimal. The larger network also benefits other properties (elasticity, resistance to solvents and chemicals). A simplified version of the model presented here is shown in Figure 3.5.3.

Some authors have noted that the best intercoat adhesion is observed when there are covalent bonds between the two layers. It is correct to presume that crosslinked molecular networks contain residual functional groups and some uncrosslinked molecules whose molar masses are lower than average. However, it is rather improbable that such residual functional groups or low-molecular fractions will react chemically with partner groups in the other coating layer. There are kinetics reasons preventing this. Even if some interfacial chemical bonds were to be formed, they would not make any major contribution to intercoat adhesion as the forming of such bonds is relatively improbable.

In this connection, it must be said that films swelled by solvents, water or chemicals generally exhibit significantly less adhesion. It is believed that swollen films are much more sensitive to mechanical impact than non-swollen films and that their molecular network is much easier to destroy. This may also be the reason for adhesion loss (see the model described above). Adhesion is tested by applying mechanical forces to the coating layer that may lead to deformation and destruction. If swelling of the film matrix is an additional risk, the tests are expanded to include the wet-adhesion test.

The intercoat adhesion of topcoat films on primer surfacers can be improved by selecting special pigments. Very finely dispersed pigments or those with a platelet-particle structure lead to better adhesion. These pigment types influence rheology, levelling and generate special surface structures that can improve the adhesion of the next layer to be applied. It is also possible that parts of such pigment particles project out of the film surface to act as an anchor for the next layer. There are restrictions on the use of such pigments when they are added to the composition, because their effect and quantity must not impair levelling, filling power, topcoat holdout, or gloss.

Precise physical tests for measuring film adhesion are difficult and expensive. For example, the pull-off test, in which a two-layer combination in the form of a free piece of film is placed between two punches which are pulled until the adhesion fails, is problematic. First, it is difficult to prepare representative free film pieces. Second, the adhesive on the punches may influence the film properties and it may exhibit poorer adhesion than the films. Therefore, more practicable methods of measuring adhesion are chosen. The most popular is the cross-cut test in which several parallel cuts are made in the paint at a distance of 1 or 2 mm apart, the distance depending on the layer thickness, and then at right angles across them. The cuts form a series of small squares. The cut part is carefully covered with adhesive tape, which is then pulled off rapidly. The measure of adhesion is the amount of damage done to the squares in the cross-cut. The number of removed film squares can be expressed as a percentage. The structure of the cut edges also provides

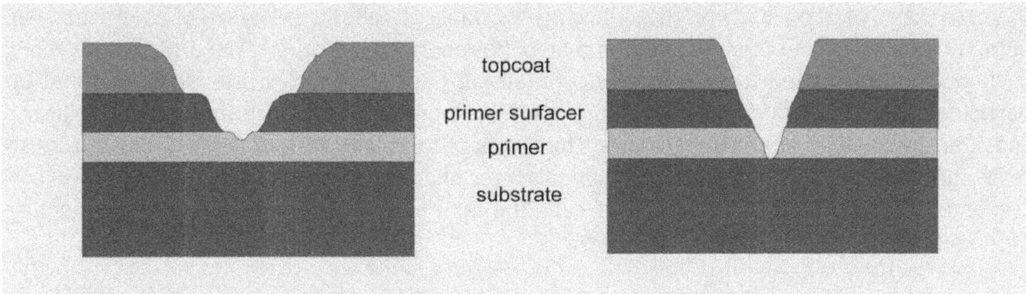

Figure 3.5.4: Different types of stonechipping

information about the quality of adhesion. The test is fairly subjective, but nonetheless highly compelling. It must be remembered that the result of the cross-cut is influenced not only by the adhesion but also by flexibility of the film.

Stonechip resistance

As mentioned above, one of the most important requirements on primer surfacers today is an ability to absorb mechanical impact on the entire coating system. Developers responded to this with stonechip primer surfacers (a better term would be stonechip-resistant primer surfacers), both solvent-borne and water-borne. Virtually all car makers employ these types of primer surfacer on their car bodies. Stonechip resistance is influenced by the properties of adhesion and flexibility. Of course, if a stone hits the surface of a speeding car, it will cause damage; that cannot be avoided. However, as far as the risk of corrosion is concerned, it is important that the damage be minimal. Different criteria are employed for assessing the damage. Some coating customers stipulate very slight damage, while allowing the damage to extend as far as the substrate. Others specify that the different layers may be compromised but that the primer, or, better still, the primer and the primer surfacer layer, must remain intact. Accordingly, customers employ various methods to assess stonechip resistance.

One method simulates stonechipping by blasting the surface of the coating system with a defined quantity of real stones, steel scrap, screws, and the like. These are propelled from a compressed-air gun or are dropped from a pipe of defined diameter and a specified height. The damage is judged on the basis of amount of material lost and the size of the damaged areas [82].

The other group of methods employs single shots. These may be falling steel balls or wedge shapes [83], or a small steel ball shot from an air-gun onto the surface of the coating layers [84]. The tests are carried out at different temperatures [85] because stonechipping is more problematic at low temperatures than at ambient. The assessment is based on the amount of material removed and the damage done. A crucial factor is whether only the upper layers are damaged over a more or less extended area or whether the damage penetrates the entire system down to the substrate (steel) over a very small area. The different types of damage are shown schematically in Figure 3.5.4.

Damage down to the metal surface poses a corrosion risk while large damaged areas in the topcoat are more conspicuous. Of course, it is best to avoid both. As mentioned above, the different automakers evaluate damage differently. The producers of primer surfacers therefore have to adapt their formulations to meet the various demands.

Why do paint layers absorb or cushion mechanical impact? Two reasons have been advanced. First, the film matrix absorbs the impact, which is transformed into kinetic and thermal energy. This occurs when the resins of the film matrix are of a pronounced elastomeric nature. However, such resins might not provide excellent filling power and topcoat holdout. It is therefore necessary to find a compromise. Impact energy can also be absorbed at interfaces. The type of pigmentation therefore has a marked effect on the stonechip resistance. It takes a great deal of energy to move pigment particles about in a flexible film matrix (for energy dissipation). The difficulty becomes particularly pronounced when extenders containing platelet particles are used that build up particle-particle interactions. To explain the effect of pigmentation in primer surfacer layers, consider the analogy of an asphalt road. The layers of the road must be flexible, but must also resist mechanical load. They are made from mixtures of gravel, sand, and asphalt. The gravel corresponds to the extenders in primer surfacers, the sand to the titanium dioxide, and the asphalt to the resin matrix (see Chapter 3.5.3 below).

3.5.3 Composition of solvent-borne primer surfacers

Whereas the earlier primer surfacers consisted of alkyds and urea resins, current solvent-borne primer surfacers contain a resin made up of saturated polyesters bearing hydroxyl groups, and melamine resins as crosslinker. Improved versions of such primer surfacers contain at least some blocked polyisocyanates as crosslinker. To boost corrosion resistance in addition to the other properties, primer surfacers may also contain some epoxy resins (see Chapter 3.4.5.1). As already indicated, the composition of pigments is crucial for optimum stonechip resistance and adhesion. In most cases, combinations of titanium dioxide and extender are chosen and, in addition, some extenders with platelet particles. Such extenders are rheologically efficient and support resistance to mechanical impact. Although it is important for primer surfacer to form smooth film surfaces, the use of levelling agents is problematic.

3.5.3.1 Saturated polyesters

Saturated polyesters [86-88] consist of polycarboxylic acids and polyols. Thus, they are in principle the simplest type of polyesters. The definition given in DIN 55958 is not very enlightening: *"Polyester resins are synthetic binders based on polyesters, whose structures bear ester groups in the molecular chain."* While saturated polyesters may be the simplest type of polyesters, it took a long time for them to be used in substantial quantities in the preparation of coatings. The first to be used were alkyd resins, i.e. polyesters modified with oil or fatty acids, which were widely used in the coatings sector. These were followed in the 1950s by unsaturated polyesters. The reason that saturated polyesters were not used more widely was that - as long as they consisted of simple diols - their solubility and compatibility with other resins were poor. However, with the advent of new raw materials for polyesters, saturated polyester resins started to find greater application in the coatings market. Then the new molecular building blocks introduced solubility and compatibility in saturated polyesters for paints. The crucial step was the introduction of neopentyl glycol (2,2-dimethyl-1,3-propanol) onto the market. To differentiate these polyesters from the widespread unsaturated polyesters and alkyd resins, they were called saturated polyesters. Today, unsaturated polyesters are only used for special applications (repair putties, see Chapter 4.3) and the use of alkyd resins is falling. However, saturated polyesters are growing strongly.

Composition of saturated polyesters

The most important polycarboxylic acids for saturated polyesters are aromatic. The dicarboxylic acids or their derivatives employed are isophthalic acid (contained in nearly all common saturated polyesters for coatings applications), phthalic anhydride, and terephthalic acid (often in the form of its derivative dimethyl terephthalate). The derivatives of a tricarboxylic acid, trimellitic anhydride, is used mainly for water-borne polyester systems (see Chapter 3.5.6.1). Aromatic polycarboxylic acids and their derivatives bestow hardness and resistance on coating layers made with polyesters thereof. However, they limit the properties of solubility, compatibility and flexibility. To compensate, the aromatic polycarboxylic acids or derivatives are mixed with aliphatic polycarboxylic acids. These would be adipic acid and, less often, azelaic acid, sebacic acid, and dimer fatty acids. More important for generating solubility are the polyols. The polyols in saturated polyesters for coatings are blends of different diols and some polyols of higher functionality. Suitable blends of diols introduce solubility, compatibility, and flexibility. The quantity and length of the diol side chains greatly influence the solubility and compatibility. Flexibility is determined by the length of the diol molecular chains; short chains confer hardness and less solubility and long chains, plasticity. Plasticity is also provided by diols bearing ether groups. Optimum solubility, hardness and some flexibility is achieved with cycloaliphatic diols. Polyester molecules having uniform structures tend to be crystalline and do not yield clear solutions. Such polyester systems therefore adversely affect levelling, smoothness, gloss and resistance. Any blend of diols will suppress crystallisation. Suitable diols for saturated polyesters are ethylene glycol, propylene glycol, the ether diols thereof, 1,4-butane diol, neopentyl glycol, 1,6-hexane diol, hydroxypivalic acid neopentyl glycol ester (HPN), dimethylol cyclohexane (cyclohexane dimethanol), 4,4'-bis-(hydroxycyclohexyl)-propane (perhydro bisphenol A). Ethoxylation of bisphenol A with ethylene or propylene oxide yields building blocks that can introduce aromatic polyol compounds into polyester molecules. The content of higher-functional polyols determines the degree of branching among polyesters. The most commonly employed triol for this is trimethylol propane.

Structure of saturated polyesters

Saturated polyesters employed in automotive coatings are crosslinkable. The functional groups for crosslinking are hydroxyl groups. The only exception is the polyesters used for powder coatings - they contain crosslinkable carboxyl groups. The hydroxy polyesters are more or less branched molecules. They have number average molecular weights of 800 to 4000 g/mol. The molecular weights are dictated by two parameters: the molecular ratio of polyols to polycarboxylic acid, and the degree of condensation. The higher the molecular weight and the higher the degree of branching the broader is the molecular weight distribution. Highly branched polyesters can therefore only be realized with relatively small average molecular weights (number average). Attempts to prepare such polyesters with high molecular weights lead to gelation [86]. That means, due to the broad molecular weight distribution, some molecules strive to become infinite in size, although the average molecular sizes are still relatively low [89]. Nevertheless, the molecular weight distribution is always narrower than the results suggested by statistical calculations. The reason is the balance maintained between polyester molecules of different size by transesterification reactions, which generate more molecules of average size. Most theoretical calculations only take account of the esterification reaction.

Polyesters for primer surfacers

Polyesters for primer surfacers are crosslinked amino resins and blocked polyisocyanates. For the crosslinking reaction, they have hydroxyl values of 80 to 140 mg KOH/g. The acid values are 10 to 20 mg KOH/g. If the quantity of hydroxyl and acid groups of polyesters is measured by titration against sodium hydroxide solution, the values are defined by that quantity of potassium hydroxide (KOH in mg) which is equivalent to the hydroxyl or acid groups contained in 1 g polyester resin. To illustrate: if a polyester has an average molecular weight of 2000 g/mol and a hydroxyl value of 112 mg KOH/g, each average polyester molecule contains 4 free hydroxyl groups. The acid groups do not participate in the aforementioned crosslinking reactions, but they catalyse the reaction between the hydroxyl groups and the functional groups of amino resins.

Since polyesters for primer surfacers must contribute to stonechip resistance, they contain sufficient quantities of building blocks that confer flexibility. However, the polyesters must not be overly soft as that would reduce topcoat holdout. Since primer surfacers are sometimes sanded, the resins must provide adequate hardness to accommodate this. Polyesters for solvent-borne primer surfacers are diluted with high-boiling aromatic hydrocarbons. Sometimes the delivery form contains small quantities of more polar solvents, such as glycol ethers and glycol ether acetates. The solids content of delivery forms are usually between 50 and 70% by weight.

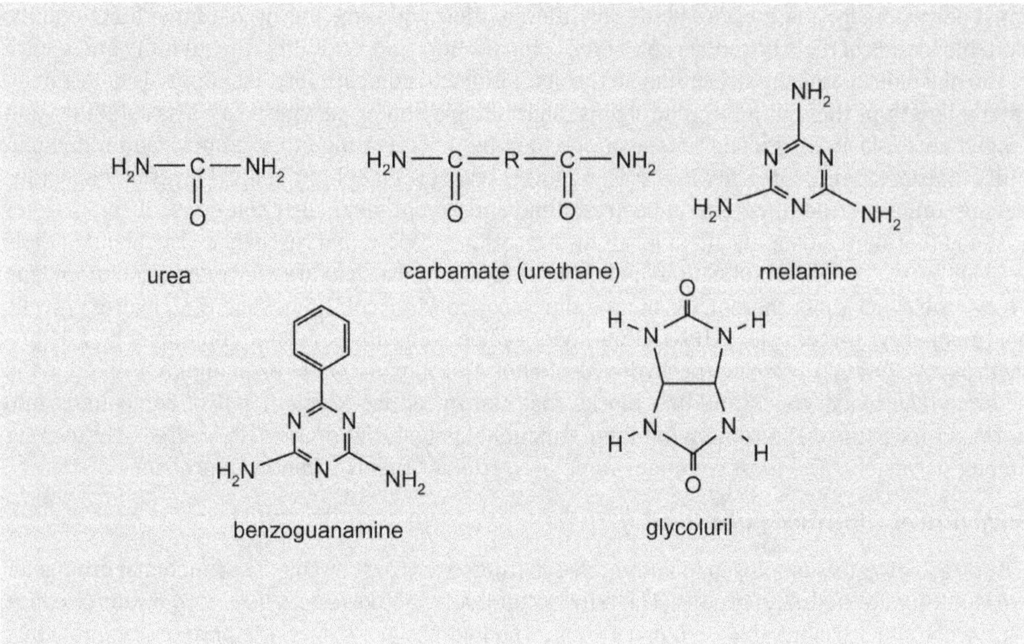

Figure 3.5.5: Basic building blocks of amino resins

Production of polyesters

The aforementioned polyesters are prepared in a melt condensation process at temperatures of 180 to 240 °C. In most cases, the reactors utilise indirect heating systems, multistage controllable stirrers, inert-gas purging, packed columns, condenser, and a separator for reaction water. Small quantities of reflux solvents are employed (e.g. 1 to 3 % xylene). The column prevents loss of polyols, which either boil below the reaction temperature or evaporate along with water vapour. Controlling the increase in temperature and the vapour-separating effect of the column and returning the condensed reflux solvent to the top of the column avoids loss of polyols, with the result that only the reaction water is entrained. The remainder of the process consists in measuring both the acid values to arrive at the degree of condensation and the viscosity to determine the increase in molecular weight. The viscosity is measured on a test solution or in melt at a defined temperature. Once the target values have been reached, the condensation process is stopped by cooling. The delivery form is then prepared, in most cases by adding inert solvents (e.g. aromatic hydrocarbons) which do not contain ester or hydroxyl groups. If the temperature is reduced to a level that esterification or transesterification reactions are very slow, other solvents can be added. Finally, the solution is adjusted to defined solids content for storage or delivery.

3.5.3.2 Amino resins

Structure of amino resins

Amino resins are not the correct chemical term for this resin group. They should properly be called "amido resins", but they never are. The raw materials for amino resins are amides: urea (diiamide of carbonic acid), melamine (triamide of isocyanuric acid or 2,4,6-trisamino-1,3,5-triazine), benzoguanamine (6-phenyl-1,3,5-triazine-2,4-diamine, glycoluril (acetyleneurea), and also some urethanes, e.g. butyl urethane or butanediol diurethane. The structures of these basic building blocks are given in Figure 3.5.5.

On account of the mobile hydrogen atoms in these amides, it is possible to add formaldehyde to afford methylol groups. Both hydrogen atoms of the primary amides can add on formaldehyde. The

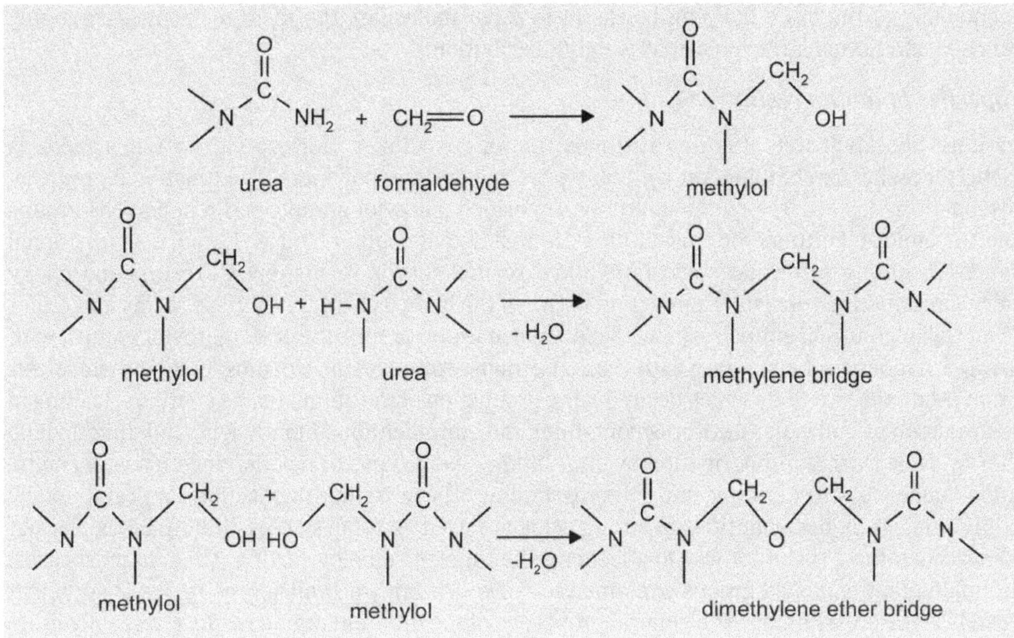

Figure 3.5.6: Formation of methylol groups and bridging elements in the preparation of amino resins

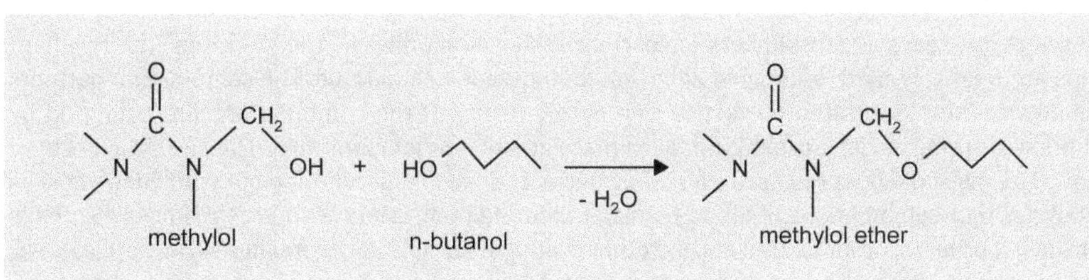

Figure 3.5.7: Etherification to modify amino resins

methylol groups are amphoteric. Mostly in acid media, they split off hydroxyl groups which form water either with the hydrogen atoms of other methylol groups or with the remaining hydrogen groups on nitrogen atoms of amides. The resultant bridging elements are methylene groups or dimethylene ether groups. The formation of methylol groups and two bridging elements is illustrated in Figure 3.5.6.

The resulting amino resins (made from urea and melamine) are highly polar compounds; they are soluble only in water and in lower alcohols, and are incompatible with most potential combination resins. They are used for glues and for laminated panels, but not for the production of coating materials. They must be modified for use in coatings. The modification consists in at least partial etherification of the free methylol groups with monoalcohols, e.g. methanol, n-butanol and i-butanol. The etherification reaction is presented in Figure 3.5.7.

Etherification confers solubility in common solvents for those amino resins as well as compatibility with most potential combination partners (resins) in coating materials. Amino resins etherified with butanols are soluble in alcohols, ketones and esters, and thinnable with aromatic hydrocarbons. Amino resins etherified with methanol are soluble in the aforementioned solvents, but are also soluble in or thinnable with water. The amino resins for coatings have relatively small molecules. They contain on average from one to five amide elements. On account of the high functionality of the basic building blocks (4 to 6 per molecule), the resin molecules are highly branched and have a large molecular weight distribution.

Properties of amino resins

The most important role of amino resins is that of crosslinker. However, a few types serve as plasticiser resins for combination with only physically drying polymers. The functional groups of crosslinker resins are free methylol groups, etherified methylol groups, and residual NH-groups from the amides. Partners for crosslinking are resins that contain hydroxyl groups as functional groups, including saturated polyesters, alkyd resins, acrylic resins, epoxy resins, and epoxy esters. Crosslinking consists in a reaction between the hydroxyl groups of these resins and either the methylol groups of amino resin, with cleavage of water, or the etherified methylol groups, with cleavage of monoalcohols. Along with these reactions, amino resins can crosslink by themselves. In that event, the reactions which occur during production of the amino resins continue, leading to the formation of molecular networks containing the aforementioned methylene and dimethylene ether bridges. During film formation by crosslinking, both types of reaction take place in combination, namely co-crosslinking, and self-crosslinking. The crosslinking reactions are catalysed by adding acids. If higher quantities of strong acids are used as catalyst, crosslinking takes place at room temperature. The other way to effect crosslinking is to employ stoving. Then, in most cases, the small quantity of acid groups contained in polyesters, alkyds, and acrylic resins is sufficient to catalyse the crosslinking. The general properties of the different amino resins are greatly influenced by their basic building blocks.

Urea resins

If urea resins serve as crosslinkers for partner resins containing hydroxyl groups, the resultant films are relatively hard, offer good adhesion, but are not weatherable. The combination partners are alkyd resins, saturated polyesters and epoxy resins. If the combinations are catalysed by addition of strong acids, crosslinking takes place at ambient temperatures. The classic acid-cured coatings consist of alkyd resins, urea resins and, e.g., an alcoholic solution of hydrochloric acid as catalyst. Other combinations of alkyd resins or saturated polyesters with urea resins are suitable for stoving primers, primer surfacers and other coating materials in general industrial applications. In such combinations, the urea resins are preferred to melamine resins as they provide better hardness, adhesion and are more stable to overstoving, but are not weatherable and are less flexible. In the past, urea resins were used instead of melamine resins because they were cheaper.

Melamine resins

Melamine resins are the most important members of the amino group of coating resins. Their molecular structure is determined by the molecular ratio of melamine (M) to formaldehyde (F) to monoalcohol (A). This ratio influences the average molecular weights and the quantities of functional groups. The type of monoalcohol plays a role, too. The highest average molecular weights results from a slight excess of formaldehyde, a low etherification degree, longer condensation reaction time and lower pH values. Melamine resins of relatively low-molecular weights result from a high excess of formaldehyde and monoalcohol (e.g. M : F : A = 1 : 6 : 6). The average number of melamine groups per molecule of melamine resin can be a little higher than one.

Melamine resins are usually produced in aqueous phase. First, an aqueous solution of formaldehyde is combined with melamine in a basic catalyst (alkali hydroxide or an amine) at temperatures of about 80 °C to yield the methylol compounds. Then monoalcohols are added and the reaction mixture is rendered acidic with small quantities of acid catalysts, leading to both molecular extension and etherification of free methylol groups. After that, the water from the formaldehyde solution and from the etherification reactions is distilled off. Finally, the delivery form is prepared by adding solvents (butanols, sometimes along with aromatic hydrocarbons) and adjusting the solids content. Some low-molecular products are delivered in a 100 % form.

The free methylol groups and the etherified methylol groups offer the possibility of reaction with the hydroxyl groups of partner resins (co-crosslinking). In these cases, either water or monoalcohols are cleaved. Etherification proceeds faster than transetherification. The reactions may be accelerated by adding acid catalysts. Co-crosslinking reactions are illustrated in Figure 3.5.8.

Melamine resins additionally react by self-crosslinking. The self-crosslinking reactions are: methylol group with NH group (1st), with cleavage of water; methylol ether group with NH group (2nd), with cleavage of monoalcohol; methylol groups with methylol groups (3rd), with cleavage of water; methylol ether group with methylol group (4th), with cleavage of monoalcohol. Self-crosslinking generates methylene groups and methylene ether groups. The reaction rate decreases in the order given. The self-crosslinking reactions are illustrated in Figure 3.5.9 (page 90).

These self-crosslinking reactions can be accelerated with acid catalysts, in the same manner as co-crosslinking. In strong conditions (strong acids, large quantity of catalyst, high stoving temperatures), some side reactions also occur, such as the cleavage of formaldehyde. Both co-crosslinking and self-crosslinking always occur during film forming of systems containing melamine resins. Co-crosslinking supports flexibility, chemical resistance, and weatherability. Self-crosslinking contributes hardness and solvent resistance. It is therefore important to control the extent of both types of reaction in order that the various demands on properties may be met. Control parameters are: the different reactivities of melamine resin and resins containing hydroxyl groups, the quantities and types of functional groups, the different blending ratios of the resins, the crosslinking temperatures and time, the type and quantity of acid catalyst. The blending ratios for polyesters or alkyds with melamine resins depend on the molecular weights and quantities of the functional groups, but range from 60:40 to 80:20 (in terms of solids).

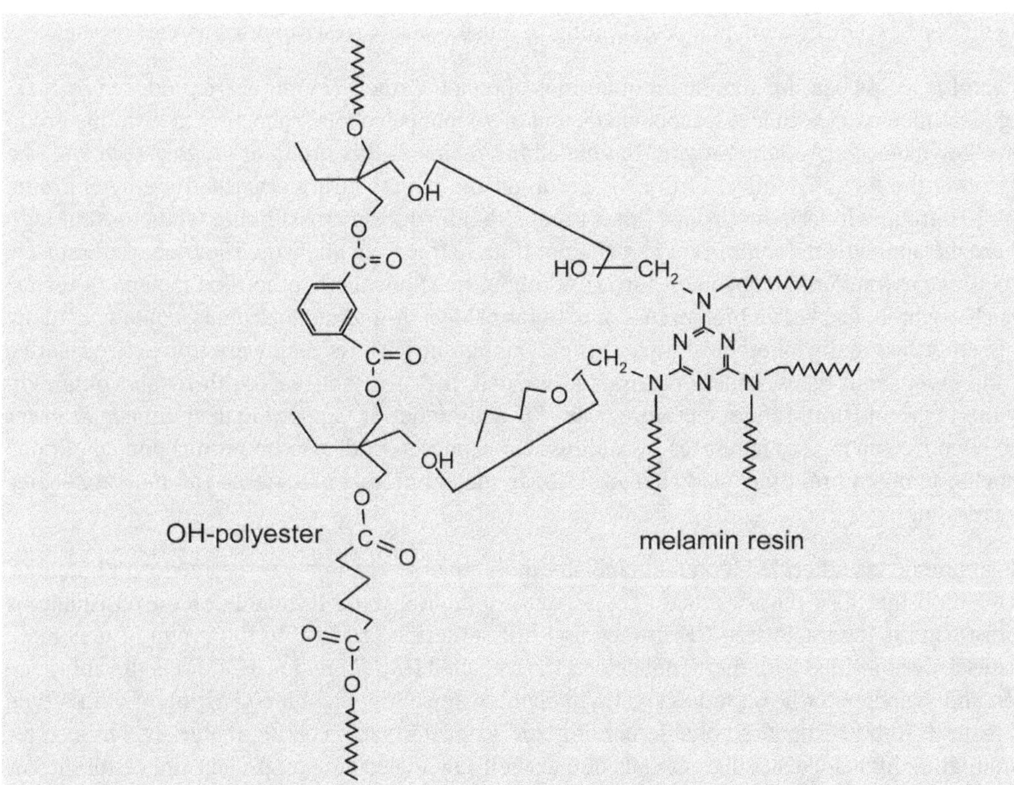

Figure 3.5.8: Possible co-crosslinking reactions

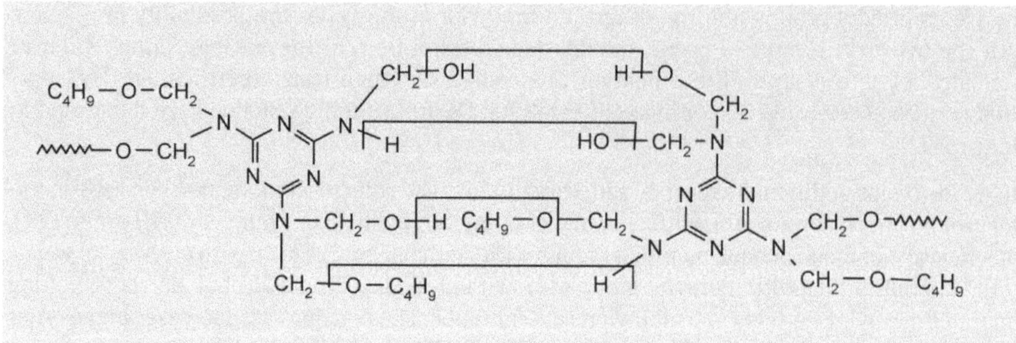

Figure 3.5.9: Self-crosslinking reactions of melamine resins

To describe the different types of melamine resins as used in automotive coatings, they are classified here into three groups; other classifications are possible, and the boundaries between the groups are fluid.

- Partly etherified melamine resins, characterised by free methylol groups
- Highly etherified melamine resins, characterised by residual NH (imino) groups
- Fully etherified melamine resins, containing just etherified methylol groups

Partly etherified melamine resins containing significant quantities of free methylol groups are produced by reaction of a high excess of formaldehyde with a slight excess of monoalcohols. Most resins of this type are high molecular. They are relatively reactive; the effective stoving temperatures (without addition of catalysts) are 120 to 160 °C, with the precise temperature depending on the degree of etherification. Self-crosslinking in these types is relatively extensive.

Melamine resins bearing significant quantities of residual free NH groups are produced by making melamine react with less formaldehyde and more monoalcohols. Some products in this group have low molecular weights. Again, the melamine resins of this group are highly reactive. The reason is the inductive effect of the NH group on the neighbouring etherified methylol group. The NH group will form methylene ether groups rapidly during crosslinking (self-crosslinking). These melamine resins contain very small quantities of free formaldehyde. They, too, generate low quantities of free formaldehyde during crosslinking reactions. Fully etherified melamine resins, which are produced with a high excess of both formaldehyde and monoalcohols, contain virtually only etherified methylol ether groups as functional groups. Those resins are unreactive relative to the other resin types. Unless catalysts are added, they will not react with resins containing hydroxyl groups until the temperature is 180 °C and higher. It is possible to significantly lower the effective stoving temperatures by adding acid catalysts. Under optimum reaction conditions (stoving temperature, time, acid catalysis), these melamine resins undergo the most extensive co-crosslinking.

Melamine resins etherified with methanol are more reactive than those etherified with butanols. In addition, they have the advantage of being water-soluble or water-thinnable. This group includes both resins that contain free NH groups and fully etherified resins. To the group of melamine resins fully etherified with methanol belong the so-called HMMM resins. HMMM is the abbreviation and synonym for hexamethoxymethylmelamine since it is the main constituent of this type of resin. HMMM resins are suitable not only for water-thinnable coating materials, but also for formulating high-solid stoving enamels due to their low molecular weight and the resultant low solution viscosity. HMMM resins are suitable for coil-coatings. Use of HMMM resins for stoving temperatures below 160 °C requires an acid catalyst.

Benzoguanamine resins

Unlike melamine, benzoguanamine can add on a maximum of just four molecules of formaldehyde to form methylol groups. The resultant resin molecules are therefore less highly branched. Also, benzoguanamine resins are at least partially or fully etherified by monoalcohols, which renders them more soluble and compatible. If benzoguanamine resins are used to crosslink polyesters or alkyds, the resultant coating films are notable for their outstanding flexibility. The reason for this is not only the lower functionality; the molecular structure also plays a role by conferring excellent wetting properties for pigments and substrate surfaces. However, benzoguanamine resins are not weatherable, and they are more expensive than melamine resins. On account of the aforementioned properties, benzoguanamine resins were used for automotive primer surfacers that offered excellent wetting properties and great flexibility. They are now being replaced in this application by blocked polyisocyanates as crosslinkers. Although these are even more expensive, they offer an even better balance between flexibility, hardness and resistance.

For the sake of completeness with regard to melamine resins, resins that play a minor role, if any, in automotive coating materials are described below.

Urethane resins

These are extremely soft resins on account of the urethane building blocks. A product made from butyl urethane and formaldehyde (molar ratio of 1 to 1) acts a polymer plasticiser [90]. 1,4-Butanediol and two moles of urea are made to react to yield butane diurethane. The diurethane reacts with formaldehyde to form methylol compounds which are then etherified with monoalcohol. The resultant resin is structurally similar to urea resins, is a suitable crosslinker, but - unlike urea resins - forms very flexible films.

Glycoluril resins

Glycoluril is prepared by making acetylene react with two moles of urea. The compound can add on maximum four molecules of formaldehyde. The resultant methylol compound is then etherified with four molecules of methanol. The etherified methylol groups have fairly low reactivity, and they need strong acid catalysts. Glycoluril resins are suitable for powder coatings [91].

3.5.3.3 Blocked polyisocyanates

As already described, blocked aromatic polyisocyanates are used for primers (electro deposition primers). For primer surfacers, however, aliphatic or cycloaliphatic polyisocyanates serve as the crosslinkers. These polyisocyanates are resistant to yellowing, and also confer a degree of weatherability on the primer surfacer layer. Neither of these properties can be achieved with aromatic polyisocyanates. Aliphatic and cycloaliphatic polyisocyanates are less reactive than their aromatic counterparts. As the stoving temperatures for primer surfacers are 160 to 170 °C, it is necessary to find blocking agents which are cleaved at these temperatures, or, more correctly, which re-urethanise with hydroxyl groups of polyesters or alkyds at these temperatures. Suitable polyisocyanates for primer surfacers are include the isocyanurate trimers of hexamethylene diisocyanate (HDI) or isophorone diisocyanate. The preferred blocking agent so far has been the methyl ethyl ketoxime. Such blocked polyisocyanates have effective reaction temperatures of 150 °C and higher. The mechanism by which these blocked polyisocyanates react is shown in Figure 3.5.10 (page 92).

Although methyl ethyl ketoxime is defined as hazardous, it is still in use. Special precautionary measures need to be adopted when it is handled. The blocking agent is not cleaved until the stov-

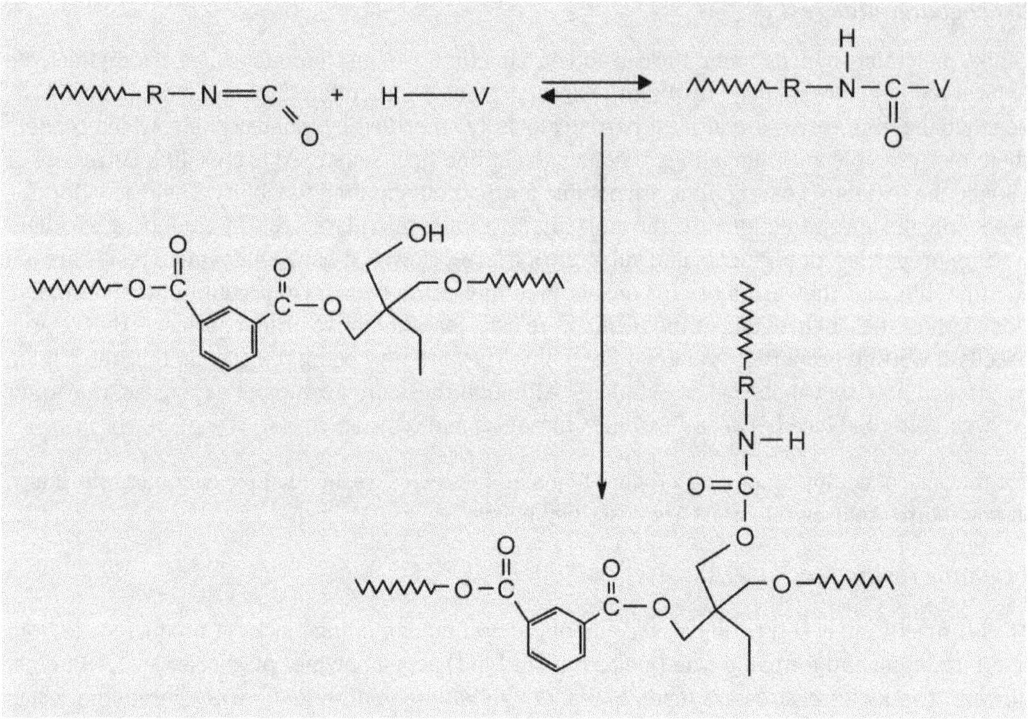

Figure 3.5.10: Mechanism by which blocked polyisocyanates react

ing process. Together with evaporated solvents, it is burned to generate heat energy (catalytic after-burning). Until now, alternative blocking agents either have other disadvantages or are more expensive than methyl ethyl ketoxime.

Unlike the complex crosslinking reaction of melamine resins, the reaction of blocked polyisocyanates with hydroxyl groups definitely yields only urethane groups. Urethane groups are much more resistant to chemicals - particularly acids - than the methylol ether groups of molecular networks containing melamine resins. As urethane groups form hydrogen bonds, they support the elastic component of the property of flexibility. More importantly, blocked polyisocyanates form molecular network structures that are different from those of melamine resins. The networks are more homogeneous, the meshes are wider, but the networks are more extended. Therefore, the potential to form flexible <u>and</u> resistant films is much higher. However, networks with wide meshes are open to diffusion; the networks can swell, for example, on treatment with solvents. Swollen networks are sensitive to mechanical impact. Although urethane groups are less sensitive to chemicals, the networks can be damaged after swelling. The solution to this problem is to combine both crosslinkers, namely melamine resins and blocked polyisocyanates, in one primer surfacer formulation. This also saves money because blocked polyisocyanates are much more expensive than melamine resins. In sum, a combination of the two crosslinkers offers a couple of advantages. These are an optimum of flexibility and hardness, and resistance to chemicals and solvents. In addition, such combinations lead to improved topcoat holdout.

It is possible to extend the elastomeric nature of primer surfacer films by adding polyurethanes. For solvent-borne primers, polyurethanes are available which are relatively low-molecular and contain blocked isocyanate end groups, which can be incorporated into the crosslinked network. On account of their soft-segment structures, those polyurethanes confer significant flexibility on the entire films.

3.5.3.4 Pigments for primer surfacers

Coloured pigments

The most important coloured pigment for primer surfacers is titanium dioxide, with the preferred shades being grey. To this end, the titanium dioxide is mixed with small quantities of black pigments (carbon black) and yellow pigments (yellow iron oxide) for neutral greys or beige greys. Those shades can be covered with topcoats that confer different special effects and colours.

There are also **coloured primer surfacers** whose colours have been matched to the colours of the topcoat systems. These primer surfacers boost the hiding power and brightness of topcoats. It is possible furthermore to reduce the layer thickness of topcoats, which sometimes is necessary due to application restrictions. The use of coloured primer surfacers entails a great deal of effort, though. First, the various coloured primer surfacers have to be stored close to the application line. Changes of primer surfacer colour require careful cleaning of the application lines. This can be compensated for by running colour campaigns. Another way of minimising the effort is to limit the number of colours for primer surfacers since then it is not necessary to produce a separate primer surfacer for every topcoat colour. Topcoat brightness can be boosted by maintaining a limited number of colours for the primer surfacers: white, light grey, dark grey, yellow, red, and black. These coloured primer surfacers are then grouped according to the various topcoat colours. Nor is it necessary for the coloured primer surfacers to be as bright as the topcoats. Hiding power can be enhanced by selecting inorganic pigments which are not as brilliant as organic pigments, but which have greater light-scattering power. The various coloured pigments are described in detail in Chapters 3.6.3.4 and 3.6.3.5.

Extenders

Extenders are a crucial ingredient of primer surfacers. They are pigment-like substances that have no pigmenting action. They are more or less fine particles of inorganic compounds that are totally insoluble in organic solvent or aqueous solutions and have low refractive indices. Since the refractive indices are at the same order of magnitude as those of the film matrix resins, extenders are colourless and transparent. Table 3.5.1 (page 94) lists the most important extenders along with their composition and characteristic values.

Accordingly, extenders are oxides, hydroxides, and salts (carbonates, sulphates, silicates) of alkaline earth and earth metals. If the products are based on natural minerals, it is important that only very pure materials be used. The products must not contain any pigmenting compounds. The ubiquitous red iron oxide (iron(III) oxide) poses the biggest risk in this regard. Minerals are obtained mostly by open-cast mining, cleaned by physical methods, ground and classified. They include natural **barium sulphate** (heavy spar, barytes), **calcium carbonate** (calcite, chalk), **calcium magnesium carbonate** (dolomite), and various **aluminium silicates** (kaolin, mica, feldspar, and wollastonite). Average particle sizes here range from 1 to 25 μm. Although there are products containing larger particles, they are not used as extenders for primer surfacers.

Silicon dioxide (silica) is one of the most widespread minerals in nature. Commercial extenders based on the minerals quartz and cristoballite are available. Quartzes confer high hardness and high abrasion resistance on coating films. Unfortunately, these properties also cause abrasion of grinding media in stirrer mills for dispersing pigments. Kieselguhr is the geological sediment of the shells of diatoms. It has a very complex structure and has numerous cavities that account for its very high specific surface area.

Unlike most inorganic coloured pigments, extenders have relatively low densities, the exception being barium sulphate.

Table 3.5.1: Extenders and the characteristic values

Extenders	Versions	Density [g/cm³]	Average particle size [µm]	Oil absorption value
barium sulphates	natural	4.0–4.5	2–15	9–19
	precipitated	4.1–4.5	0.7–3.0	11–20
aluminium silicates	kaolin	2.1–2.6	0.8–25	30–68
	mica	2.8	1.5–50	37–75
	feldspar	2.6–2.7	7–25	20–32
	talc	2.7–3.5	0.8–25	20–60
	synthetic	1.9–2.6	0.015–5.0	35–160
other silicates	wollastonite	2.85	3.5–25	80–90
calcium sulphate		2.5–2.9	10–12	15
calcium carbonates	chalk	2.7	1.0–10.0	15–20
	calcite	2.7	0.8–25	15–25
	dolomite	2.7–2.9	2–25	12–22
	synthetic	2.7	0.02–1.0	22–56
aluminium oxide		2.4	0.7–15	–
aluminium hydroxide		2.1–2.4	0.1–2.2	60
silicium dioxide	cristobalite	2.35	3.5–22	20–30
	kieselguhr	2.4–2.5	2–10	80–115
	quartz	2.6–2.7	2–25	20–30
	silica gel	2.1	2–10	80–300
	precipitated	1.9–2.1	0.008–10	180–260
	pyrogenic	–	0.007–0.2	120–350

Synthetic extenders are mainly precipitated from aqueous solutions. Given the right precipitation conditions, very fine particles can be prepared. This group also includes products which have average particle sizes (primary particles) below 10 nm (e.g. nanoscale silicon dioxide in pyrogenic silica and silica gel). Silica gel (colloidal silica) is precipitated from solutions of alkali silicates with acids. The resultant products have silanol groups on their surfaces and contain occluded water. Other precipitated silicas are prepared from solutions of alkali silicates under neutral or basic conditions. There are also products that have very fine particles. Pyrogenic silica is prepared by flame hydrolysis of silicon tetrachloride. This process yields very fine particles (nanoparticles).

The particle sizes listed in Table 3.5.1 refer to the average diameter of primary particles. Table 3.5.1 lists oil values [92]; these refer to the quantity of oil (linseed oil) needed to wet 100 g of a pigment and form a homogeneous paste. The requisite quantity depends not only on the particle size but also on the surface structure of the particles. The value stands for the specific surface area of an extender. For example, due to the highly structured surface, the oil value for kieselguhr is ten times that of barium sulphate, which has a spheroidal particle structure, even though the two extenders have nearly the same average particle size.

The extenders chosen for formulating primer surfacers have to be resistant to chemicals; these are barium sulphate, aluminium silicates, and silica. Less suitable are calcium carbonates and calcium sulphate as they are soluble in acids. Those extenders which have very small particles (precipitated and pyrogenic silicas) and those which consist of platelet particles, such as kaolin, mica and talc, influence the rheological behaviour of the paint material. They give rise to intrinsic viscosity, and so are used in limited quantities to avoid lowering the application solids and adversely affecting application behaviour (levelling, film gloss). They benefit primer surfacer formulations by preventing settling of dispersed pigments and sagging of paints applied to vertical surfaces.

In the past, as their name suggests, extenders were used make materials stretch further, primarily to reduce the overall material costs. It was subsequently discovered that a combination of pigments of different particle sizes, used in defined pigment volume concentrations, imbues primer surfacer films with a great capacity to absorb mechanical impact energy. It is believed that the various particles alter their relative positions within the film matrix on impact. The impact energy is transformed into kinetic energy (i.e. is dissipated). This idea forms the basis for the aforementioned analogy between a primer surfacer layer and an asphalt road. Figure 3.5.11 conveys an impression of a combination of barium sulphate extender and titanium dioxide with average diameters of 1 µm and 200 nm respectively. The weight ratios are approximately 1:1, while the pigment volume concentration in the film is about 20 %.

There is evidence that pigments with platelet particles help to absorb much more energy. However, as mentioned above, such extenders are not widely used due to the negative influences on viscosity, levelling, and topcoat holdout.

3.5.3.5 Solvents

The bulk of solvents for primer surfacers are medium- and high-boiling aromatic hydrocarbons (alkyl-substituted aromatics with boiling ranges of 155 to 175 °C and 180 to 210 °C). The solvents are relatively inexpensive. Saturated polyesters of isophthalic acid, melamine resins, and blocked polyisocyanates require solvents that are more polar. These include butanols, which are also present in the delivery form of melamine resins. With regard to the stoving conditions for film formation by primer surfacers, it makes sense to also add some high-boiling polar solvents that have nearly the same evaporation rate as the aromatics. That ensures that the balance of solvent polarity is maintained throughout the film-forming process, including the crosslinking phase in the oven. Glycol ethers and glycol ether esters serve as high-boiling, more-polar solvents. Solvents

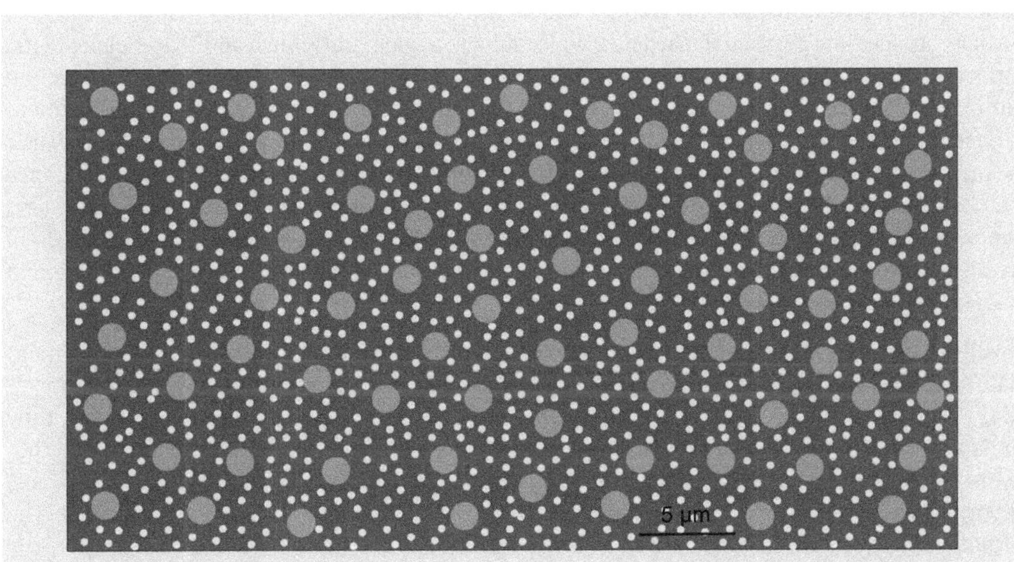

Figure 3.5.11: Distribution of titanium dioxide and barium sulphate extender in a primer surfacer layer

bearing hydroxyl groups, such as butanol and the glycol monoethers, stabilise the primer surfacers that contain melamine resins. The hydroxyl groups influence the equilibrium of the reaction between the methylol ethers and the polyester hydroxyl groups to the extent that premature crosslinking is avoided.

3.5.3.6 Additives

Theoretically, the aforementioned extenders with very small particles (colloidal and pyrogenic silicas), which are added in small quantities to the primer surfacer formulations, belong to the additives group. The bentonites are also pigment-like additives. Bentonite is a meta-silicate with layer-like crystal lattices (montmorillonite mineral $Al_2[(OH)_2Si_4O_{10}]\cdot n\ H_2O$), which can occlude alkaline and alkaline earth cations in the lattice and absorb water, a fact which leads to swelling of the crystal particles. There are also types of bentonite which are modified with organic compounds and can swell in solvents. All these products influence the viscosity of paints by increasing the intrinsic viscosity. Such pigment-like additives are therefore employed to prevent pigment settling (which is crucial when high-density barium sulphate is used) and, in addition, film sagging on vertical parts of objects between application and stoving. However, as mentioned above, high quantities of such additives would lower the solids content in the application state and interfere with levelling.

There are also organic additives that have rheological effects. These exhibit complex solubility behaviour. Such products will form gels if the concentration is high enough, in which case they are called thickeners. The thickening effect stems from greater interaction between the additive molecules rather than the additives and the solvent. The nature of the solvent is important, of course. Different organic thickeners are available for polar and non-polar solvents, and also for aqueous solutions. Organic thickeners are effective even in very small concentrations. Their effectiveness is therefore not just due to molecular interactions; it is believed that they build up an associative structure with the molecules in paints. Different classes of compounds can serve as organic thickeners (e.g. segmented polyurethanes, alkylated polyethers, and hydrogenated castor oil).

Wetting agents are additives which support the wetting of pigments during dispersing and later stabilise the pigment dispersion. Wetting agents are important for pigments that have complex particle surfaces and very small particles, which tend to agglomerate. Such additives belong to the surfactants group. Most suitable are alkaline earth salts of aliphatic carboxylic acids (e.g. calcium salt of 2-ethylhexanoic acid). It is believed that the metal cations are arranged on the pigment surface (mainly inorganic pigments) and the more hydrophobic acid groups are oriented towards the organic film matrix. The polarity of the pigment surface, the resins and the solvents influence the choice of additive. Besides the salts of monocarboxylic acids, sulphonates, phosphoric acid esters, polyurethanes, polyethers, polyamides, and acrylates with polar groups are used. There are also pigments and extenders which are treated with such compounds during the production process, and are therefore dispersed more easily.

Levelling agents are surface-active compounds. They are dilutable in resin solution or in paint materials. However, during film formation by physical drying, the additives become incompatible, and float to the surface of the film. On account of their very low surface tension, they spread over the entire film surface, forming a very thin layer. The interfacial tension of this thin layer creates a levelling effect, and the surface of the entire system becomes smooth. The most important compounds for levelling additives are modified silicone oils and polyacrylates. The formation of a very thin layer of additive on the surface of the coating film poses a risk since the following layer has to adhere to the surface of the previous layer very well. This is possible only if the very thin additive film is readily absorbed by the following paint material. That capability is especially important for primer surfacer layers if they are to be topcoated. Topcoats, too, are tested under this aspect in case they have to be used for refinishing (line repairs). Polyacrylates are less surface-active than silicone oils, but are also less sensitive to over-coating. Therefore, polyacrylates are preferred for formulating primer surfacers. Silicone oils are surface-active to an extent depending on the type and degree of modification. They are described in more detail in Chapter 3.8.3.7.

3.5.3.7 Formulations

Table 3.5.2 shows the composition of a typical OEM primer surfacer [93].

A combination of a flexibilised saturated polyester, melamine resin, and blocked polyisocyanate (ratios by weight: 65:20:15) leads to excellent stonechip resistance. Optimum application behaviour is controlled with solvents in a balanced ratio of low boiling non-polar (xylene) and polar (n-butanol and isobutanol) solvents, medium-boiling non-polar (Aromatic 100) and polar (butyl glycol

Table 3.5.2: Composition of an OEM primer surfacer

Pos.	Component	Solids wt-%	Formulation wt-‰
01	saturated polyester (65 % in aromatic 100. BDGA. isobutanol. xylene = 77 : 11 : 6 : 6)	99.1	152.4
02	barium sulphate	182.9	182.9
03	titanium dioxide, rutil	121.9	121.9
04	carbon black	0.6	0.6
05	thickenner	4.3	7.6
06	dispersing agent, soy lecithine	1.5	1.5
07	methoxy propyl acetate		30.5
08	n-butanol		9.1
grind on required fineness, than add:			
09	saturated polyester (65 % in aromatic 100, BDGA, isobutanol, xylene = 77 : 11 : 6 : 6)	99.1	152.4
10	blocked polyisocyanate (75 % in ar. 100) isocyanurate-trimer + methyl ethyl ketoxime	45.8	61.0
11	melamin resin (71 %), butanol-etherified	61.0	85.9
12	poly acrylate (25 % in Xylene)	0.5	2.1
13	butyl diglycol acetate		36.6
14	butyl glycol acetate		30.5
15	pine oil		24.4
16	n-butanol		36.6
17	xylene		64.0
	sum	616.7	1000.0

characteristic values

application solids (15' 165 °C)	61,2 %
viscosity (DIN 4/23 °C)	25 sec
viscosity (ISO cup 23 °C)	38 sec

pos.

01 + 09:	Setal 1671 SS-65 (Nuplex)
02	Blanc Fixe F (Sachtleben)
03	Kronos 2059 (Kronos)
04	Printex 200 (Degussa)
05	Borchiset UZ (Borchers)

pos.

06:	Nuosperse 657 (Elementis)
10:	Desmodur BL 3175 (Bayer)
11:	Setamine US 132 BB-71 (Nuplex)
12:	Modaflow (Cytec-Monsanto)

acetate) solvents and high-boiling non-polar (pine oil) and polar (butyl digol acetate) solvents. Of course, a wetting agent and anti-settling agent also belong in the pigment dispersion. The levelling agent is added to finish the formulation. Application of the primer surfacer to primed test panels is followed by flash-off and stoving for 15 minutes at 165 °C. The resultant film thickness is 35 µm. The Erichsen indentation test yields a figure of more than 9 mm, the hardness (Persoz) is 149 s, and stonechip resistance is ranked 2 on the Ford scale.

3.5.4 Application

Primer surfacers applied by electrostatic high-performance spray guns are typically applied at a flow rate of 240 ml/minute, and a bell speed of 33,000 per minute. For pneumatic application by high-performance spray guns, 4.5 bar and a flow rate of 350 ml/min are needed. The conveyor belts move at speeds of between 2 and 10 m/min, and typically at 4.5 m/min. Nowadays, application is nearly fully automated. After a short flash-off zone, the car body is transported into the stoving oven for 15 to 30 minutes at 160 to 170 °C. There is usually little need for corrective measures after cooling. However, it is common practice to clean the surface of the primer surfacer before the topcoat is applied. A genuinely exotic cleaning machine comprising a rotating wheel covered with ostrich feathers is used for this. Figure 3.5.12 shows the principle work flow in the primer-surfacing segment of an OEM application line.

Some car producers lower the stoving temperatures for primer surfacers to 130 - 140 °C (low-stove primer surfacers). There are melamine resins available which are sufficiently reactive at such temperatures; in addition, acid catalysts can be added to accelerate crosslinking. However, it is much more difficult to achieve optimum crosslinking of blocked polyisocyanates at these temperatures. Of course, the requisite stonechip resistance can only be expected if the polyisocyanates become involved in the crosslinking. There are other products capable of supporting the required flexibility, e.g. polyurethanes, but they are not really a substitute.

In this connection, it must be pointed out that it makes sense for the stoving temperatures to decrease with the number of applied layers. If low-stove primer surfacers (stoved at 130 to 140 °C) are covered with topcoats which also are stoved at 130 to 140 °C, the primer surfacer is exposed to the same temperature again, when it has already been crosslinked in the previous step. The development of low-stove primer surfacers is not so easy. The product needs to be sufficiently reactive at low temperatures, but it must also be stable if it is stoved at that temperature again. It must not decrease in flexibility, nor become discoloured or lose other properties. In addition, the material has to offer overlapping stability at ambient temperatures, even though it is more reactive.

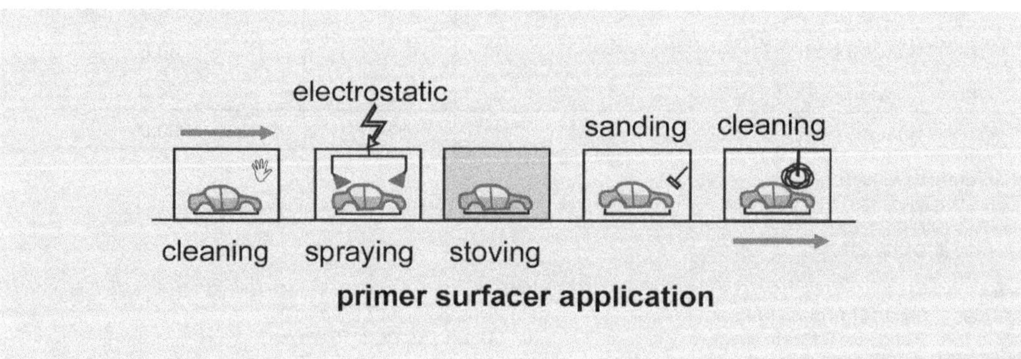

Figure 3.5.12: Process flow in a primer-surfacing segment

3.5.5 *High-solid primer surfacers*

Common solvent-borne primer surfacers have solid content at application viscosity of 55 to 60 % by weight. In comparison to other solvent-borne coating systems, the application solids are relatively high. Nevertheless, there have been studies aimed at increasing the application solids content of primer surfacers and decreasing the quantity of volatile organic compounds (VOCs) in application processes. This was only an intermediate step in the development of water-borne primer surfacers and ultimately powder primer surfacers. The result was an increase in application solids to more than 65 % by weight. Success was achieved by lowering the molecular weight of the polyester resins and using HMMM resins as crosslinker. Blocked polyisocyanates are less suitable for increasing solids content on account of their solution viscosity. The absence of blocked polyisocyanates meant that the end properties were not comparable to those of conventional primers. This was one reason to proceed to the next step, the development of water-borne primer surfacers.

3.5.6 *Water-borne primer surfacers*

Currently available water-borne primer surfacers contain approx. 45 % by weight application solids, and between 6 and 12 % by weight organic solvents. Even so, this constitutes a significant saving over the VOC levels of common or high-solid solvent-borne primer surfacers. The technology of water-borne primer surfacers is established mainly in Europe, and particularly in Germany. In Europe in 2006, about 60 % of primer surfacers for OEM application were water-borne. The global figure is only about 20 %. Assembly plants that use water-borne primer surfacers outside Europe are mainly the transplants of European car producers.

3.5.6.1 *Resins for water-borne primer surfacers*

It is theoretically possible to use nearly all types of resin in water-borne systems. Consequently, unlike solvent-borne primer surfacers, water-borne primer surfacer resins contain water-thinnable polyesters and melamine resins. Blocked polyisocyanates, too, are suitable for water-borne systems and are incorporated mainly by an indirect method. Polyurethane dispersions are additionally used in some water-borne primer surfacers.

Water-thinnable polyesters

Only a few polyesters are actually soluble in water. These resins have the disadvantage of remaining sensitive to water and moisture after film formation. For this reason, recourse is made to the aforementioned method of introducing small quantities of hydrophilic groups into polyester molecules, which are mainly hydrophobic. The hydrophilic groups are ions, mainly anions. Such anions offer the possibility of preparing colloidal aqueous solutions that contain relatively large particles. The anions are mainly oriented on the surface of the colloidal particles, acting as carrier groups in aqueous phase. They are mostly prepared from carboxyl groups which are neutralised by bases. However, sulphonic acids, phosphonic acids, or acid groups of partly esterified phosphoric acid may act as carrier groups.

Doping with carboxyl groups is preferred. There are different ways to introduce free carboxyl groups into polyester molecules:

- addition of cyclic anhydrides of carboxylic acids
- partial condensation of polycarboxylic acids
- diene addition of unsaturated carboxylic acids

The different ways of incorporating carboxylic acids into polyester molecules are described by chemical formulas in Figures 3.5.13, 14 and 15.

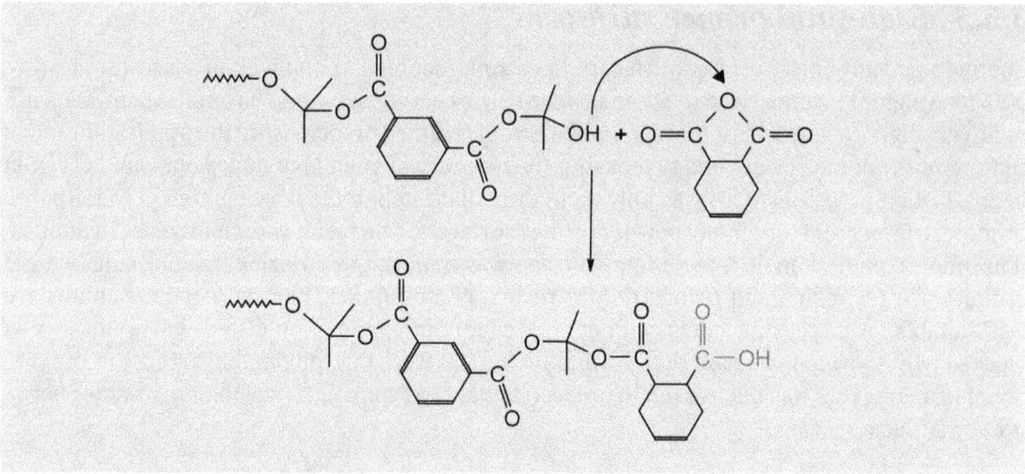

Figure 3.5.13: Incorporation of carboxylic groups by adding cyclic anhydrides

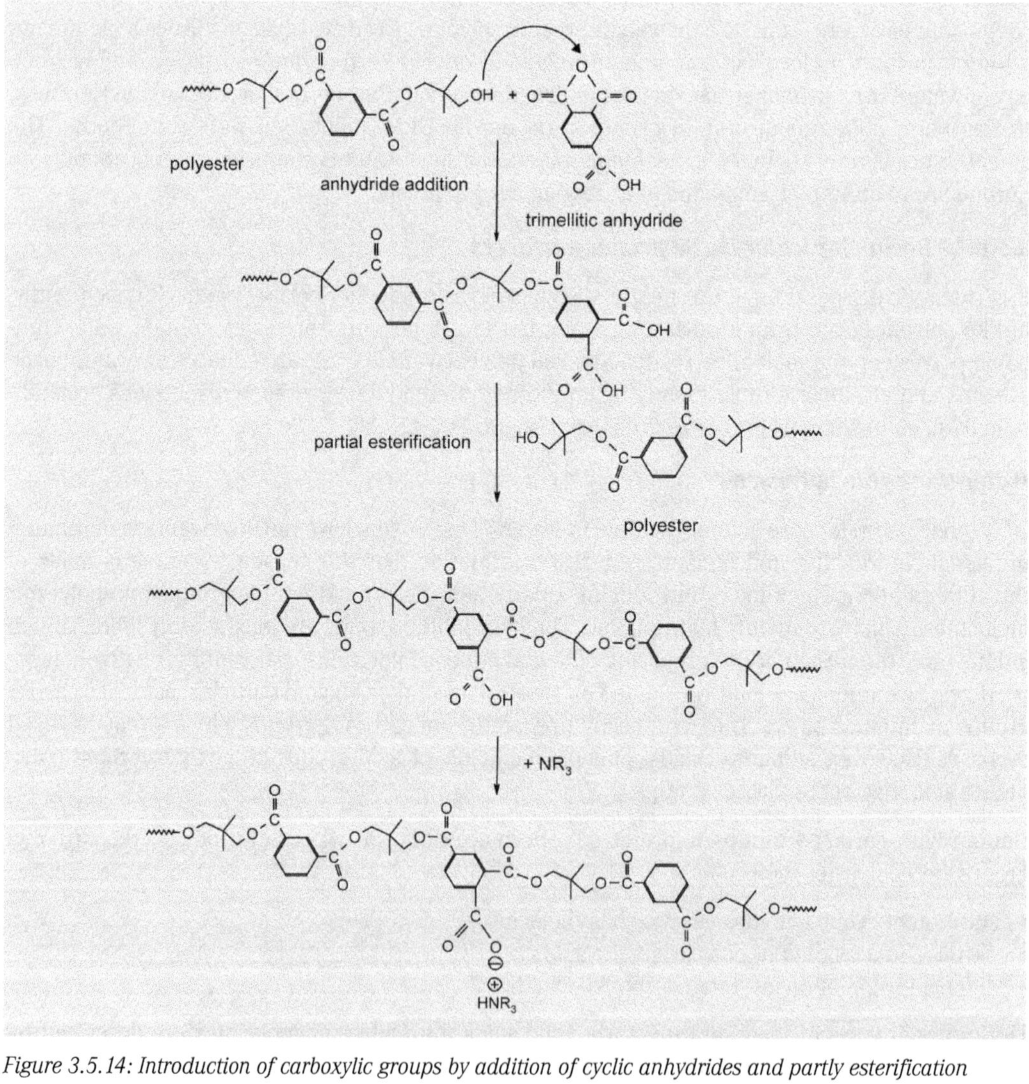

Figure 3.5.14: Introduction of carboxylic groups by addition of cyclic anhydrides and partly esterification

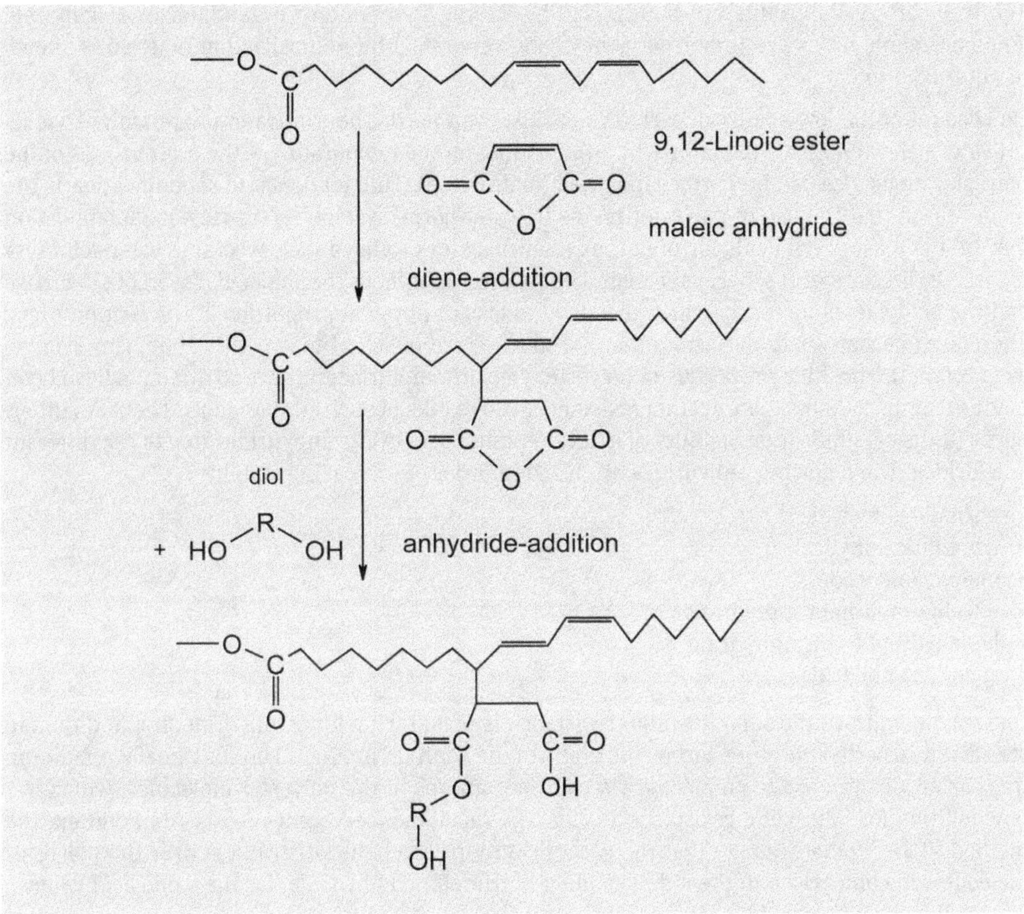

Figure 3.5.15: Incorporation of carboxylic groups by diene addition of unsaturated acid

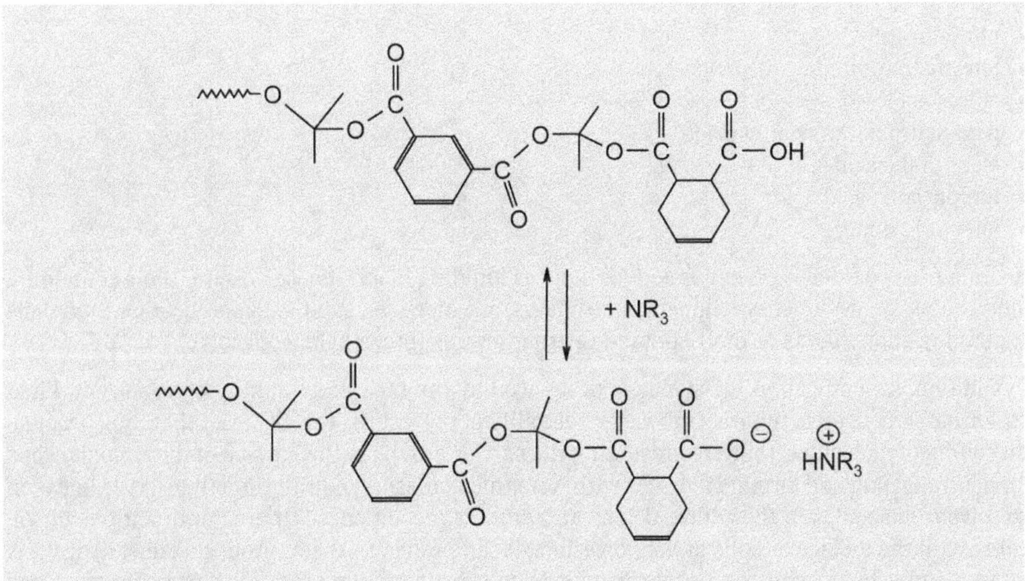

Figure 3.5.16: Neutralisation reaction

The free carboxylic groups are neutralised by adding inorganic or organic bases to generate anions. Organic bases are preferred if they can evaporate. The neutralisation reaction is shown in Figure 3.5.16.

On account of the low acidity of carboxylic acids, even partial neutralisation with amines yields pH values above 7.5. This explains why - depending on the temperature - the ester groups of the polyester molecules can be partly saponified (hydrolysed). This tendency to saponification is the main reason that the use of polyester resins in water-borne systems is restricted. Saponification may be the reason that acids forming the anionic carrier groups are cleaved and water-solubility is lost. Or, if esters of the polyester main chains are saponified, the molecular weights decrease and the acid values increase, which may lead to loss of physical properties of the coating films. There are alternatives to polyesters that are much more resistant to saponification. However, so as to retain the positive properties of polyesters for primer surfacers, several attempts have been made to improve their saponification resistance. The results of such studies show: There are differences in the saponification stability of addition products of cyclic anhydrides due to the different anhydrides. The following anhydrides are listed in order of increasing stability:

- phthalic anhydride
- trimellitic anhydride
- maleic anhydride
- tetrahydrophthalic anhydride
- hexahydrophthalic anhydride
- succinic anhydride

One way to improve the saponification resistance is to add trimellitic anhydride first and then to statistically esterify one other carboxylic group of the addition product. The statistically remaining free carboxylic groups are connected via two ester groups in the polyester molecules, which is a precondition for improving resistance to cleavage reactions (see Figure 3.5.14). In addition, the method [94] has the advantage that the carboxyl groups are better distributed over the polyester molecules in comparison to the simple adduct of trimellitic anhydride, for the same acid values.

It is also possible to rank polycarboxylic acid contained in the polyester chains by their saponification resistance. The following list shows the polycarboxylic acids or corresponding derivatives in order of increasing saponification resistance:

- phthalic anhydride
- tetrahydro phthalic anhydride
- adipic acid
- hexahydro phthalic anhydride
- isophthalic acid
- terephthalic acid
- fatty acid dimers

A similar list could also be generated for diols as building blocks for polyesters. Diols containing long chains, or, even better, aliphatic side chains, are more resistant to saponification than diols containing short chains or diols containing ether groups (polyethylene glycols).

In addition, saponification resistance is influenced by the type and quantity of co-solvent. First, co-solvents are used to improve the water solubility of polyesters. It is believed that they promote un-coiling of molecules, thereby presenting the carrier groups to the surface of the colloidal coils for neutralisation and forming solvates with water molecules. An equilibrium then exists between co-solvent molecules in the colloidal coils and in the aqueous phase. If greater quantities of solvent are in the molecular coils of the polyesters, it can protect the ester groups against hydrolysis by water molecules, and the polyester molecules remain stable. From all these aspects, butyl glycol (ethylene glycol monobutyl ether or butyl cellosolve) has proved to be the best co-solvent.

It is also possible to employ small quantities of solvents which are incompatible with water. Such solvents are placed in the polyester molecule coils and are also able to protect the ester groups against saponification. The quantity of such solvents is restricted because of their lack of a mediator role with the aqueous phase, which is important for optimum film forming.

Furthermore, the type and quantity of neutralisation agent influence the saponification resistance. Suitable neutralisation agents are N,N-dimethyl ethanolamine (DMEA), diisopropanol amine (DIPA), triethyl amine (TEA), aminomethyl propanol (AMP) and ammonia.

Water-borne polyester solutions often exhibit very strange and unusual viscosity behaviour when thinned with water. During thinning, the viscosity rises significantly at first, unlike the case for all organic solutions. After that, as thinning continues, the viscosity drops very rapidly to specific values. This anomalous viscosity behaviour is sometimes termed a “water hill”. The process is reversible, so it is not the effect of an inversion reaction (observed in emulsions). The water hill may cause problems, first in the handling of water-borne polyesters in paint production, and second during film forming, where such anomalies may interfere with levelling. The position and height of the water hill are influenced by the polyester building blocks, the type of neutralisation agent, the degree of neutralisation and type and quantity of co-solvent.

Water-thinnable melamine resins

As mentioned above, all melamine resins which contain methanol as an etherification alcohol are water-soluble or at least water-thinnable. These include HMMM resins and melamine resins containing NH groups. Combining water-borne polyesters with these types of melamine resins decreases the viscosity effect of the water hill. Crosslinking of HMMM resins in water-borne coating systems also necessitate the use of acid catalysts to ensure effective crosslinking at temperatures of 120 to 150 °C. Catalysts are available for water-borne systems; they consist of amine salts of strong acids, which, of course, are water-soluble.

Blocked polyisocyanates

To meet the requirement for excellent stonechip resistance in water-borne primer surfacers as well, it makes sense to use blocked polyisocyanates as crosslinkers for those systems, too. Doping of blocked polyisocyanates with carboxylic groups is relatively difficult. In addition, even if it is successful, compatibility with water is often worse. However, it is possible to introduce blocked polyisocyanates into aqueous phase through using other water-soluble resins as a carrier for those types of crosslinker. Preparation consists in mixing the blocked polyisocyanate and the carrier resin, e.g. polyester containing carboxyl groups, in an organic phase, neutralising the carrier groups and then transferring the mixture to aqueous phase. Another possibility is to partially block the polyisocyanates. The residual free isocyanate groups are then made to react with hydroxyl groups of water-soluble polyester and so are incorporated chemically into the entire system. After that, the product of the reaction is transferred to the aqueous phase by the conventional method.

Polyurethanes

Another way to introduce elastomeric character into water-borne primer surfacers is to add aqueous polyurethane dispersions. These consist at the molecular level of soft segments that contain hydroxyl groups (polyethers, polyesters), diisocyanates, hydroxy carboxylic acids, which ultimately form the anionic carrier group, and chain extenders, which are most often primary and secondary polyamines but may also be polyols (e.g. trimethylol propane). Different molecular weights are achieved by varying the molecular quantities and ratios of the named components. If an excess of polyols is chosen for chain extension, the results are polyurethanes with relatively low molecular weights and with terminal and lateral hydroxyl groups, which are available for crosslinking reactions.

In addition to chain extension, blocking agents may be used. The results are low-molecular polyurethanes with terminal blocked isocyanate groups, which are also available for crosslinking. The equivalent weights of potential isocyanate groups, in this case, are much higher than in the case of blocked polyisocyanate adducts. Such polyurethanes serve as components for water-borne primer surfacers (see formulation examples below). Polyurethanes are also important ingredients of water-borne basecoats. The description of this resin class is therefore continued in Chapter 3.7.5.

3.5.6.2 Formulations for water-borne primer surfacers

The chosen example of a water-borne primer surfacer [95] contains a water-thinnable polyester, polyurethane dispersion, and melamine resin. Pigments for water-borne primer surfacers are principally the same as for solvent-borne primer surfacers. In the example given, these are titanium dioxide, carbon black, barium sulphate, and talc.

The **water-thinnable polyester** [96] contains isophthalic acid, a dimer fatty acid (trimer content 18 % by weight), trimellitic anhydride, 1,6-hexanediol, and a special modification comprising an adduct of two moles of epoxy resin (EEW 185 g/mol) and one mole of dimer fatty acid (trimer content less than 2 % by weight). The composition is shown in detail in Table 3.5.3. In the first step, the isophthalic acid, dimer fatty acid and the 1,6-hexanediol are esterified nearly to completion. Then trimellitic anhydride is added and esterified until the acid value is 67.7 mg KOH/g. After cooling, the polyester is diluted with a small quantity of butyl glycol. The epoxy adduct is then added. The final acid value is 40.9 mg KOH/g. N,N-Dimethyl ethanolamine is added for neutralisation. The resultant product is transferred to aqueous phase; the solids content is adjusted with water to 35 % by weight. Owing to the content of relatively hydrophobic building blocks, this polyester is relatively resistant to saponification.

The polyurethane dispersion [97] contains a soft segment, which is a polyester prepared by making two moles of 1,6-hexanediol and one mole of neopentyl glycol react with two moles of adipic acid (the acid value is less than 1 mg KOH/g, number average molecular weight is 555 g/mole). The diisocyanate is 4,4'-diisocynanato biscyclohexylmethane (H_{12}MDI). Water solubility is introduced via dimethylol propionic acid. Chain extension and doping with lateral hydroxyl groups are effected with trimethylol propane (TMP). The reactions are carried out in a ketone process solvent that is water-compatible. The resultant polymer is neutralised with N,N-dimethyl ethanolamine (DMEA) and transferred to aqueous phase. Then the process solvent is distilled off. The solids content is adjusted to 40 % by weight with deionised water, and the pH value is adjusted to 7.2 by adding small quantities of DMEA. The melamine resin is an HMMM type [98].

Production of the primer surfacer consists in mixing the pigments and extenders (titanium dioxide, carbon black, barium sulphate, and talc) with some of the aforementioned polyester, and adding defoamer, some more DMEA and water. The products are premixed and then dispersed on a sand mill, until the particle size (as measured by the Hegmann method) is less than 12 µm. The let-down portion consists of the residual polyester, polyurethane dispersion, and melamine resin. Some deionised water is used to adjust the solids content. The composition of the water-borne primer surfacer is given in Table 3.5.4.

The water-borne primer surfacer has a solid content of 52 % by weight, it contains just 4.8 % by weight of solvent, the pH value is between 7.8 and 8.0, and the viscosity is 120 s (DIN 3511, Ø 4 mm, 20 °C equivalent to ISO cup 460 sec). For spray application, the material is thinned down to 25 s. The primer surfacer is sprayed electrostatically onto electro primed panels. The flash-off conditions are 10 min at 23 °C plus 10 min at 80 °C. The stoving conditions are 20 min at 160 °C. The resultant film thickness is 35 µm. The primer surfacer is then topcoated. Final assessment includes levelling, topcoat holdout, adhesion, and stonechip resistance. Since the primer surfacer formulation contains a sufficient quantity of plasticising building blocks, very good results are achieved in the stonechip tests (VDA method).

Table 3.5.3: Polyester for water-borne primer surfacer

n = m/M	Component	M	m (example)	wt-‰
3,749	hexandiol-1,6	118	442,4	371,4
0,294	dimer fatty acid	567	166,6	139,9
1,110	isophthalic acid	166	184,3	154,7
1,389	trimellitic anhydride	192	266,7	223,9
0,152	adduct of epoxy resin (EEW 184) and dimer fatty acid (2 : 1)	1422	216,0	181,3
	sum		1276,0	1071,3
4,718	water	18	84,9	71,3
	yield (AN = 40,9)		1191,1	1000,0

average molecular weight [g/mol]	M_n	2029
acid number [mg KOH/g]	SZ	40,9
degree of branching [mol/kg]	v'	1,55
OH-number [mg KOH/g]	OHN	101,3

Table 3.5.4: Composition of water-borne primer surfacer

Pos.	Components	nfA wt-%	Formulation wt-‰
01	polyester (35 %)	6.3	18.0
02	defoamer		0.2
03	deionised water		4.3
04	DMEA		0.1
05	titanium dioxide, rutil	11.2	11.2
06	blanc fixe	11.0	11.0
07	talc	1.3	1.3
08	furnace black	0.1	0.1
dispers on ≤12 µm, then add:			
09	polyester (35 %)	2.1	6.0
10	polyurethane dispersion (40 %)	16.0	40.0
11	HMMM resin (100 %)	4.0	4.0
12	deionised water		4.0
	sum	52.0	1000.0

pH-value: 7.8–8.0
viscosity 120" (DIN 4/20 °C) or 460" (ISO cup 20 °C)

pos.		pos.	
01 + 09	polyester 1, example of EP 0339433	07	"Talkum" NT (GFR)
02	"Surfynol" 50 % in ethylene glycol (Air Products)	08	"Furnace" black 101 (Degussa)
05	"Kronos" 2310 (Kronos)	10	polyurethane dispersion 1, example of EP 0339433
06	"Blanc Fixe" F (Sachtleben)	11	"Cymel" 301 (Cytec)

3.5.7 Powder primer surfacers

While powder coatings are widely established in the market for general industrial coatings, their share of the automotive coatings market is relatively small. Of course, the greatest advantage of powder systems is the elimination of volatile organic compounds (VOCs) during application. Globally, powder primer surfacers account for less than 10 % of total primer surfacer consumption. The bulk of it is used in the USA [99], where nearly 30 % of primer surfacers for OEM coatings are powder primer surfacers. In Europe the quantity is only about 2 % [100].

3.5.7.1 Resins for powder primer surfacers

The resins for powder primer surfacers for OEM coatings are the same as those which were used for industrial powder coatings because the experiences gained with those resins were positive. These resins are combinations of aromatic epoxy resins (see Chapter 3.4.5.1) and polyesters containing carboxyl groups.

The **epoxy resins** are prepared from bisphenol A and epichlorohydrin. Of course, they have to be solids. The epoxy equivalent weights (EEW) are 500 to 900 g/mol and the softening temperatures are 80 to 110 °C. Although these resins have low molecular weights, and the softening temperatures are relatively high, a fact which benefits use in powder coatings. Another advantage is that the melt viscosity decreases relatively quickly with rise in temperature. That boosts levelling in powder coatings.

Polyesters containing carboxyl groups for powder primer surfacers are prepared by the same methods used for the production of water-borne polyesters (see Chapter 3.5.6.1). However, by contrast, they must have building blocks that confer higher glass-transition temperatures or high melting temperatures. Such building blocks are aromatic and cycloaliphatic polycarboxylic acids, aliphatic diols with relatively short molecular chains, cycloaliphatic diols and, in special cases, ethoxylated bisphenol A. Small amounts of building blocks with long aliphatic chains may be added to optimise levelling during the melting process for film forming. Doping with carboxylic

Table 3.5.5: Composition and characters of a typical polyester resin for powder primer surfacer

n = m/M	Component	M	m (example)	wt-‰
8.370	propylene glycol	76	636.12	262.8
2.000	dipropylene glycol	134	268.00	110.7
10.000	terephthalic acid	166	1660.00	685.7
1.105	trimellitic anhydride	192	212.16	87.6
	sum		2776.28	1146.9
19.751	water	18	–355.52	–146.9
	yield (AN = 57)		2420.76	1000.0

average molecular weight [g/mol]	**M_n**	3910
acid number [mg KOH/g]	**SZ**	57
degree of branching [mol/kg]	**v'**	0.46
OH-number [mg KOH/g]	**OHN**	~ 0

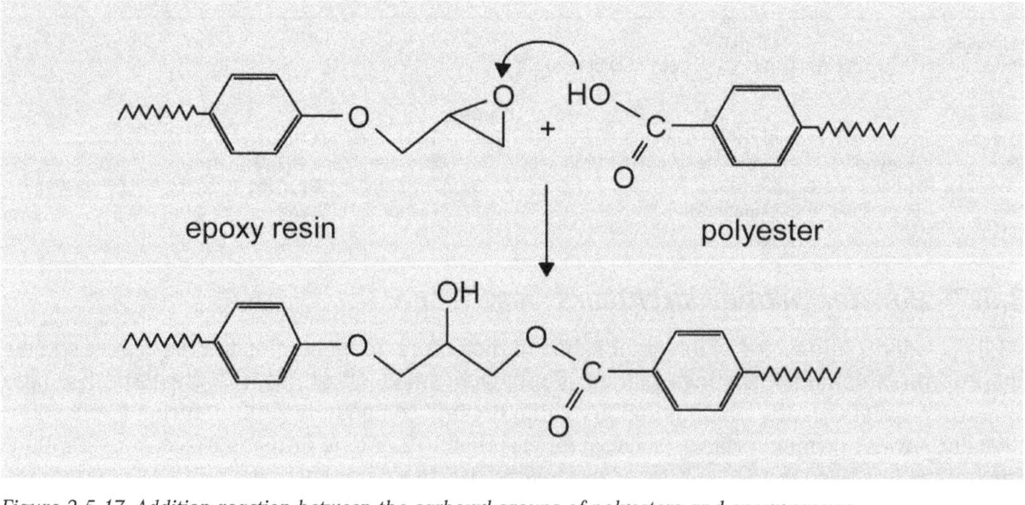

Figure 3.5.17: Addition reaction between the carboxyl groups of polyesters and epoxy groups

groups takes place via anhydride addition (see Figure 3.5.13) or by partial esterification of polycarboxylic acids or a combination of both (see Figure 3.5.14). Diene adducts of unsaturated dicarboxylic acids are unsuitable because of the low softening or melting temperatures.

Suitable polyesters for powder primer surfacers have softening temperatures of 80 to 100 °C, acid values of 30 to 80 mg KOH/g, and number average molecular weights of between 1500 and 3000 g/mol. Since the solubility of these polyesters is not important, the polyester may exhibit a tendency to crystallise. This tendency raises the softening temperature, but decreases the melt viscosity just after softening; both are ideal for promoting flow and levelling during film forming. However, the tendency to crystallise may not impair the compatibility of polyesters and epoxy resins. Table 3.5.5 shows the composition and characteristics of a typical polyester resin for a powder primer surfacer [101].

Crosslinking between the epoxy resin and the polyester containing carboxyl groups is an addition reaction between epoxy and carboxyl groups. The underlying principle is given in Figure 3.5.17.

The addition reaction has the advantage of not producing cleavage products that could possibly lead to film defects, such as blisters or craters. This is particularly important for thick film layers. Analysis of the stoichiometric conditions of the reaction reveals that some side reactions must occur. The reaction presented above fails to adequately describe the formation of networks of very high molecular weight. It is conceivable that the hydroxyl groups of the epoxy resin, too, react with the epoxy groups (self-crosslinking), and that the carboxyl groups of the polyester have a catalytic effect. Furthermore, other commonly employed catalysts, metal salts, phosphines, or phosphonium salts and especially tertiary amines might affect not only the reaction between the carboxyl and the epoxy groups, but also the reaction between the hydroxyl and the epoxy groups. The stoving conditions for powder primer surfacers are 160 to 180 °C. Although epoxy resins and appropriate polyesters, due to the high glass-transition temperatures, contain building blocks that form hard films, the films are surprisingly flexible. The reason for this is surely that the crosslinking reaction leads to wide-meshed, extended molecular networks.

3.5.7.2 Formulation and production of powder primer surfacers

The pigments employed for powder primer surfacers are a combination of titanium dioxide and extenders, similar to those for solvent-borne and water-borne primer surfacers. The usual pigment volume concentration is lower than in the liquid primer surfacers. Small quantities of pigment-like additives (e.g. pyrogenic silica) are selected to influence the viscosity. The levelling agents chosen are non-polar polyacrylates. Since these are liquids, it is first necessary to prepare an intermediate product where the polyacrylates are melted together with epoxy resin. The final

Table 3.5.6: Composition of a typical powder primer surfacer

Pos.	Components	Formulation wt-‰
01	polyester resin. containing carboxyl groups	35.5
02	epoxy resin	23.6
03	titanium dioxide, rutil	30.1
04	barium sulphate	10.0
05	levelling agent, acrylic polymer	0.6
06	flow agent, benzoin	0.3
	sum	1000.0

pos.		pos.	
01	Crylcoat 344, SZ 50 mg KOH/G (UCB)	04	EWO (Alberti)
02	D.E.R 633 (Dow)	05	Acronal 4 F (BASF)
03	Kronos 2310 (Kronos)		

solid product is called a master-batch. The reaction catalysts are those already described above, with solid products preferred. Most patents specify the addition of benzoin, which obviously is an excellent flow additive for powder coatings. Table 3.5.6 shows the composition of a typical powder primer surfacer [102].

Powder primer surfacers are made by grinding the resins to a defined particle size (mostly to 1 to 2 mm average particle size) and then mixing them together with pigments extenders and additives in a dry-mixer. The dry mixture is fed to an extruder where it is homogenised, and the

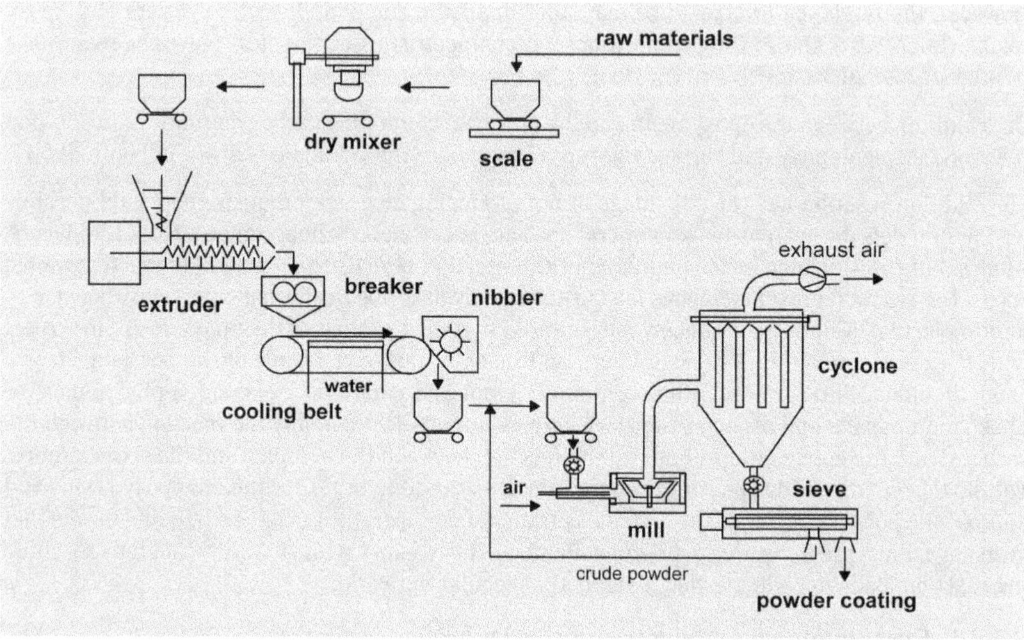

Figure 3.5.18: Production process for powder primer surfacer

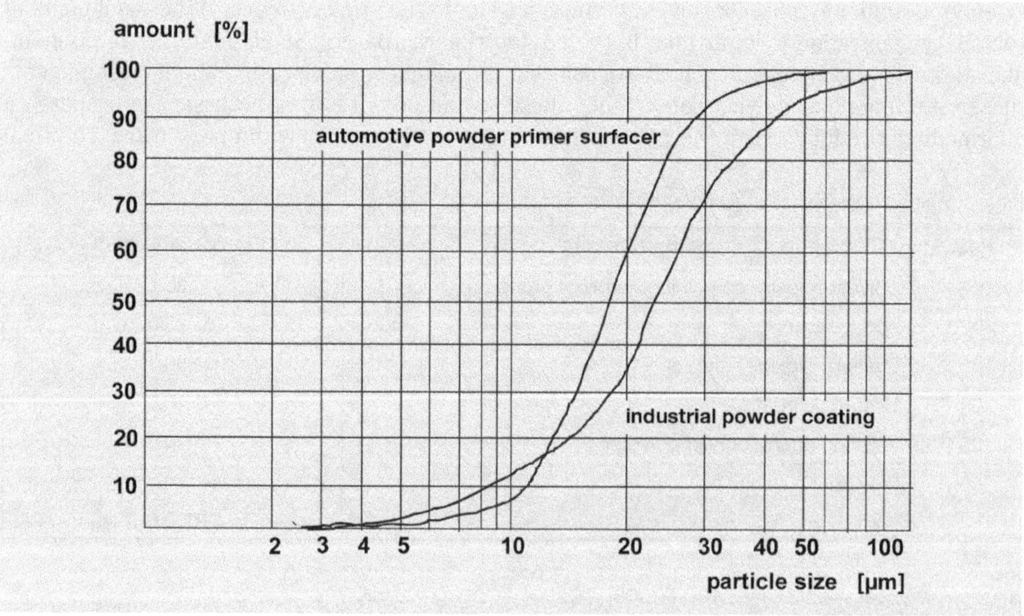

Figure 3.5.19: Particle size distributions of different powder coatings

pigment particles are dispersed and wetted effectively. The extrusion temperature is between 80 and 110 °C. Extrusion must not cause the resins to start crosslinking. The extruder product is fed over a cooling belt and chopped in a nibbler. Then the broken material is ground intensively by a counter-current mill. The coarse material is separated on screens and fed back to the mill. The super-fine fraction is separated on a cyclone and fed back to the extruder. The entire process is presented in Figure 3.5.18 [103].

The particle size for powder primer surfacers must be as low as possible for the sake of optimum flow and levelling at the usual layer thickness. Figure 3.5.19 compares the particle size distributions of a powder primer surfacer and a powder coat for general industrial application [104].

3.5.7.3 Application

Application consists in first converting the powder primer surfacers to an aerosol (a powder dispersed in air). To this end, the powder is fluidised in a feed container by compressed air. The aerosol is fed to an electrostatic spray gun, where the powder particles are electrically charged. There are two ways to do this. The first takes the form of either corona charging, in which the aerosol is fed along electrodes, which may be acicular or cylindrical capacitors in the spray gun (internal charging), or charging the powder particles as they pass electrodes outside the spray gun (external charging). The applied potential depends on the method and ranges from 10 to 100 kV/cm. The second consists in feeding the aerosol through a pipe made from special plastic materials; charging takes place by friction of the particles inside the pipe. The method is called tribo-charging, and it can be boosted by incorporating additives into the powder. The charged particles are distributed as they exit the gun from rotating application bells, much in the manner of liquid materials (see Chapter 3.3). The charged particles follow the electrical high voltage field and are deposited on the object, which is earthed. On account of their charge, the particles adhere to the objects very well, and so transportation of the object to the stoving oven does not pose a problem. Film forming in the oven takes the form of melting and crosslinking. Figure 3.5.20 shows the principle behind the application of powder primer surfacers.

Powder primer surfacers are usually applied in film thicknesses of 60 to 70 μm. That is much higher than the film thicknesses of liquid primer surfacer materials, which are 30 to 40 μm. The reason is that such film thicknesses are necessary for providing primer surfacer layers with adequate levelling. High film thicknesses are an advantage as they create an optimum barrier effect and boost - if sufficiently smooth - topcoat holdout. High film thicknesses of powder coatings are free from film defects, such as blisters. However, the disadvantages are the higher weight of thick layers and higher material consumption (higher costs). Consequently, a great deal of effort is being expended on devel-

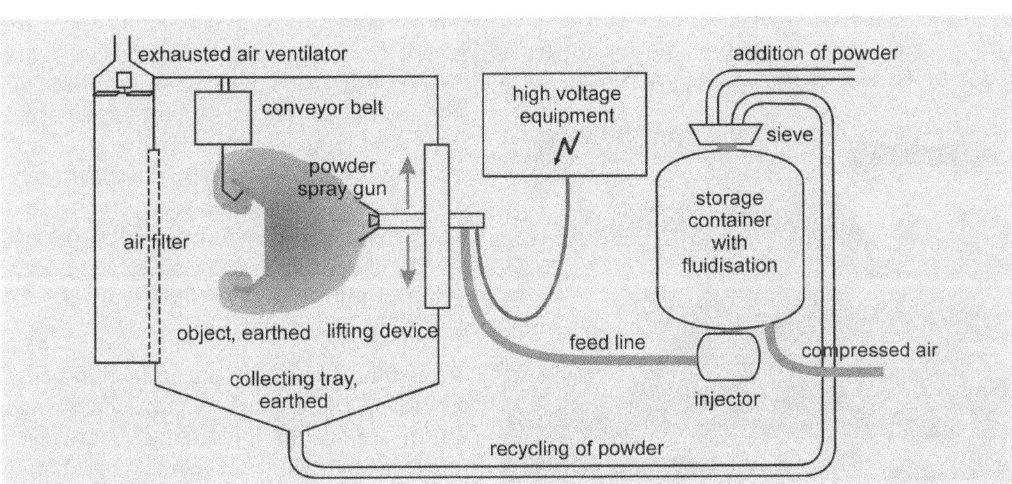

Figure 3.5.20: Principle behind the application of powder primer surfacers

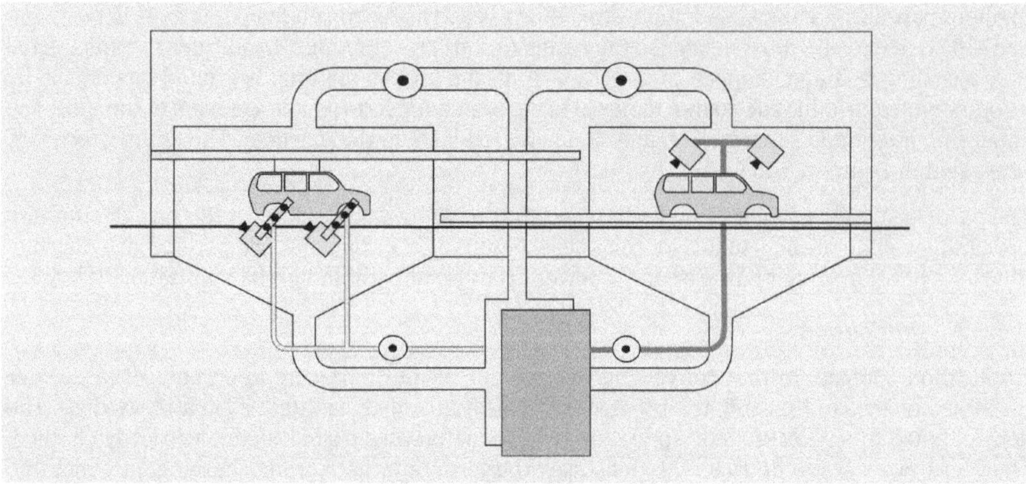

Figure 3.5.21: Two-stage application method for powder primer surfacers

oping powder primer surfacers that are applied at lower film thicknesses yet retain adequate levelling and smoothness. The main reason for the need to apply powder primer surfacers in a film layer greater than 60 µm is the particle size distribution. Since powders contain particles with diameters around 50 µm, optimum film forming to yield smooth surfaces requires higher film thicknesses than that. Thus, these studies have focussed on reducing particle size and producing a narrower particle size distribution. The processes which have been developed are very complex and expensive. However, there are limitations to this approach. Smaller particles are less amenable to fluidisation; they do not form mobile aerosols due to strong particle-particle interaction. Furthermore, additives that boost fluidisation do not work perfectly as they impair other film properties.

Unlike electrostatically applied liquid primer surfacers, electrostatically sprayed powder primer surfacers offer a primary material transfer efficiency of just 50 % by weight (the figure for other methods is up to 85 %). This means that only half of the powder is applied to the object during spraying. However, again unlike liquid primer surfacers, the overspray can be recycled and returned to the application. The resultant transfer efficiency for the whole application process, including recycling is as much as 95 % by weight, which exceeds that of all other application methods. However, recycling poses a risk of contamination. Once again, unlike liquid paints, powder coatings are not made in a way that lends itself to removal of contamination. To circumvent this problem, special application methods have been developed for powder primer surfacers. As with other types of coating, there are two application steps. In the first, the powder is sprayed onto the lower part of the car body (bottom group). In the second, the upper parts, which are more visible, are coated. Fresh powder is used for the second step only. All powder collected for recycling is fed to the first step, i.e. the coating of the lower parts of the car body. If a potential contamination means that the primer surfacer film is not so perfect in that area, this is not visible. The underlying principle is shown in Figure 3.5.21.

Figure 3.5.22: Car body in powder primer surfacer spray booth.

Coloured powder primer surfacers are also available. It must be borne in mind that changing the colours of powder coatings in the application line is much more complicated than with wet paints. For a better impression of the application of a powder primer surfacer, see Figure 3.5.22 [105].

3.6 Topcoats

3.6.1 Development of topcoats for automotive coatings

The first boost to automakers' productivity came with the advent of topcoats based on cellulose nitrate and plasticisers. These topcoats dried fast enough to meet the needs of on-line conveyor production (see Chapter 3). The main disadvantage was their low weatherability. Gradual improvements were made by replacing them with combinations of alkyds and cellulose nitrate. However, continually rising demands meant that research had to continue, too. The introduction of **stoving enamels** based on alkyds and amino resins achieved both goals, namely fast drying and adequate weatherability (as demanded at that time). The first stoving enamels contained urea resins as crosslinkers. However, melamine resins offered even better weatherability. Even though these were more expensive than urea resins, they were nonetheless incorporated into automotive topcoat formulations. Initially, alkyd resins acted as plasticising resins. They consisted of unsaturated fatty acids, with the oil sources being cottonseed oil, soy oil, sunflower oil, and more particularly dehydrogenated castor oil. When alkyds are modified with unsaturated fatty acid, they can dry by crosslinking with atmospheric oxygen. For a long time, it was thought that, in stoving enamels based on the aforementioned combination, the amino resins and alkyd resins crosslinked separately at elevated temperatures in the oven, and that the alkyd was the plasticising component of the combination. Surprisingly late, in the mid-1950s, it was discovered that co-crosslinking also occurs between them. As co-crosslinking yielded optimum film properties, the resins were developed so as to promote the participation of co-crosslinking in film formation: the hydroxyl values of the alkyds were increased. Alkyds modified with unsaturated fatty acids were therefore replaced by saturated fatty acids. These were non-yellowing and offered better weatherability. The saturated fatty acids at that time were based on coconut oil, mainly the short chain moieties of the fatty acids of coconut oil. Later, when the linear saturated fatty acids of coconut oil were used increasingly in the production of surfactants and high performance lubricants, their place in alkyds was increasingly taken by synthetic fatty acids. Such fatty acids have branched alkyl chains and are prepared mainly by oxo synthesis from different olefins. The alkyds modified with such monocarboxylic acids are notable for their excellent weatherability and lack of yellowing. They offer an optimum balance of hardness and flexibility. However, in terms of pigment wetting and application behaviour, the long-chain linear fatty acids have some advantages. Thus, the alkyds in optimised topcoat formulations contain synthetic fatty acids and some quantities of long-chained natural fatty acids.

Meanwhile, in the USA, development work took a different direction. Topcoats based on cellulose nitrate were initially replaced by systems based on high-molecular polyacrylates. Thermoplastic polyacrylates, like topcoats based on cellulose nitrate, dried by physical means only. They conferred better weatherability and did not yellow. When the effects of crosslinked systems were recognised, the logical step was to replace the thermoplastic, drying polyacrylates by acrylates that had lower molecular weights and contained functional groups for crosslinking. Since the late 1950s, topcoats in the USA had been based on acrylic resins containing hydroxyl groups and melamine resins as crosslinker. They offered excellent weatherability, did not yellow, and were additionally resistant to solvents and chemicals. Their weatherability paved the way for their use in other markets. In Europe, topcoat makers persisted with alkyds because their weatherability now matched that of acrylic resins. More crucially, alkyd systems bestowed a much better appearance than acrylic systems. By appearance here is meant the totality of levelling, smoothness, filling power (topcoat holdout), and gloss.

The solids content (application viscosity) of such topcoats containing alkyds and melamine resins varied from 45 to 55 % by weight, the precise figure depending on the colour. The relatively high application solids meant that the pressure to economize on solvents and to eliminate volatile organic compounds (VOCs) was less than for conventional solvent-borne basecoats, which had much lower application solids. Nevertheless, trials were conducted with a view to launching topcoats with higher application solids onto the market. Several resin groups were involved: alkyds, saturated polyesters, and acrylic resins. For topcoats based on high-solid alkyd resin and HMMM resin as crosslinker, the solids content varied from 55 to 65% by weight, again the precise figure depending on the colour. Another way to achieve such high-solids was to use low-molecular polyisocyanates as crosslinkers. Of course, that meant dealing with two-component coatings. However, in contrast to clearcoats, where two-component coatings had become established (see Chapter 3.8.3.4), the topcoats posed problems due to the need for frequent colour changes.

Another development was an improvement in resistance to chemicals, mainly acids. The resultant topcoat systems employed hybrid crosslinking, the crosslinkers being a combination of melamine resins and blocked polyisocyanates [28]. Finally, the positive experiences gained from the use of basecoat-clearcoat systems for effect topcoats were then applied to topcoats without effect matter.

To distinguish between topcoats with and without effect matter, those without were called solid colour topcoats or straight shade topcoats. All the definitions convey the impression that the colour and brightness of such topcoats do not change with the angle of view. Thus, the effect basecoats, which were introduced into the automotive coating market in the late 1960s, were followed in the 1980s by **solid colour basecoats**. The combination of solid colour basecoats and clearcoats greatly improved gloss, filling power, smoothness, weatherability, resistance to various chemicals and mechanical impact. At first, the solid colour basecoats contained solvent-borne resins. As the solids content of such basecoats was much lower than that of the common one-layer topcoats, solvent emissions (VOCs) increased significantly. This was something of a backward back given that solvents were also emitted from the clearcoats. Even back then, emissions reductions and lower VOC levels were major goals. It thus made sense to formulate solid colour basecoats as water-borne systems. That was the only way to introduce solid colour basecoats onto the market without increasing the VOC values. Optimum application behaviour was more difficult to achieve for **water-borne solid colour basecoat** than for effect basecoats. To provide optimum hiding power and brilliance of colour, a great many colours require relatively high film thicknesses, much more so than is the case for effect basecoats. Water-borne basecoats readily blister (popping) if they are applied in thick layers and over-coated with solvent-borne clearcoats. In addition, because water evaporates more slowly than most organic solvents, sufficient levels of rheological additives (thickeners) must be added, as otherwise the thick film layers will sag on vertical car body parts. Nevertheless, most car makers in Europe have replaced the one-coat solid colour topcoats with water-borne solid colour basecoat and clearcoat systems. Another advantage is that it is no longer necessary to change lines or line conditions during application to suit the colour or effect. All colours and effects are applied in the same manner and with the same clearcoat. Only large vehicles, such as trucks, buses, and train wagons, are still coated with one-layer solid colour topcoats.

Powder topcoats play only a subordinate role [106] in automotive coating.

3.6.2 Requirements on topcoats

As solid colour topcoats are the upper layer of the automotive coating system, their primary task is to protect the entire coating system and substrate against all external influences. They also have to convey an appealing impression consisting of colour, gloss, and smoothness. The various requirements may be classified into three groups:

- application behaviour
- appearance
- resistance

Of course, there are interactions between these groups of requirements: optimum application behaviour is a precondition for optimum appearance. Optimum resistance ensures an enduring appearance over the product lifecycle. The colour impression of a topcoat is generated by absorption and reflection of light. It is determined by the concentration and distribution of pigment particles. The concentration of pigment particles in the film matrix is limited. Once the specific packing density (PVC or pigment volume concentration) of pigment particles in the film matrix is exceeded, the hiding power no longer increases and other film properties such as gloss, smoothness, and resistance to solvents and chemicals are impaired. Of course, scattering and, more importantly, absorption vary with film thickness [10].

3.6.2.1 Application behaviour

To provide adequate hiding power, topcoats must be applied in sufficiently thick layers. The dry film thickness ranges from 40 to 45 µm. Application conditions must be rendered conducive to such film thicknesses. Sufficient quantities of paint must be applied in the time allowed by the line speed. The still wet paint layers must not sag on vertical car parts. In the drying phase (flash-off) and at the start of stoving, the solvents must evaporate uniformly. They must not cause blisters or other film defects. Smooth, glossy surfaces must be formed.

Optimum coating conditions on line are achieved with electrostatic spray equipment and high rotation bells. Topcoats may also be applied with high-efficiency, pneumatic spray guns. The goal is to maximise transfer efficiency.

Sagging is avoided by incorporating rheological additives into the paints. Sagging resistance is tested by spraying a wedge-shaped paint layer onto a panel perforated with holes of defined diameter. The panels are coated, flashed-off and stoved vertically. The holes indicate where sagging first occurs. The dry film thickness at which only minor sagging at the holes (e.g. less than 1 cm) occurs is determined. The target is set to a film thickness 30 % higher than the average application layer thickness [107].

Popping is the formation of blisters due to the sudden evaporation of solvent or reaction products (a boiling process on small film spots). These defects are observed particularly when evaporation occurs at a stage when the surrounding film has already achieved higher viscosities through evaporation and crosslinking reactions. The popping tendency is also determined with the aid of a wedge-shaped paint layer [108]. The application conditions are chosen so as to promote popping. Parameters that influence popping are, in addition to layer thickness, the amount applied per time (transfer masses), the air temperature, the air-fall velocity (the rate of air circulation in the flash-off zone) and finally the temperature curve in the stoving oven. Popping may also occur when air is introduced into the paint material during the production process, either in dissolved form or as micro-foam, and then escapes during film forming. Therefore, the specific application conditions which vary from one application line to the next, must be taken into consideration for specific paint formulations. Thus, the composition of topcoats of the same colour may vary with the application line on which they are used. Parameters that can be used to influence the popping tendency are solvent composition, application solids content, application viscosity, and the reactivity of the resins for crosslinking, including the type and quantity of catalysts. Since reactivity is very high, the film viscosity increases rapidly in the first stoving phase. If, after that, further reaction products are cleaved, they may give rise to blisters (reaction blisters).

Levelling, too, is significantly influenced by the application conditions. For optimum levelling, the spray particles must wet the surface perfectly, spread over the surface and flow together to form smooth, homogeneous films. Evaporation of solvents in the flash-off zone and at the start of stoving creates convection currents in the paint layer which may impair the film smoothness. These processes differ in vertical and horizontal film layers. Nonetheless, the entire topcoat film must exhibit the same degree of levelling. Such processes are counteracted through the judicious choice of type and quantity of rheological additives.

During stoving, the viscosity of the film drops as the temperature rises. However, with progressive evaporation and the onset of crosslinking, it rises again. At this stage of the stoving process, the temperature is constant. In the phase described, the viscosity curve forms a "valley". The shape and size of the "viscosity valley" is crucial to levelling. If it is very deep, the paint layer may penetrate into the surface of substrate and may later reproduce the structure of the substrate (so called telegraphing of structure), which is deleterious to the filling power. In addition, if the viscosity is very low, sagging may take place. Conversely, if the "viscosity valley" is very shallow or the viscosity increases too quickly (through evaporation or crosslinking reactions), convection currents are "frozen" at an early stage, flow ceases, and a relatively rough structure is generated. Several trials aimed at accurately measuring the viscosity of films have been performed [109], with a further goal of enabling levelling to be analysed on the basis of measurable values. However, the process is too complex and the throughput time is too short to enable reproducible viscosity values to be obtained. Formulation experts must be capable of interpreting the results for variations in paint composition with a view to optimising levelling properties. Thus, development work continues after their results have been tested on model applications. Since levelling is also influenced by several parameters with complex interactions, visual assessment is necessary for development to proceed. It is also possible to measure and classify the roughness of a paint surface by pertometer analysis [110]. Parameters affecting levelling also include the application conditions: the type of spray gun, the gun settings (quantity of fed paint per time, quantity of feed air and regulation air, spray time), air-fall velocity in the booth, flash-off temperature and duration, temperature curve and duration in the oven. Of course, the topcoat system must be adapted to these specific conditions. Usually the application solids and the application viscosity are default parameters and cannot be varied. Finally, the topcoat system has to be adapted by varying the composition. The optimisation parameters in this area are the choice of optimum solvent combination, characteristic resin values (molecular weight and molecular weight distribution, crosslinking reactivity), and the choice of additives. Additives that can influence the levelling properties are wetting agents, surfactant additives (levelling additives in the narrower sense), and in some cases antifoaming additives and rheological additives. The rheological additives introduce intrinsic viscosity into the topcoat material. The type and quantity of such additives must be balanced very well because they have a dual role: to prevent sagging and to generate surface structure so as to create an aesthetically appealing surface effect (see definition of filling power in Chapter 3.5.2).

Appearance

The goal of the statements above is not to generate very planar surfaces. As already mentioned, the human eye is positively receptive to a structure with small amounts of long waves and nearly no short waves. The technical terms for these impressions are filling power and topcoat holdout. These properties are the most important components of the term "appearance", which is the sum of several parts. They describe the subjective impression gained by the human eye of optimal coverage of a substrate by a coat of paint that is smooth and sufficiently thick. As mentioned above, not only are the topcoat properties responsible for meeting the demands for optimum appearance, but the primer surfacer also has to play its part in generating filling power and topcoat holdout (see Chapter 3.5.2). It is very difficult to define the different impression parameters by physical measurements.

Gloss also plays a part in overall appearance. Gloss is the reflection of light from a surface. The gloss of a paint surface therefore depends primarily on the refractive index of the film matrix. The higher the refractive index, the higher is the gloss. Gloss is measured with the aid of reflectometers under defined reflection angles (20° and 60°). The measurement is expressed in terms of a given standard in percent (%) [111]. Since the refractive index of a paint film can vary considerably and, for example, also depends on the colour, the readings can also vary substantially. However, the human eyes can compensate for such differences.

To better capture the impression gained by the human eye, measurement of the so-called **distinctness of image** (DOI) was introduced, which is the sharpness in reproduction of a surface reflection. The measured parameters are the mirror effect and the contrast of the mirror image. The measurement is carried out visually by observing the mirror image of a black-and-white-chequered specimen panel; or digitally by the camera-focusing principle. The DOI is quoted as a number [112]. DOI measuring is also important for assessing the surface properties of OEM clearcoats (see Chapter 3.8.2.1).

Gloss and DOI are influenced by the micro structure of topcoat surfaces. The lower the quantity of structures with wavelengths half the size of visible light, the higher is the gloss and the greater is the DOI. Such surfaces are achieved through optimum levelling. However, the most important precondition is that the pigment particles are well wetted and incorporated perfectly into the resin matrix of the film. The pigment volume concentration must therefore be much lower than the theoretical critical value, which is the maximum volume packing density. The pigments must be dispersed perfectly during production of the paint. And they may not agglomerate during storage or application of the paint or during film forming. In addition, the resin components must be compatible in solutions and in films. If resins are incompatible, this may give rise in the films to domains whose interfaces may differ in optical density and generate lower gloss and DOI values.

Until now, car producers and their customers requested high-gloss finishes, and car washes offer high-gloss polishing after washing. Now, there is also talk of offering cars that have matt surfaces. Matt surfaces are usually generated by adding sufficient quantities of pigment-like additives (e.g. colloidal silicas), which generate a micro structure on the surface that scatters visible light. Or, the topcoats deliberately contain an incompatible resin combination that creates the matt surface.

Reproducible colours are absolutely essential. First, customers like their cars to come in the colour which they selected from colour samples. More important for the colour impression, however, is that the colours match if parts of the car are coated separately and then assembled afterwards. In addition, if car body damage has to be repaired, the colour of the repair part must be a very good match for the existing colour. The human eye is relatively sensitive to differences in colour, but even here there are individual differences. There are a couple of methods for ensuring that the colours match very well. First, the pigment producers are asked by the paint producers to guarantee that pigments are always available in the first standardised characterisation as regards chemical composition, type and quantity of surface treatment, particle size and particle size distribution. Of course, that is an important precondition for faithful reproduction of different colours. The paint producers themselves are responsible for producing the optimum pigment dispersions, which must be capable of storage without re-agglomeration. Pigments that have not been dispersed very well have lower colour strength. When pigments re-agglomerate, the number of dispersed particles decreases, which also leads to lower colour strength and lower colour brightness. Since most of the suitable coloured topcoats consist of more than one pigment, it is important that the pigments be mixed perfectly and be reproducible in order that the same colour may be matched across all production batches. A great deal of effort and colour adjustment goes into ensuring reproducibility for all batches. First, the colour is analysed by measuring the reflection of a standardised light on the surface of a test panel (the so-called colour location). The readings are compared with a standard, which is prepared from the first or the previous batch. If the colour does not match very well, there must be means provided for adjusting the colour by correcting the batch. This corrective action is called **tinting**. It is carried out by adding tinting pastes, which are standardised pigment dispersions of single pigments (see Chapter 3.6.3.6). A number of computer programs are available for calculating the type and quantity of tinting pastes needed to adjust the colour. The computer programs were developed to render the tinting process efficient which, in the past, took a lot of time as it was done visually by tinting specialists. It must be remembered that the application process may also influence the resultant colour. Even the heat capacity of different parts of the car body may influence the colour as pigments may reagglomerate differently during film forming and crosslink at different stoving temperatures. It is therefore necessary to install a colour management system to guarantee the reproducibility of colours by controlling all the steps from production, delivery, storage and application.

3.6.2.3 Resistance properties

Besides the requirements on appearance and colour, of course, the topcoats have to protect the entire coating system against external impact. To fulfil these requirements, topcoats are optimised with regard to various resistance properties.

The first is protection against **mechanical impact**, such as stonechipping, deformation, abrasion, and scratches. Although primer surfacers are particularly designed to absorb the energy of stonechipping, the topcoats are also called on to provide some of that protection. Consequently, topcoats must adhere well on the primer surfacers and withstand some deformation, without cracking or loss of adhesion.

Resistance to mechanical impact is tested by applying the entire coating system to test panels (panel material corresponding to that of the body) and then performing adhesion [113] and various impact tests. The most important mechanical resistance is the stonechip type. The various methods of testing stonechip resistance have already been described in Chapter 3.5.2. During a car's life time, sand and dust may impinge on the finish, causing abrasion and scratches. Scratches are observed particularly after a car has passed through a car wash.

Topcoats must be resistant to **chemicals,** by which is meant a wide variety of substances, including acids (in practice, mainly sulphuric acid from the car battery) and bases. Topcoats must also resist solvents - essentially fuels, but also including high octane fuels which may contain alcohol; and brake fluids, which are really aggressive. Resistance to natural rosins as the excretions of greenfly, which can damage coating surfaces by dissolving and swelling them is also needed. Of course, the coatings must be stable to aqueous solutions of surfactants, which are used for all cleaning processes. Even water (deionised water) can swell coating layers at elevated temperatures. As already mentioned, swollen coating layers are sensitive to mechanical impact. Last, but not least, the requirements for optimum resistance were the reason that one-layer topcoats were replaced by solid-colour basecoat clearcoat systems (two-layer systems) in which the colour function is assigned to the basecoat and the other functions -mainly resistance - are allocated to the clearcoat.

A very important requirement is **weatherability**. The primary concern here is resistance to sunlight. The UV content of sunlight can destroy molecular bonds in organic molecules, which in our case are contained in the resins. Its action is boosted by elevated temperatures, high humidity and airborne pollutants, e.g. sulphuric acid and nitric oxides (from acid rain) and also products of the reactions of ozone (e.g. hydroxyl free-radicals). All these influences have a major impact on the properties of clearcoats (see Chapter 3.8.2). In the case of solid-colour topcoats, the influences are highly specific. If pigment particles are not perfectly embedded, moisture may penetrate along the interfaces of pigment and film matrix and cause swelling. Besides the purposes of conferring hiding power and colour brightness, there is a further reason that pigment particles need to be wetted perfectly. As already described, coloured pigments absorb electromagnetic waves of specific wavelengths, including UV wavelengths. However, in case of topcoats, the selected pigments must be resistant to UV light. It is an advantage, as absorption of the UV light protects the lower layer of the coating structure. Some pigments induce damage to film matrix through catalytic interactions of particle surfaces and the light (photocatalysis). The surfaces of such pigments need to be modified by doping with other compounds or other treatment methods.

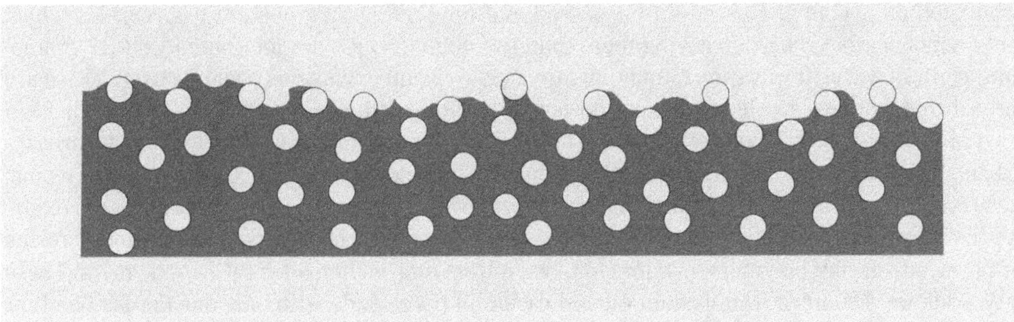

Figure 3.6.1: Schematic description of matting and chalking of topcoat films

Symptoms of surface damage due to decomposition by light and other weather influences are matting and chalking. **Matting** is the local destruction of the film surface to form a microstructure which scatters light, creating the matt effect. As a rule, matting is compensated by polishing. **Chalking** results when the process continues and the film matrix is destroyed such that pigment particles are released. When this happens, the free pigment particles can be wiped off, like chalk from a blackboard. Figure 3.6.1 shows a schematic diagram of matting and chalking of films.

The choice of resins for bestowing optimum weatherability on topcoats is totally different from that for clearcoats. For topcoats, it makes sense to choose aromatic compounds for the resins (alkyds). Aromatic compounds – together with some pigments – absorb UV light and in this way protect the entire coating system against destruction by UV light. Unlike the case for clearcoats, there is no risk of the formation of check-cracks (stress cracks) as the interfaces of pigment and resin matrix absorb the energy that forms those cracks. It is not necessary to add UV absorbers to pigmented topcoats as absorption occurs anyway. However, it does make sense to add free-radical scavengers, which benefit the weatherability of topcoats, mainly for obtaining optimum gloss retention. The effect of UV absorbers and free-radical scavengers is described in detail in the Chapter 3.8.2.

3.6.3 Composition of OEM topcoats

Resins for OEM topcoats are combinations either of alkyds and melamine resins or of acrylic resins containing hydroxyl groups and melamine resins. The resistance of topcoats to chemicals can be improved by incorporating blocked polyisocyanates as at least partial crosslinker. The colour impression is generated with inorganic and organic coloured pigments, including pigments for white and black. The most important additives for topcoats are wetting agents and dispersion agents, rheological additives and levelling agents. It is also possible to add reaction catalysts for crosslinking.

3.6.3.1 Alkyd resins

Alkyd resins are polyester resins. Unlike the saturated polyesters (see Chapter 3.5.3.1), they contain monocarboxylic acids in addition to polycarboxylic acid and polyols. The actual definition of alkyd resin is [114]: *"polyester resins prepared by polycondensation of polycarboxylic acids, polyfunctional alcohols, and oils or fatty acids."*

Development of alkyd resins

Alkyd resins were the outcome of attempts to produce hard resins by combining vegetable oils (triglycerides) with polyesters. Some vegetable oils were well-known binders that could form films by crosslinking with atmospheric oxygen (oxidative drying). Since the drying process takes a long time, attempts were made to accelerate drying properties, mainly by adding binders which boosted physical drying, e.g. rosin, phenolic resins and finally polyesters. Such preparations were initially formulated on drying oils based on linseed oil and tung oil. Drying oils are triglycerides which contain monocarboxylic acids having more than two double bonds per molecule (e.g. 9,12,15-linolenic acid). However, it was soon found that fatty acids of semi-drying oils could be used. These are derived from cottonseed oil, soy oil, sunflower oil, safflower oil, and also tall oil (fatty acids from wood produced during paper production). These fatty acids are mixtures of various monocarboxylic acids, the most important ingredient being the 9,12-linoleic acid. As components of alkyd resins, these fatty acids, which have a much higher molecular weight than the oil, confer optimum oxidative-drying properties. Alkyd resins containing non-drying fatty acids served as plasticiser resins for physically drying polymers, e.g. for cellulose nitrate and – later – for PVC copolymers. These fatty acids were sourced from peanut oil and castor oil (fatty acids with only one double bond) or coconut oil (with saturated fatty acids). A semi-synthetic oil was produced by dehydrogenation of castor oil (DCO or dehydrogenated castor oil). The fatty acids thereof consist mainly of 9,11-linoleic acid. Like the fatty acids of tung oil, the conjugated double bonds of DCO fatty acids do not take up high quantities of oxygen by addition. Instead, film formation by crosslinking takes the form of polymerisation reactions (1,4-polymerisation), which is accelerated at elevated temperatures. DCO fatty acids were therefore preferred for formulating stoving enamels. Initially, such grades of alkyds were combined with amino resins to plasticise them. It was later found that alkyds, mainly those which contain sufficient quantities of free hydroxyl groups, can crosslink with amino resins (co-crosslinking), and that the resultant films offer an optimum balance of hardness and flexibility and are also resistant to solvents and chemicals. With that knowledge, it was only a small step to the development of topcoats containing alkyds that were modified with saturated fatty acids that contain significant quantities of hydroxyl groups, and contained melamine resins for crosslinking. Originally, the sources of those fatty acids were coconut oil. Nowadays, acids from coconut oil are used for the production of biodegradable surfactants (e.g. for cosmetic articles), high performance lubricants (e.g. for jet engines) and cooling media (e.g. for transformers). Synthetic fatty acids thus became more important for alkyds. Synthetic fatty acids are aliphatic, branched monocarboxylic acids made from various olefins by the oxo synthesis. Alkyds based on them are distinguished by excellent hardness, adequate solubility, yellowing resistance, and weatherability. For pigment-wetting properties and optimum application behaviour (surface wetting, no popping, levelling, and flow), it makes sense to combine the synthetic fatty acids with linear fatty acids that bear long aliphatic chains and also small quantities of double bonds.

Structure of alkyds

As already mentioned, alkyd resins for topcoats consist of polycarboxylic acids or their derivatives, polyols, and monocarboxylic acids. The most important polycarboxylic acid derivative is phthalic anhydride. The specific esterification behaviour of phthalic anhydride makes it possible to prepare alkyd resins with relatively high average molecular weights. The polyols for these alkyds are triols (trimethylolpropane, glycerol) and tetrol (pentaerythritol). Combinations of polyols and diols are occasionally used, e.g. pentaerythritol and propylene glycol (average functionality of three). Suitable monocarboxylic acids (synthetic fatty acids) are isooctanoic acids, e.g. 2-ethylhexanoic acid, isononanoic acids, e.g. 3,5,5-trimethylhexanoic acid, and isodecanoic acids, e.g., 2,2,4,5-tetramethylhexanoic acid. Since 2,2,4,5-tetramethylhexanoic acid contains a tertiary carboxyl group, the extent of esterification by standard processes is very low. Therefore, the tertiary carboxylic group is made to react with epichlorohydrin and the epoxy group is re-formed by reaction of the chlorohydrin with sodium hydroxide (see preparation of epoxy resin in Chapter 3.4.5.1). It is easy to incorporate the resultant glycidyl ester into alkyd resins [115]. The fractions of monocarboxylic acids bearing long aliphatic chains are fatty acids from coconut oil, as well as unsaturated fatty acids from soy oil, dehydrogenated castor oil, and tall oil. Some alkyds also contain aromatic monocarboxylic acids, e.g. benzoic acid or p-tert.-butylbenzoic acid, which increase the hardness and the physical drying properties of the resins thereof. For greater flexibility, alkyd resins are modified with aliphatic dicarboxylic acids, e.g. adipic acid, instead of phthalic anhydride.

Alkyd resins are produced in a one-step process starting with all three raw materials (phthalic anhydride, polyols, and monocarboxylic acids). The first – and fastest – reaction is the addition of polyol to the phthalic anhydride, which forms an ester group and generates a free carboxyl group. This carboxyl group esterifies more slowly with the remaining hydroxyl groups of polyols than do the free carboxyl groups in monocarboxylic acids. Only then does the free carboxyl group of the phthalic monoester react with hydroxyl groups, to yield chain extension and high molecular weights. In addition, the growth of larger molecules is controlled by transesterification reactions, which take place at the same time as the esterification reactions. The equilibrium between esterification and transesterification reactions yields increasing amounts of molecules of average molecular size. However, the most important influence is the fact that monocarboxylic acids reduce the functionality of polyols *in-situ*, which mainly led to relatively small molecular weight distributions. In the past, monocarboxylic acids were defined as chain stoppers. However, in reality, modification with monocarboxylic acid offers the possibility of obtaining resins which are branched and achieve very high average molecular weights without gelling. As already mentioned, that is not possible with saturated polyesters. High branching of saturated polyesters limits the average molecular weight of products attainable. Figure 3.6.2 shows the basic reactions leading to the formation of alkyd resins. Naturally, the reactions do not take place one after the other. The order shown relates the different velocities of the reactions, which all start together but involve very different consumption rates over time.

Alkyd resins are made by fusion at temperatures of 180 to 240 °C. The process is controlled by removing the water of reaction as effectively as possible with an entraining agent. Reflux solvents are added to support water evaporation. The progress of the reaction is checked by measuring

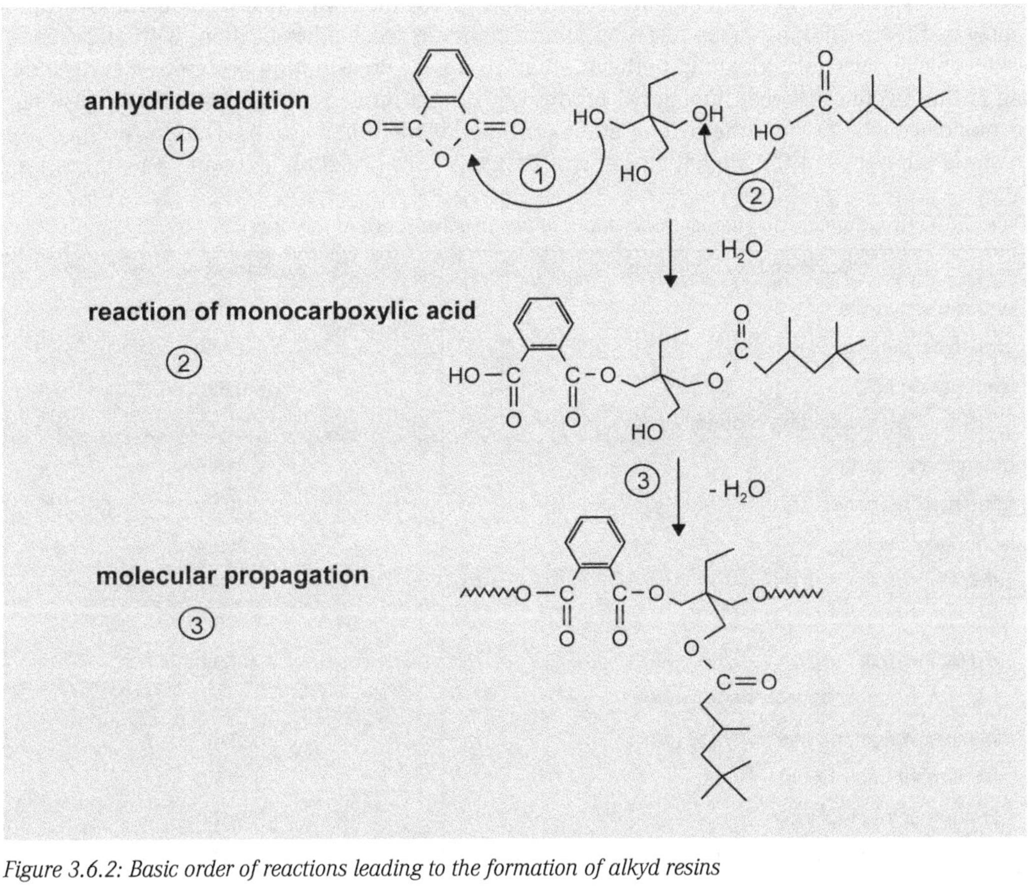

Figure 3.6.2: Basic order of reactions leading to the formation of alkyd resins

the acid number, which is related to the degree of condensation, and the viscosity of the melt or of a test solution, which depends on the molecular weight of the resin. The attainable molecular weight is determined by the molar ratio of polyols to polycarboxylic acids and the degree of polycondensation (conversion rate). In general, alkyds have number average molecular weights of 1500 to 8000 g/mol. Typical alkyds for OEM topcoats have lower average molecular weights in the region of 1500 to 3000 g/mol. Relative to the degree of branching, the molecular distribution is relatively small. The molecular dispersivity, i.e. the quotient of the weight-average molecular weight and the number-average molecular weight, describes the width of the molecular distribution, and lies between 3.5 and 5.5 for the alkyds discussed here. The alkyds for stoving enamels normally have hydroxyl values of 75 to 140 mg KOH/g, which is optimal for crosslinking with melamine resins. Acid values are a measure not only of the degree of condensation, but also of the functional groups in the finished alkyd resin. These do not participate directly in the crosslinking reactions, but exert a catalytic effect on the reactions; the hydroxyl values are mainly 10 to 20 mg KOH/g. Table 3.6.1 describes the formulation and characteristic values of a model alkyd resin [116] that is also suitable for topcoats containing phthalic anhydride, trimethylolpropane and isononanoic acid, without any special modifications. Delivery forms of such alkyds are solutions in aromatic hydrocarbons (e.g. xylene or Aromatic 100), which may additionally contain small fractions of more polar solvents (e.g. n-butanol or propylene glycol monoethers). The solids content lies between 50 and 70 % by weight.

3.6.3.2 Melamine resins

As already described, melamine resins in stoving coatings react via their functional groups, both with the hydroxyl groups of partner resins (co-crosslinking), and with themselves (self-crosslinking). That is also true of combinations of alkyd resins and melamine resins in automotive OEM topcoats. Co-crosslinking takes place by etherification or trans-etherification, with cleavage of water or monoalcohols, yielding methylol ether groups as crosslinking bridges. Self-crosslinking is the reaction between functional groups on the melamine resins, with cleavage of water or monoalcohols, to yield methylene ether or dimethylene ether groups. Due to the increase in steric hindrance of the aliphatic chains of fatty acids, co-crosslinking of alkyd resins is less

Table 3.6.1: Composition and characteristic values of model alkyd for OEM topcoats

Building blocks. moles	Triol-type
phthalic anhydride	1.000
trimethylol propane	1.050
isononanoic acid	0.630
building blocks, wt-‰	
phthalic anhydride	413.6
trimethylol propane	393.2
isononanoic acid	269.4
sum	1076.2
water	76.2
yield (AN = 19.0)	1000.0
characterisitc values	
molecular weight, number average [g/mol]	2457
acid number [mg KOH/g]	15.0
OH-number [mg KOH/g)	100
Viscosity, 60 % in xylene [mPa·s, 23 °C)	5500

effective than with saturated polyesters containing comparable quantities of hydroxyl groups. It should be remembered that co-crosslinking boosts flexibility and weatherability; in contrast, self-crosslinking supports hardness and solvent resistance. The goal is to achieve the greatest balance of these properties. Once the types of alkyd resin and melamine resin are defined, optimisation takes place by varying and testing different mixing ratios of both resins. The usual mixing ratios for alkyd resins and melamine resins range from 60:40 to 75:25. The optimum mixing ratio is influenced by the quantity of functional groups on both resins, the alkyd hydroxyl groups, and the molecular weight of the melamine resin, which is directly related to the quantity of functional groups. If low-molecular melamine resins are used, the optimum quantities are lower. Figure 3.6.3 shows the co-crosslinking reaction between an alkyd resin and a melamine resin.

A typical melamine resin for an automotive topcoat contains melamine, formaldehyde and n-butanol in the molar ratio of 1:4:2.5 [117]. Thus, it still contains residual NH groups, which take part in self-crosslinking by reacting with methylol groups or etherified methylol groups to form methylene groups. In addition, the methylol groups and etherified methylol groups react by co- and self-crosslinking. The average molecular size of the melamine resin is defined by the average number of melamine molecules per resin molecule, and here is 2.2.

Film forming takes place at temperatures of 130 to 145 °C, maintained for 15 to 20 minutes. Including the heat-up phase, the total stoving time is 20 to 30 minutes. These conditions require relatively reactive melamine resins. As already mentioned, the crosslinking reactions of melamine resins are accelerated by catalysts. For the given conditions and the melamine resins usually chosen, the carboxyl groups of the alkyds (acid values) are sufficient for catalysis. However, less reactive melamine resins, for example HMMM resins, require external acid catalysts. Sulphonic acids are used for this purpose. The resultant films are adequately hard and flexible. They are highly weatherable and relatively highly resistant to chemicals and solvents. Combinations of alkyd resins and melamine resins confer better wetting, levelling, and gloss than saturated polyesters or acrylic resins containing hydroxyl groups.

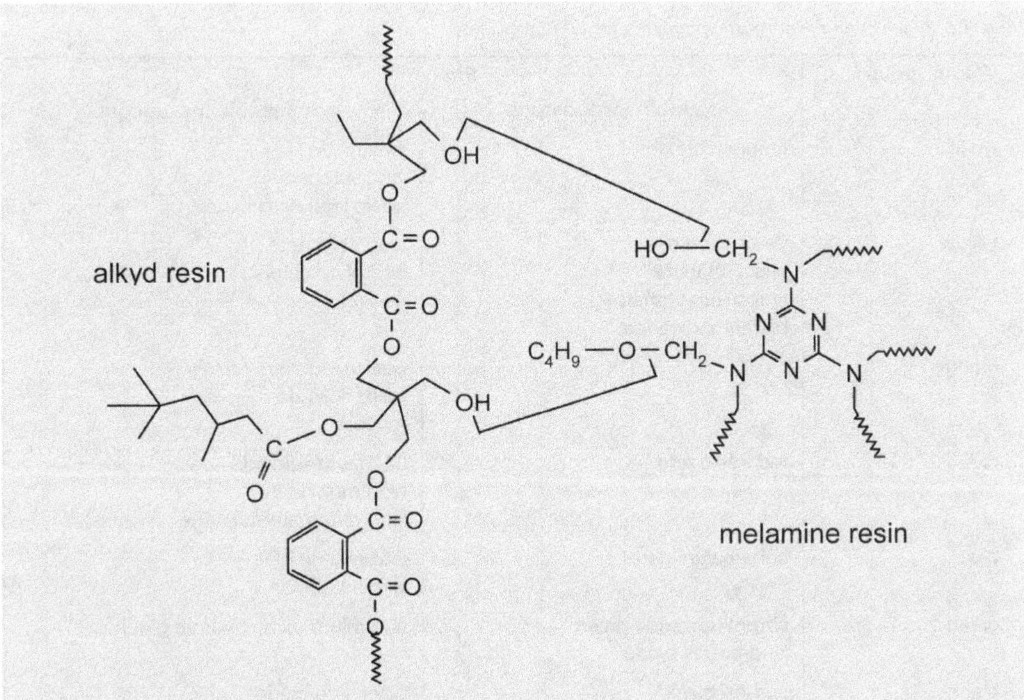

Figure 3.6.3: Co-crosslinking reaction of an alkyd resin and a melamine resin

3.6.3.3 Other resins

Some producers of automotive OEM topcoats employ hydroxy acrylic resins crosslinked by melamine resins to enhance weatherability. Acrylic resins have a reputation in the USA and the Far East for offering superior weather resistance to alkyd resins and polyester resins. Outdoor weathering tests on topcoats based on acrylic resins and alkyd resins containing saturated fatty acids have failed to show significant differences after several years of exposure to the Florida weather. Some alkyd topcoats achieve much better gloss retention than formulations based on acrylic resins. The explanation is not that the polyester chains of the alkyd resins are the weakest link in the film matrix that is exposed to the sunlight and the elements. Rather, the weakest link in the topcoat films is the methylol ether bridges of melamine resins, which both topcoats series contained. Methylol ethers are sensitive to acid attack, and therefore also to environmental influences, such as acid rain. Topcoats containing at least partially blocked polyisocyanates as crosslinker instead of melamine resins [28] are much more resistant to chemicals, i.e. to acid rain. These results have been extensively exploited in the formulation of OEM clearcoats (see Chapter 3.8.3.5).

3.6.3.4 Inorganic pigments

The choice of inorganic pigments for OEM topcoats is mainly influenced by their physiological compatibility. Topcoats therefore no longer contain pigments comprising lead, cadmium, or chromates. In addition, pigments with low tinting power or with low resistance to chemicals or low weatherability are excluded from topcoats. Table 3.6.2 lists the different inorganic pigments by colour in two columns according to their suitability for pigmenting OEM topcoats.

Titanium dioxide

Titanium dioxide is the most important and commonest white pigment. The white pigments employed in the past (zinc white, lithopone, white lead) are no longer used. Titanium dioxide is also used in numerous other applications (plastics, ceramics, papers, textiles, catalysts, glasses, cosmetics, UV absorbers). Apart from offering outstanding resistance to solvents and chemicals,

Table 3.6.2: Selection of inorganic pigments for OEM topcoats

Colour group	Pigments	
	suitable for topcoats	not suitable for topcoats
white	titanium dioxide	zinc oxide zinc sulphide basic lead carbonate
yellow	yellow iron oxide nickel titanate chromium titanate bismuth vanadate	lead chromate cadmium sulphide
orange		basic lead chromate lead molybdate red lead
red	red iron oxide	cadmium selenide lead molybdate lead chromate-sulphate
blue	cobalt aluminate	ultramarine blue Berlin blue
green	chromium oxide green zinc-cobalt oxide	chromium oxide hydrate green
black	carbon black	black iron oxide iron titanate

the key reason for its ubiquitous use is its total physiological harmlessness. Titanium dioxide occurs in three crystalline lattice structures known as anatase (tetragonal) rutile (tetragonal), brookite (orthorhombic). Titanium dioxides with anatase and rutile structures are suitable for technical applications. At 700 °C, anatase is converted into rutile. Rutile is the most stable form of titanium dioxide and has the highest atom packing density. The melting temperature of titanium dioxide is about 1800 °C. The absorption spectrum changes above 400 °C, at which temperature its colour becomes yellow; this process is reversible below 400 °C. At temperatures above 1000 °C, titanium dioxide splits off one atom of oxygen per molecule to form Ti^{3+}-ions. Titanium dioxide is amphoteric, but is highly resistant to chemicals. It is soluble only in concentrated sulphuric acid, in hydrofluoric acid and, at very high temperatures, in alkali melts. Titanium dioxide absorbs near-UV light. The absorption maxima are 385 nm for anatase and 415 nm for rutile. Rutile is therefore somewhat yellower than anatase. However, due to its significantly greater resistance to light and weathering, rutile is preferred to anatase where topcoats are concerned. Titanium dioxide as such is virtually colourless. However, due to a high refractive index (rutile: 2.70), fine particles of titanium dioxide scatter visible light very efficiently. The optimum particle size for titanium dioxide as a white pigment is about 200 nm.

Titanium is the 9th most abundant element in the earth's crust. It is widely distributed but mostly occurs in a number of minerals. The most important minerals for mining of titanium dioxide are ilmenite (iron titanate), its weathering product leucoxene, and the mineral rutile. Rutile mineral is mainly found in secondary mineral deposits (sands). Today, titanium dioxide is produced in the **chloride process**, which has almost completely superseded the earlier **sulphate process** [118]. Most of the mined minerals contain only small quantities of titanium and so must be dressed. Significant quantities of iron compounds are often associated with titanium in ores. The strong colour of iron requires that it has to be removed as extensively as possible. Titanium dioxide pigments may contain no more than 50 ppm red iron oxide. In the chloride process, the dry mineral or the dressed forms are heated with petroleum coke and chlorine at a temperature of 700 to 1200 °C to yield titanium tetrachloride and other chlorides. The titanium tetrachloride distils, leaving the other chloride compounds behind. Cooling to below 300 °C causes the residual quantity of iron(III) oxide to precipitate. The titanium tetrachloride is condensed at 0 °C. It can be cleaned by partial reduction and a second distillation step. The pure titanium tetrachloride is oxidised at 900 to 1400 °C in excess oxygen to yield fine particles of titanium dioxide and chlorine (which is recycled). Addition of aluminium chloride (up to 5 % per mole) affords very fine particles of titanium dioxide. This is then calcined in a rotary furnace and post-treated. The latter consists in adding small quantities of aluminium oxide, silica, or other colourless oxides for immobilisation on the surface of the titanium dioxide particles. The purpose of post-treatment is to improve the weatherability of titanium dioxide. Untreated titanium dioxide can react with atmospheric moisture. This reaction is promoted by UV light (absorption maximum at 415 nm) and yields hydroxyl and hydroperoxide free-radicals and Ti^{3+}-ions. The pigment turns grey. A greater disadvantage is the reaction between the free-radicals and the surrounding polymer matrix. Destruction of the resin molecules leads to matting and chalking (see Chapter 3.6.2.3). It can be largely avoided by post-treating the pigment surfaces and choosing resins which form films that resist moisture diffusion. To facilitate dispersion, the pigments are treated with organic compounds that improve wetting by acting like surfactants. The light-scattering properties of titanium dioxide are also exploited in pigments that mostly absorb rather than scatter visible light. Some pigments, e.g. organic violets and phthalocyanine pigments, look black in pure dispersions. They develop bright colours only in combination with white. Finally, all pastel colours are based on titanium dioxide.

Iron oxide pigments

Minerals containing high quantities of iron oxides served as pigments in prehistoric times. The mineral goethite contains α-iron III oxide hydrate (FeO[OH]). The iron(III) oxide hydrates are greenish to brownish yellow, or ochreous to orange coloured. Haematite consists of α-iron(III) oxide (Fe_2O_3) in a corundum structure; maghaemite is the γ-form in a spinel structure. Iron(III) oxides range from light red to violet red. The mineral magnetite consists of iron II/III oxide (Fe_3O_4) in a spinel structure and is black. Where iron oxides contain manganese oxides, they are brown. Down to the present, natural iron oxides have been mined and used as pigments for coating materials and artists' colours. However, the iron oxides for OEM topcoats are exclusively synthetic in order that bright colours and high resistance may be obtained. The raw materials are iron and steel scrap and iron sulphate as the by-product of the synthesis of titanium dioxide, iron(III) chloride as the by-product of steel pickling, and other synthesis by-products containing a significant quantity of iron. Iron oxide pigments are produced in a number of processes. The iron oxides may be dissolved in sulphuric acid and then, through addition of alkalis or alkaline carbonates and with the pH still below 7, iron(III) hydrates and iron(III) oxide hydrates can be precipitated under oxidative conditions. The precipitation conditions (temperature between 10 and 90 °C) influence the crystal structures of the products. The second method (Penniman process) avoids the formation of large quantities of alkaline sulphates. A thinned solution of iron II sulphate is treated with small quantities of alkalis which cause nucleation of iron(III) hydrate crystals. Scrap is then added and, under oxidative conditions, the iron is transformed into iron(III) oxide hydrate. The third method produces iron oxides as the by-product of the reduction of nitro-aromatics to amino-aromatics (e.g. aniline) with scrap iron, which is transformed into iron(III) oxide hydrate. The iron oxides vary from yellow, to red and black iron in accordance with the production conditions and addition of other salts. The water-rich products are calcined, washed, and dried. Yellow iron oxides must be treated at low temperatures only. Above 180 °C, they are transformed into red iron(III) oxide, with cleavage of water. Red iron(III) oxide is stable up to 1200 °C. The higher the calcination temperature, the larger are the particles and the colours change from light brick-red to a violet red.

Iron oxide pigments are physiologically safe. The TLV value for the inert pigment dust is 6 mg/m^3. Yellow iron oxide has an acicular structure and is 50 to 200 nm wide and 300 to 800 nm long. The pigment particles are sensitive to high shearing energies (mainly during grinding processes). The anisotropy of the yellow iron oxide particles leads to metameric effects. However, yellow iron oxides with spherical particles are also commercially available. In topcoats, yellow iron oxides create colours ranging from light to dark ochre. No brilliant yellow is available, but the colour strength is relatively high. Of course, the pigment is not resistant to acids or elevated temperatures.

The particle sizes of red iron oxides lie between 100 and 800 nm. The colours vary from light brick-red to dark violet-red. Furthermore, the tinting strength and the hiding power decrease. It is not possible to generate brilliant red colours with iron oxides. Red iron oxide is very stable to chemicals and high temperatures. Red iron oxide is also suitable for corrosion protection primers. Black iron oxide has particle sizes of between 100 and 600 nm. On account of its low tinting strength, mainly relative to carbon black, black iron oxide is not used in topcoats.

Rutile mixed-phase oxides

Rutile mixed-phase oxides are rutile (titanium dioxide) pigments containing a specific quantity of other metal oxides in the crystal lattice. The heavy metal oxides are incorporated into the lattice structure of the colourless titanium dioxide where they generate specific colours. The modification is carried out by adding the other oxides during the calcination process. The colour effect stems from transfer of d-shell electrons of the heavy metal ions to the basic lattice of the titanium dioxide. The likelihood of such doping depends on the charge number of the heavy metals and the diameter of the ions. The basic lattice of the crystal is transferred to the structure of the added oxide, and the charge differences of ions can be compensated for by mixing heavy metal ions of different charge. The mixed-phase oxides group includes nickel-titanium-yellow and chromium-titanium-yellow. Both pigment types contain significant quantities of antimony.

The ions of nickel-titanium-yellow for example contain 3.3 % nickel, about 13.9 % antimony(V) ions and 78 % titanium ions. It has the composition $Ti_{0.85}Sb_{0.10}Ni_{0.05}O_2$. Nickel-titanium-yellow produces greenish yellow colours, which are more brilliant than the yellow iron oxides, but do not produce the bright colours of lead chromates. Chromium-titanium-yellow contains 3.0 % chromium(III) ions, about 13.9 % antimony(V) ions and 78 % titanium ions. Its composition is $Ti_{0.85}Sb_{0.30}Cr_{0.05}O_2$. The pigments have light and dark ochreous colours; they are more brilliant than the yellow iron oxides. The rutile mixed-phase oxides offer outstanding resistance to chemicals and elevated temperatures, and are totally physiologically safe.

Bismuth vanadate

Bismuth vanadate ($BiVO_4$) is prepared by precipitation when bismuth nitrate and sodium vanadate react in dilute nitric acid. Adding sodium hydroxide solution leads to the formation of crystals which are either monoclinic or tetragonal in accordance with the conditions. Precipitation of bismuth vanadate is modified with small quantities of phosphates, which improve weatherability and resistance to acids. Unlike the yellow iron oxide pigments, which produce light and dark ochreous colours, bismuth vanadate pigment coatings lead to very brilliant, pure yellow colours. The remission of the yellow part of the visible light (deep colour saturation) is comparable to that of pure cadmium yellow. The first bismuth vanadate pigments yielded only greenish yellow colours, but red-shifted yellow pigments are now also available. The colour strength is better than that of lead chromate pigments. Bismuth vanadate pigments are weatherable and resistant to most chemicals. They offer good hiding power. They are largely physiologically safe, although the dust may attack lung tissue. The pigments are therefore delivered in granulated form. The availability of bismuth vanadate pigments closes the gap in the colour space (brilliant yellows with scattering properties and good hiding power), which resulted from the exclusion of lead chromates and cadmium yellows due to their toxic reactions.

Cobalt pigments

There are mixed-phase oxides with spinel structure that contain cobalt. The spinels are a class of compounds in which the ions are arranged in a cubic close-packed crystal lattice. Spinels contain a mixture of oxides of bivalently and tetravalently charged metals ($Me^{II}Me^{IV}O_4$). The spinels class derives its name from the colourless mineral spinel, which consists of magnesium aluminate ($MgAl_2O_4$). If the magnesium in the lattice of the mineral spinel is replaced by cobalt, the result is a blue mixed oxide. The pigment of that compound, **cobalt blue**, contains not only cobalt, but also some zinc. The average formula of the compound is $Co_{0.65}Zn_{0.35}Al_2O_4$. Cobalt blue pigments yield brilliant red-shifted blues, good hiding power and high colour strength. More greenish blues are generated by replacing some of the aluminium ions with chromium ions.

Cobalt green has an inverse spinel structure in which the aluminium ions are replaced by mixtures of cobalt and zinc ions. Brilliant green pigments result if the magnesium ions of the base compound (spinel) are replaced by nickel ions. Then, the average formula of cobalt green is $Ni[Co_{0.5}Zn_{0.5}TiO_4]$. Cobalt green generates brilliant greens with good hiding power.

All spinel pigments offer excellent weatherability and are very resistant to chemicals and high temperatures. They are totally physiologically safe. Nevertheless, for green topcoats, a combination of titanium dioxide and phthalocyanine pigments is preferred for automotive topcoats.

Chromium oxide green

In contrast to the chromate pigments (lead and zinc chromates), chromium oxide green is practically physiologically safe. Chromium(III) oxide forms orthorhombic crystals with the structure of corundum. Its melting temperature is very high, its chemical resistance is outstanding, and the particles are very hard. However, the high hardness may cause abrasion during preparation and application of paints containing chromium oxide pigments. Chromium oxide is prepared by reduction of with alkali chromates or dichromates. The reducing agents are sulphur or carbon, but ammonia can also reduce the chromates. For example, chromium oxide is prepared by making sodium dichromate react with ammonium sulphate at elevated temperatures (see Equation 3.6.1).

Equation 3.6.1: Preparation of chromium oxide

$$Na_2Cr_2O_7 \cdot 2\,H_2O + (NH_4)_2SO_4 \longrightarrow Cr_2O_3 + Na_2SO_4 + 6\,H_2O + N_2$$

Of course, it is essential to take precautions with the raw materials and by-products of the toxic chromium(VI) compounds. Chromium oxide green offers excellent weatherability, and is resistant to chemicals, water, and high temperatures. The remission maximum is 535 nm, with a smaller peak occurring at 410 nm. The refractive index is 2.5. On account of the specific remission spectrum, chromium oxide pigments are used in camouflage paints. Although chromium oxide pigments generate good hiding power, the colours are not brilliant, but rather are matt olive greens. The pigments are therefore only used for special applications.

Carbon black

Most carbon blacks are prepared by the so-called furnace process. In this, a mixture of heavy oil (hydrocarbons) and natural gas is sprayed into a ceramic-lined furnace and combusted at temperatures of about 1900 °C. The mixture undergoes partial oxidation, with the gas providing almost all the combustion energy. Mainly, the oil generates soot. The reaction products are separated by spraying with water. The soot is separated from the water dispersion by filtering. The soot particles are very fine. Analysis of particles in electron beam microscopes show particle sizes (for the primary particles) of 5 to 10 nm. On account of their high surface energy, the soot particles agglomerate very well. The agglomerate may form particles up to 150 nm long, some of them with a fractal structure. Figure 3.6.4 show a TEM image [119] of typical particles of a carbon black pigment.

Carbon black pigments absorb nearly all visible light. Due to their high propensity to form agglomerates, a great deal of effort goes into preparing the pigment dispersion. The surfaces of most carbon black particles contain acidic carboxylic groups, a fact which must be borne in mind for the dispersing process (especially water-borne systems). The efficiency of dispersion critically determines the colour strength. The target is to produce a deep bluish black.

3.6.3.5 Organic pigments

There are large numbers of organic pigments available. The pigment classes covered here are those which play a role in automotive topcoats and basecoats. For this application, the pigments must disperse easily and be compatible with the resin combination employed for the topcoats. The pigment dispersions must be stable during paint storage, application, and film forming (including stoving). This is essential for obtaining reproducible colours. A vital property of pigments for automotive topcoats is their fastness or resistance to various impacts. The pigments must not be soluble in the aromatic or polar solvents employed in the topcoats. The goal here is to avoid bleeding (migration of pigment into other paint layers), and efflorescence (crystallisation on the

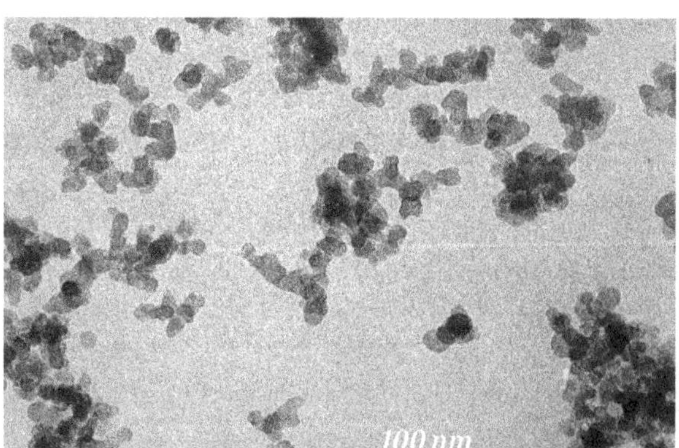

Figure 3.6.4: TEM image of a carbon black pigment (furnace carbon black FW 200)

surface of the coating layer). Of course, the selected pigments must be extremely weatherable. The only way to assess weatherability is to test the entire coating film (film matrix), including the influences of resins and crosslinking. The pigments must not react with UV light or the by-products of environmental pollution. Photo-catalytic effects triggered by the surface of the pigment particles must not lead to destruction within the binder matrix of the coating film. In this regard, it is essential to test the weatherability of organic pigments in blends with white pigments (titanium dioxide) - there must be no fading. The pigmented films must be resistant to chemicals. Since organic pigments create colour mainly by absorbing visible light, and not by scattering it, particle sizes should be as small as possible. The smaller these particles are, the greater is their colour strength (tinting strength), brilliance and transparency. However, their hiding power is diminished. Very small particles are less resistant to solvents and migration. They are difficult to disperse efficiently and may influence the rheological behaviour of the paint, and this may adversely affect flow and levelling during application. It is necessary to strike a compromise on all the influences and to balance the properties. As already mentioned, it is very important to be able to reproduce the colour impression of a pigment very well. The colour of a pigment is influenced not only by the chemical composition, but also by the crystal structure, the particle size and particle size distribution, the surface treatment, and by its degree of distribution in the paint material (dispersion efficiency) and film matrix (where flocculation should be avoided or controlled).

Organic pigments are classified first by their colour and second by their chemical structure. All available and described pigments are listed in the Colour Index (C.I. Index) [120], where numbers are used to denote the colour and chemical composition. A description of pigment classes suitable for automotive topcoats and basecoats is provided below. Chemical structures of typical examples of the various classes are also presented. The general properties of the classes have also been interpreted from descriptions provided in manufacturers' technical datasheets for the pigment classes [19] and as such are only generalities.

Azo pigments

Azo pigments are prepared by coupling reactions between aromatic diazonium compounds and various aromatic partner compounds. They include pigments which produce very brilliant and pure red colours. Unfortunately, the azo pigments have poor weatherability and some show a tendency to bleed. Since certain red colours are required, it is not possible to dispense with the entire pigment class. Accordingly, recourse is made to those members which offer the maximum resistance properties. These include the Naphthol AS pigments, which contain aromatic diazonium compounds and aromatic amides of hydroxynaphthoic acid by way of coupling component (see Figure 3.6.5)

Pigments of this kind cover the range from yellowish to bluish red.

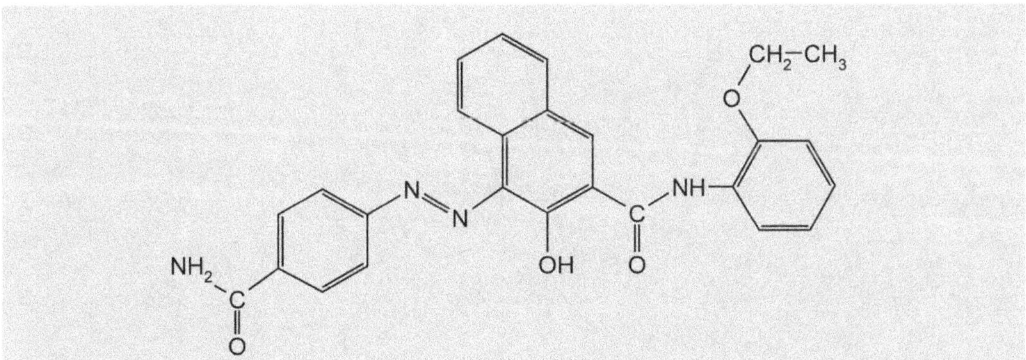

Figure 3.6.5: Naphthol AS pigment (C.I. Pigment Red 170, e.g. "Novopermred" F 3 RK 70)

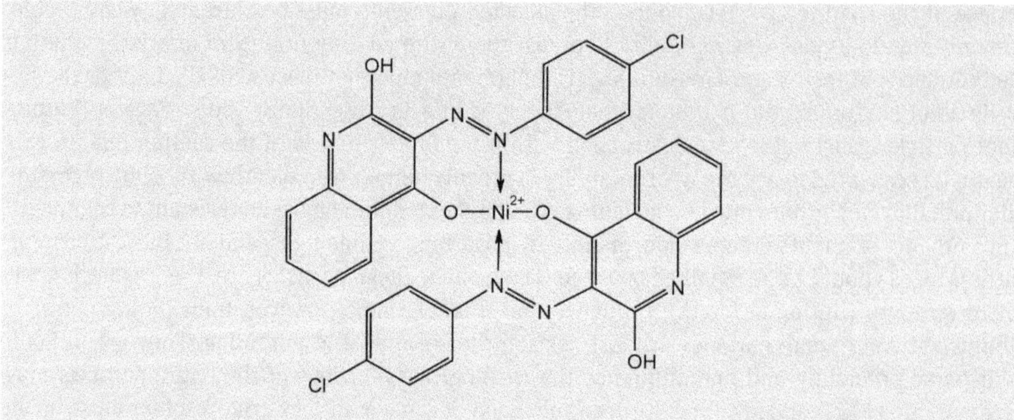

Figure 3.6.6: Azo metal complex pigment (C.I. Pigment Green 10, e.g. "Bayfast" Yellow Y 5688)

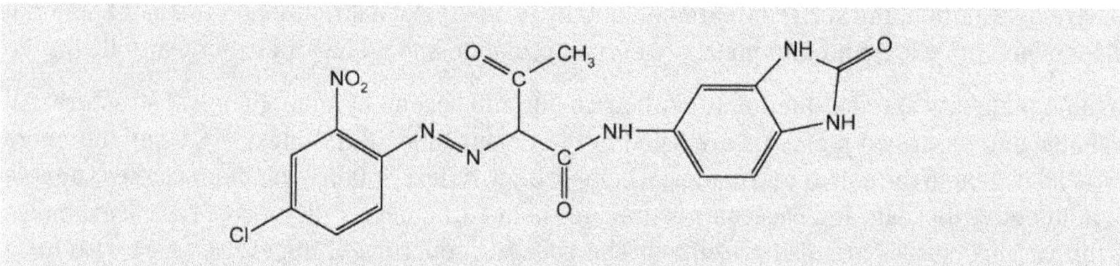

Figure 3.6.7: Benzimidazolone pigment (C.I. Pigment Orange 36, e.g. "Novopermorange" HL 70)

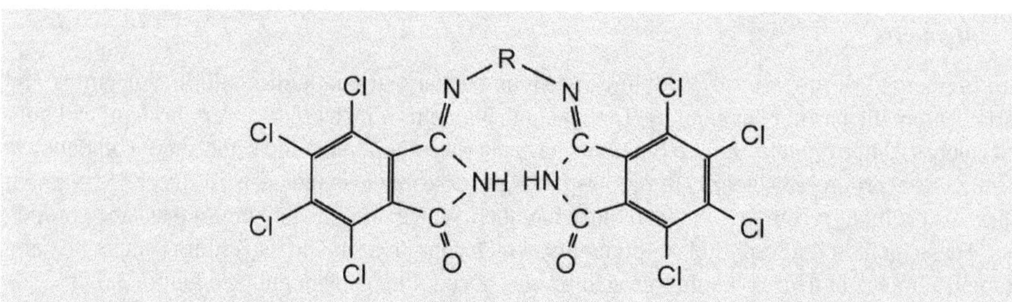

Figure 3.6.8: Isoindolinone pigment (C.I. Pigment Yellow 110, e.g. Irgazine Yellow 2 RTL)

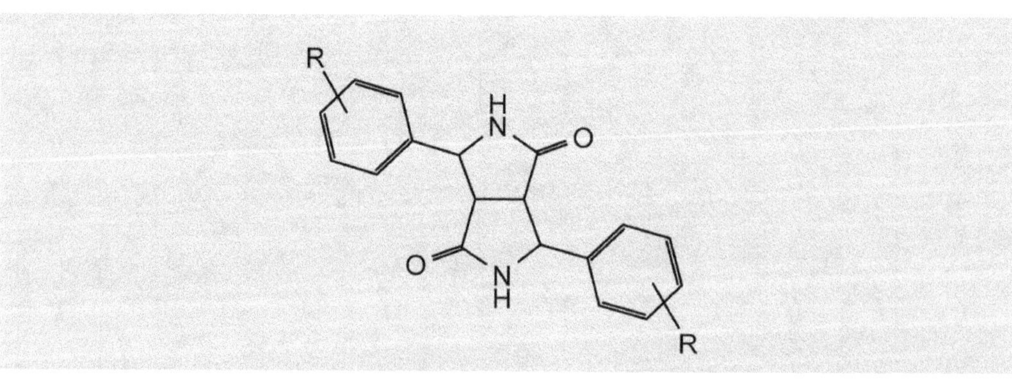

Figure 3.6.9: Diketopyrrolopyrrole pigment (C.I. Pigment Red 254, e.g. "Irgazine" DPP Red BO)

Azo metal complex pigments

On account of their molecular structure, azo metal complex pigments are more resistant to solvents and migration effects than other azo pigments. However, they are not as weatherable or resistant to chemicals. This class contains pigments which generate greenish yellow colours (see Figure 3.6.6 for an example).

Benzimidazolone pigments

Coupling aromatic diazonium compounds with acetylacetamides of benzimidazole leads to yellow and orange pigments. These have high brilliance (colour saturation) and high tinting strength. Their fastness properties are very good and they offer acceptable weatherability (see Figure 3.6.7 for example).

Isoindolinone pigments

Isoindolinone pigments produce reddish yellow colours. The most important members of this class are prepared by reaction of two moles of 3-imino-tetrachloroisoindolinone with one mole of an aromatic diamine. The isoindolinone pigments are resistant to solvent and chemicals and are relatively weatherable (see Figure 3.6.8 for example).

Diketopyrrolopyrroles

Particularly resistant are diketopyrrolopyrrole pigments, which produce colours ranging from yellow and orange to ruby red. They are prepared by making substituted benzonitriles react with succinic esters. They have particularly excellent brilliance and tinting strength and offer outstanding resistance to solvents, migration and weathering (see Figure 3.6.9 for example).

Quinophthalone pigments

Quinophthalone pigments are composed of complex ring systems containing chlorinated diketone of isoindole, quinoline, and chlorinated phthalone. These yellow pigments have good hiding power, possess excellent fastness properties and are highly weatherable. To improve wetting properties for the production of pigment dispersions, some types are post-treated (see Figure 3.6.10 for example, page 130).

Anthanthrone pigments

Anthanthrone pigments are formed by dimerisation of 1-amino-naphthalene-8-carboxylic acid in which the amino groups are cleaved and the resultant dicarboxylic acid is cyclised to yield a six-membered ring system. The pigments in this class cover different red shades and have high colour strength and outstanding weatherability. However, their overcoatability is not as good (see Figure 3.6.11, page 130 for example).

Quinacridone pigments

Quinacridone pigments are prepared by dimerisation of succinic diesters, which react with two moles of aromatic amines. At elevated temperatures, the reaction product condenses to form five-membered rings, which are ultimately converted into an aromatic compound by oxidation. Quinacridone pigments cover colours from bluish reds to violets. They offer medium to good colour strengths and excellent resistance to light, solvent and migration. They are extremely weatherable, particularly in blends with white (see Figure 3.6.12, page 130 for example).

Perylene pigments

Naphthalene and ethylene are made to react to yield acenaphthalene, which can be oxidised to 1,8-naphthalene carboxylic acid. In alkaline melt, the amide of this acid dimerises. This basic

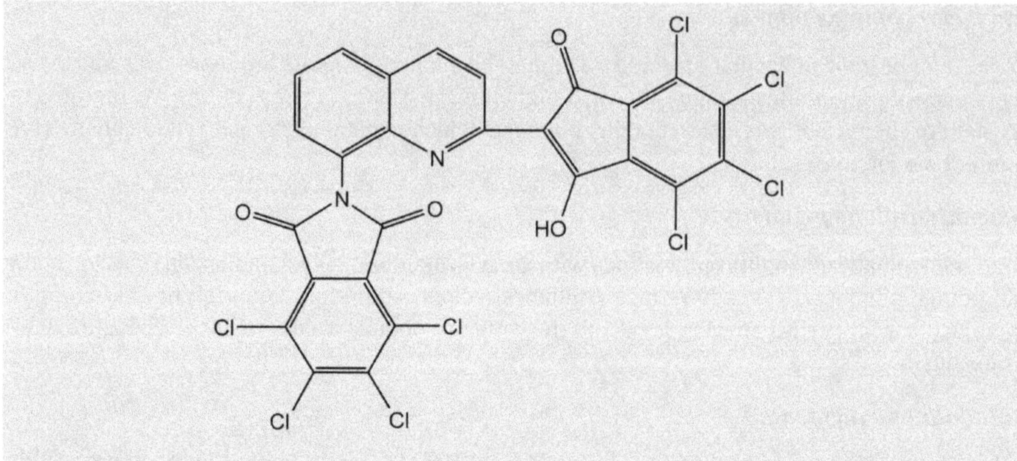

Figure 3.6.10: Quinophthalone pigment (C.I. Pigment Yellow 138, e.g. "Paliotol" Yellow 0962 HD)

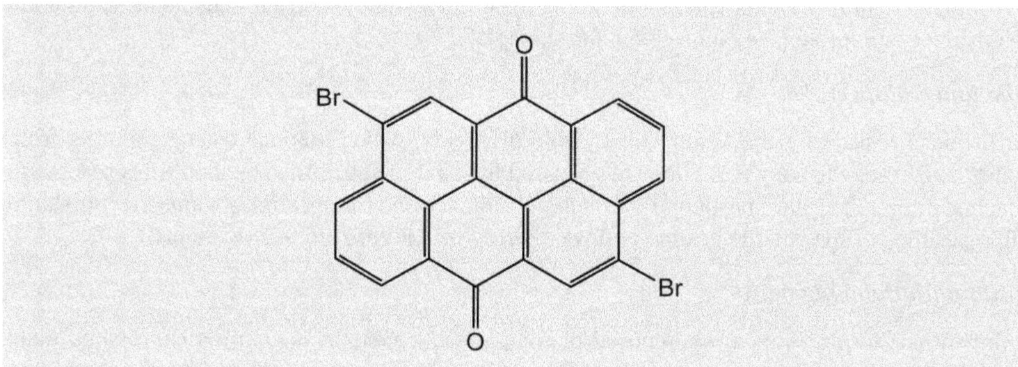

Figure 3.6.11: Anthanthrone pigment (C.I. Pigment Red 168, e.g. "Hostaperm" Scarlet GO)

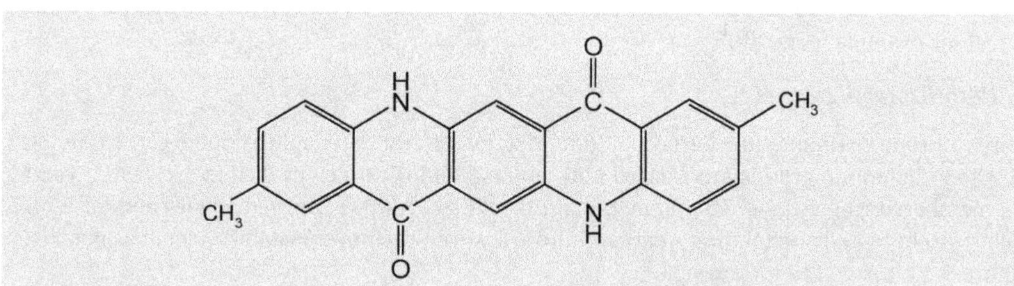

Figure 3.6.12: Quinacridone pigment (C.I. Pigment Red 122, e.g. "Hostaperm" Pink E)

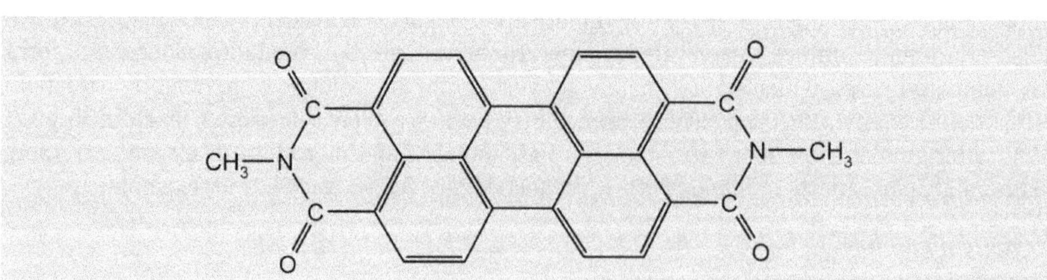

Figure 3.6.13: Perylene pigment (C.I. Pigment Red 179, e.g. "Paliogen Marrone" L 3920)

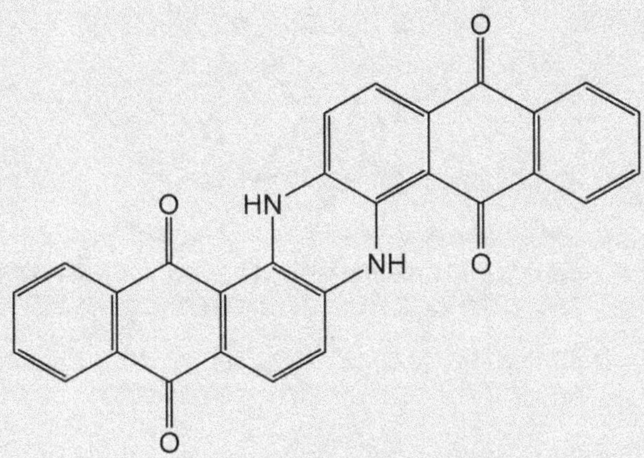

Figure 3.6.14: Indanthrone pigment (C.I. Pigment Blue 60, e.g. "Paliogen" Blue L 6470)

dianhydride of a tetracarboxylic acid reacts with mono-amines to form various diimides, the perylene pigments. Perylene pigments produce yellowish and brownish red colours. They are relatively highly resistant to solvents and have excellent migration stability. Weatherability and resistance to various chemicals are either good or excellent (see Figure 3.6.13 for example).

Indanthrone pigments

Indanthrone pigments result from the dimerisation of 2-aminoanthraquinone in an alkaline melt. The α-modification is a blue pigment of high colour strength, which covers mainly the reddish blue colours. Indanthrone pigments offer outstanding resistance to chemicals, solvents, migration, and weathering. However, indanthrone pigments are more expensive than the phthalocyanine pigments (see Figure 3.6.14 for example).

Phthalocyanine pigments

The reaction of phthalodinitrile with copper salts yields intensely blue phthalocyanine pigments. These contain porphyrin, which is also present in haemoglobin or chlorophyll. The pigments have different crystal structures, and there are also pigments which do not contain a central metal ion. Modification of the aromatic rings with halogen atoms affords green pigments; chlorine produces bluish greens, while bromine yields more yellowish greens. All phthalocyanine pigments have very high colour strength and offer outstanding resistance to many chemicals, solvents, and migration. They particularly offer outstanding weatherability. Many different grades are commercially available. Some grades are post-treated to improve the wetting properties for dispersion processes (see Figure 3.6.15, page 132 for example).

3.6.3.6 Pigmentation of topcoats

As already mentioned, the colour strength and the hiding power of pigments depend on the concentration of pigment particles in the film matrix and on the film thickness. For topcoat application, the equipment and the process (line speed) are set up so as to achieve film thicknesses of 40 to 45 μm (as a rule).

The concentration of pigment particles in the film is described by the pigment volume concentration (PVC; naturally, weight fractions are not used on account of the different densities). Increasing the PVC increases the hiding power and colour strength. The hiding power increases rapidly at first and then less and less. If the pigment particles are very small, the curve of hiding power versus PVC passes through a maximum and then declines. The PVC can influence the application behaviour. The reason is that higher quantity of pigments generates particle-particle interactions,

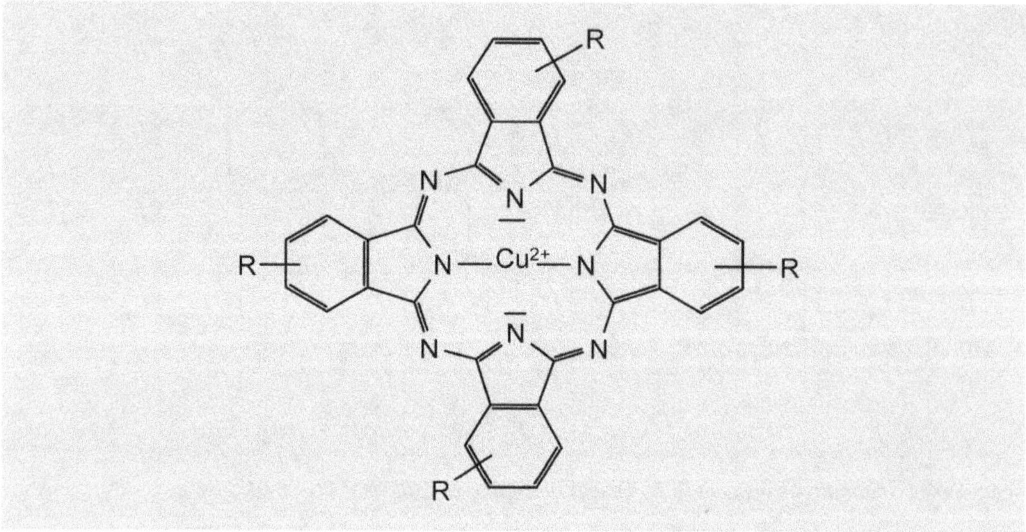

Figure 3.6.15: Phthalocyanine pigment (C.I. Pigment Blue 15, e.g. "Heliogen" Blue L 7101 F)

which lead to increasing viscosity and intrinsic viscosity behaviour. The application viscosity of topcoats is determined by measuring at low shear rates (efflux viscosimeter DIN, ISO, Ford cups and others), which may be closely related to shear rates which occur during the first steps of film forming (flow and levelling). Since different pigments not only generate different hiding power and colour strength, but also different viscosity behaviour, the compositions (by weight) of topcoats also vary with regard to achieving optimum application behaviour <u>and</u> optimum colour properties. Figure 3.6.16 shows typical compositions of topcoats of different colours, along with different ratios of binders, pigments, and solvents. All the colours must achieve the same requisite application viscosity (here: 25 s, DIN cup 4 at 23 °C equivalent to ISO Cup 96 s at 23 °C). The PVC is calculated from the quantity and density of the resin combination of alkyd resin (1.20 g/cm³), melamine resin (1.30 g/cm³) and the density of the various pigments (see Table 3.6.3). Note that the pigment density quoted refers to that of the substance and not the pigment powder.

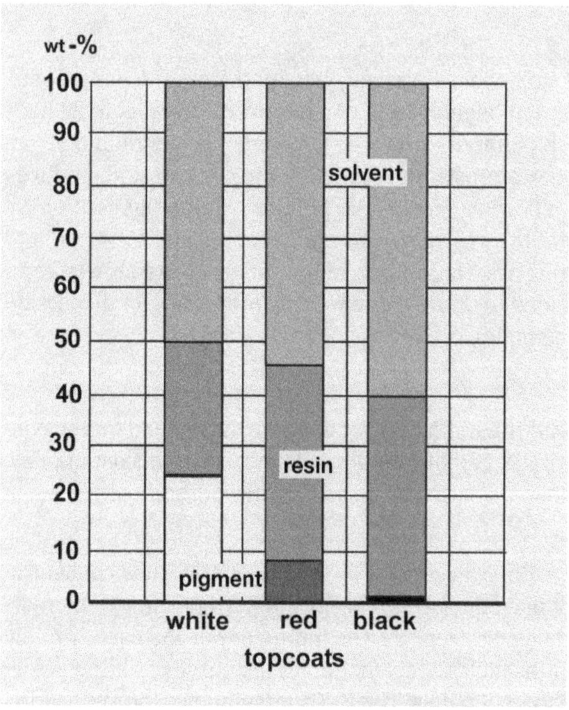

Figure 3.6.16: Typical compositions of topcoats of different colour

The goal is to build up a topcoat series on the basis of a uniform binder combination. Also, the ratio of the resin partners is the same for all colours of the topcoat series; in the example of alkyd versus melamine resin 7:3. Under the described conditions, the application solids vary, for the same application viscosity, from 45 % by weight for black, via red with 50 %, to white with 55 %. The optimum PVC values for most of the organic pigments are between

8 and 12 % per volume. Pigments with very high light absorption, such as phthalocyanine pigments and carbon blacks, have PVC values of just 1 to 3 % by volume. Pure white topcoats have PVC of about 20 % by volume.

Most commercialised topcoat colours contain mixtures of several pigments. Some topcoat formulations contain as many as five different types of pigment. Nevertheless, the pigments are dispersed individually. The advantage is that the dispersion process is highly characterised for the specific dispersion behaviour of each pigment. Therefore, not only the composition of pigment paste (ratio of pigment, binder, and solvent) is adjusted, but so too are the conditions for the dispersion process. The latter is influenced by the type of milling equipment, the dispersion medium, the batch volume, the through-put velocity, the dispersion time, and the final constants of the dispersion. Pigment pastes contain as much pigment as possible, e.g. up to 60 % by weight for titanium dioxide, 15 to 20 % for organic red pigments, and 5 to 10 % for finely dispersed carbon black, together with alkyd resin and solvent. There are alkyd resins which are specially modified for preparing pigment dispersions. For example, they may contain specific quantities of unsaturated fatty acids to improve the wetting properties: It is additionally possible to add wetting agents as additives for problematic pigments. Besides affording the opportunity to specify the composition and dispersion process for a single pigment, preparing pigment pastes for just one pigment enables standardised pigment pastes to be prepared. A series of standardised pigment pastes of well-defined colour strength and PVC allows accurate colour definition, mainly for the purpose of reproducibility. It is also possible to calculate the amount of pigment paste that must be added for tinting (see Chapter 3.6.2). Nowadays, producers of topcoats have computer programs to perform analyses, define compositions and adjust required colours. They start by measuring and defining the requisite colour. The system must be able to calculate the composition of the formulation, based on the standardised pigment paste series, and also accommodate addition of let-down resins (to allow for the quantity of alkyd resin already incorporated by pigment pastes), solvents, and additives (batch preparation). Adjustment consists in photometrically measuring the colour of the topcoat batch, comparing the result with the requisite colour, and calculating the quantity of pigment pastes needed for adjustment. The production process requires that pigment pastes be available in adequate quantities - in terms both of number to cover all possible colours and of quantity to cater for all possible batch sizes. The pigment pastes must offer excellent storage stability and may not change their dispersion state or their colour strength. Furthermore, it must be remembered that, when specific quantities of pigment pastes are added for

Table 3.6.3: Specific densities of pigments

Pigments	Densities [g/cm³]
White	
titanium dioxide. rutil	4.0 – 4.2
Yellow	
yellow iron oxide	3.7 – 4.1
nickel titanium yellow	4.4 – 4.6
chromium titanium yellow	4.3 – 4.5
bismuth vanadate	5.5 – 5.6
Ni-Azocomplexes	1.6 – 2.0
anthrapyrimidines	1.5 – 1.6
isoindolinones	1.8 – 2.1
Red	
red iron oxide	4.5 – 5.2
naphthol AS-reds	1.4 – 1.5
perylenes	1.5 – 1.7
thioindigos	1.7 – 2.1
chinacridones	1.5 – 1.7
dioxazines	1.5 – 1.9
Blue	
cobalt blue	3.8 – 4.3
indanthrene blue	1.4 – 1.6
phthalocyanine blue (Cu)	1.6 – 1.7
phthalocyanine blue	1.4
Green	
chromium oxide green	5.0 – 5.2
cobalt green	4.9 – 5.5
phthalocyanine green	1.9 – 2.9
Black	
carbon blacks	1.8

adjustment purposes, the corresponding quantities of let-down components must be adjusted in order that a change in of film properties which depend on the optimum composition may be ruled out. This can be circumvented by preparing tinting coats. Tinting coats are topcoats containing only one pigment, as well as all the other ingredients in the right combinations necessary for ensuring topcoat quality. As tinting topcoats are also standardised, all requisite topcoat colours can be prepared from calculated recipes and adjustments, but without the need to add let-down components after tinting. For some colours, this process does not make sense. Pastels, for example, are always produced by starting with titanium dioxide dispersion and adding tinting pastes.

Solvents

As shown in Figure 3.6.16, topcoats contain 45 to 55 % by weight of solvents when applied. Solvent for conventional automotive OEM topcoats comprise blends of large quantities of aromatic solvents and small quantities of more polar solvents. The aromatic solvents have boiling temperatures above 150 °C (e.g. Aromatic 100). The polar solvents are mainly butanol (a solvent for melamine resins), as well as monoethers of propylene and ethylene glycol, and the corresponding ether-esters. Sometimes, high-boiling solvents are added, albeit in small quantities to improve levelling, e.g. monoethers of diethylene glycol. The addition of such solvent also supports the compatibility of resins in the film until the onset of crosslinking and also acts to enhance gloss.

Solvents selected for topcoats have to fulfil all other general requirements on solvents, including physiological harmlessness and, possibly, low costs. Although aromatic solvents are the subject of environmental debate, restrictions do not extend as far as replacing them entirely. They are still preferred if, besides offering optimum application behaviour, they are low in costs. If it becomes necessary to replace aromatic solvents, Europe would not introduce other solvents, but rather would switch over to water-borne systems. A significant quantity of solvent gets into the formulation via the delivery forms of the resins and additives. Specific solvents are added at the let-down stage of the topcoat after the pigment pastes have been mixed. In addition, it is usual to deliver topcoats in somewhat higher viscosities for dilution with thinners to the application viscosity on the line.

3.6.3.8 Additives

Additives used for topcoats are wetting agents, or other dispersing additives, rheological additives, levelling agents and catalysts.

Wetting and dispersion agents

The wetting and dispersing agents already described for primer surfacers (see Chapter 3.5.3.6) are also suitable for topcoats. The product classes are: derivatives of fatty acids, sulphonates, and esters of phosphoric acid, phosphatides, polyurethanes, polyethers, polyamides, and polyacrylates. The polymers contain special groups that have a high pigment affinity. Such products are also used for post-treating pigments (surface layers).

Rheological agents

The most important requirement on rheological additives for topcoats is to be totally neutral in colour. They are added to prevent pigment settling and to avoid paint sagging on vertical parts during flash-off and the first phase of film forming in the oven. They must also generate special surface structures to influence levelling. However, they must not generate too much structure or reduce the gloss. Application tests are always required for finding suitable rheological agents. The most important examples of such additives are pyrogenic silica [121] and crystalline ureas [122].

Levelling agents

The most important levelling agents are modified silicone oils, which are described in more detail in Chapter 3.8.3.7.

Catalysts

Acids serve as catalysts for accelerating crosslinking reactions between hydroxyl-containing alkyds and melamine resins. The catalysts chosen for topcoats are sulphonic acids, phosphoric acids (and their esters with residual acid groups), maleic acid, or their monoesters prepared from maleic anhydride and monoalcohols (e.g. butanol). When reaction catalysts are used, storage stability can be improved by recourse to ammonium salts of the acids. Such ammonium salts of, e.g., sulphonic acids are also suitable for water-borne paints.

Light-protection additives

While UV absorbers are essential for clearcoats, they are no longer used in pigmented topcoats since the pigment particles adequately absorb the UV light. However, there are some pigment particles whose surfaces can act as catalysts for photolysis in the binder matrix under the influence of UV light. This can lead to destruction of the upper parts of the film, and ultimately to matting and chalking. For example, titanium dioxide has to be post-treated with additive oxides to avoid this photocatalysis, which induces destruction of the film matrix.

Formulation example

Table 3.6.4 shows the composition of a white automotive OEM topcoat [123]. The resins contain a combination of an alkyd resin, modified with saturated fatty acid and a relatively reactive melamine resin. In addition, the formulation contains a pigment-wetting agent, modified silicone oil

Table 3.6.4: Example of a white automotive OEM topcoat

Pos.	Components	Solid wt-‰	Formulation wt-‰
01	alkyd resin. modified with saturated fatty acids (70 % in xylene)	132.3	189.0
02	dispersion agent		1.8
03	titanium dioxide, rutil	301.0	301.0
04	xylene		66.0
05	aromatic 150		28.2
dispers on fineness <5 µm, then add:			
06	alkyd resin, modified with saturated fatty acids (70 % in xylene)	56.7	81.0
07	alkyd resin, modified with saturated fatty acids and crystalline urea (67 % in xylene / n-butanol 96/14)	94.5 4.0	147.0
08	melamine resin (70 %), butanol-etherified	95.2	136.0
09	silicone oil		1.0
10	Depanol J		38.0
11	n-butanol		30.5
	sum	683.7	1000.0

values (application):

solid (15' 165°C)	68.4 %
PVC	20.4 %
alkyd/melamine resin	75 : 25

pos.		pos.	
01 + 06:	Setal 84 XX 70 (Nuplex)	07	Setal 90084 BX 67 (Nuplex)
02	SER-AD FA 501 (Condea Servo)	08:	Setamine US 138 BB-70 (Nuplex)
03	Kronos 2310 (Kronos)	09	Byk 301 (Byk)
06	Solvesso 150 (Exxon)		

by way of levelling agent, and a quantity of crystalline urea by way of rheological additive. The crystalline urea results from the reaction of two moles of benzylamine with one mole of hexamethylene diisocyanate. Preparation consists in precipitating the crystals in some of the alkyd resin under high shear.

The application viscosity is adjusted with a thinner that contains xylene, a high-boiling aromatic solvent (e.g. Aromatic 100) and butanol (mixing ratios are 75:20:5). The application viscosity is 28 s (DIN cup 4 at 23 °C equivalent to ISO cup 108 s). For tests, the thinned topcoat is pneumatically sprayed onto primed panels and flashed-off briefly. Stoving conditions are 30 minutes at 135 °C.

3.6.4 Application

On assembly lines, topcoats are sprayed pneumatically from high-efficiency spray guns. However, application by automatic electrostatic spray equipment, which offers much superior transfer efficiencies, is more common. The usual stoving conditions are 15 to 30 minutes at 145 to 130 °C. The resultant layer thicknesses of the topcoats are 35 to 45 μm.

3.6.5 Topcoats with reduced VOC emissions

As already mentioned, the application solids of conventional solvent-borne topcoats is 45 to 55 % by weight. It is possible to increase the application solids to 55 to 65 % by weight. As a result, the quantity of solvents that evaporates from a white topcoat with a dry film thickness of 40 μm decreases from 55.7 g solvent/m^2 to 35.6 g solvent/m^3, representing a reduction of 36 % on the original level. This calculation is based on average densities of 1.3 g/cm^3 for the resins, 0.87 g/m^3 for the solvents and 4.0 g/m^3 for the pigment. It makes no allowance for transfer efficiency. Remarkably, the transfer efficiency of high-solid paints is in theory lower than that of low-solids paints. The reason is that the overspray of high-solid paints contains more film forming material.

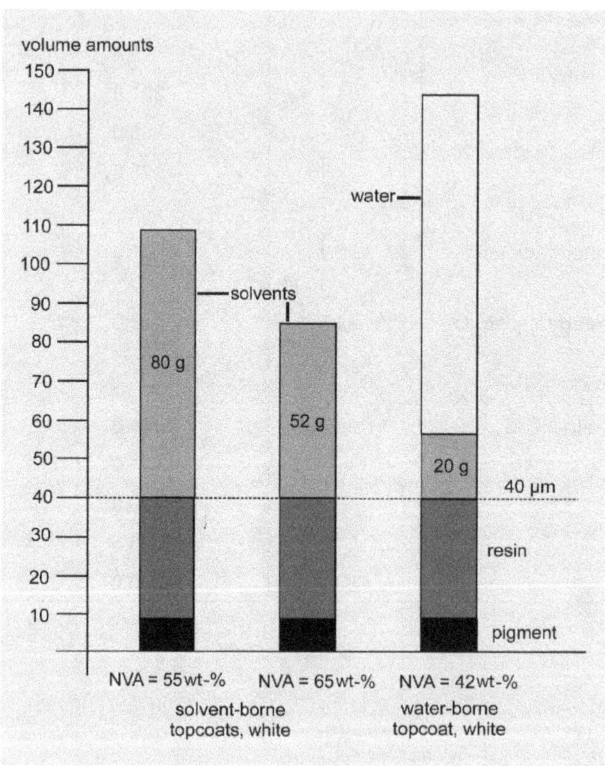

Figure 3.6.17: Volume ratios and VOC content of conventional, high-solid and water-borne topcoats

Finally, the use of water-borne topcoats yields a greater reduction in emissions of volatile organic compounds (VOCs). Figure 3.6.17 compares the volume ratios of binder, pigment, solvents (with weight quantities in figures) and water in white topcoats that yielded dry films 40 μm thick.

The weights of the topcoats shown in Figure 3.6.17 are expressed as weight ratios in Table 3.6.5.

The calculation is based on the values used in the previous calculation. The pigment density is 4.0 g/cm^3, average density is 1.3 g/cm^3 for resins, 0.87 g/cm^3 for solvents, and 1.00 g/cm^3 for water. However, for the co-solvent in the water-borne topcoat, the density was taken as 0.94 g/cm^3.

Table 3.6.5: Weight amounts and VOC of conventional, high-solid, and water-borne topcoats

Components	Conventional topcoat	High-solid topcoat	Water-borne topcoat
wt-%	NV = 55 %	NV = 65 %	NV = 42 %
resins	31.1	36.7	23.7
pigments	23.0	28.3	18.3
solvents	45.0	35.0	10.0
water	0.0	0.0	48.0
sum	100.0	100.0	100.0

3.7 Basecoats

3.7.1 Development of basecoats

In the early 1960s, the general public fell in love with sporty cars. Part of this appeal was provided by the colour and special effect coatings. The first effect paints were topcoats (stoving enamels) that contained aluminium pigments, which conferred a metallic-like appearance. Such effect topcoats are now called one-coat metallics. The resin combination for these topcoats contained hydroxy acrylic resins, which were already commonly employed for stoving enamels in the USA, and melamine resins for crosslinking. To create the special effects, these topcoats were loaded with as much aluminium pigment as possible. However, high quantities of aluminium pigments lower the resistance to chemicals, mechanical impact (abrasion) and weathering. The most important way to judge the resistance of metallic topcoats was the Kesternich test in which the test panels were exposed to high levels of humidity, elevated temperatures and sulphur dioxide. In Europe, such stoving enamels were more commonly based on alkyd resins containing saturated monocarboxylic acids. These offered a superior appearance to acrylic systems but, due to the content of aluminium pigment, they were limited in their resistance to weathering and chemicals. The resistance problems were overcome by protecting the aluminium-based topcoats with a clearcoat. In Europe, the clearcoat contained alkyds, which offered less yellowing, and melamine resins for crosslinking. The clearcoat protected the metallic effect layer (pre-coat) against chemicals, mechanical impact and weathering. In addition, the two-layer systems presented an opportunity to increase the quantity of aluminium pigment to create an appealing metallic effect.

Most car makers ran their coating processes in three steps: primer, primer surfacer, and topcoat. They were not prepared to use four steps for a two-layer topcoat system comprising metallic pre-coat and clearcoat. The four-step process was employed only on some highly prestigious cars. The first solution employed by producers who did use four-layer systems consisted in passing cars with effect coatings through the topcoat application line twice, first for the metallic pre-coat, and then for the clearcoat. This was a very intensive process. Line productivity fell 50 %. Cleaning the line between metallic pre-coat application and clearcoat application proved much more expensive than the cleaning involved in changing colours of conventional topcoats. Initially, when there were only a few cars requiring effect topcoats, the additional effort was worth it. However, in the late 1960s, demands for cars with effect colours increased significantly, even though such cars were more expensive than cars with conventional topcoats. To circumvent the effort required to meet the rising demands for cars with effect topcoats and to avoid production bottlenecks, car makers asked paint producers to develop efficient application processes. The paint industry proposed a process for applying a metallic pre-coat and clearcoat that required just one stoving process after application of both coats. This process, known as "wet-on-wet" application was not new.

The most important property of the "wet-on-wet" application was that the pre-coat, which was only allowed to dry physically, had to withstand the clearcoat just after the pre-coat had been applied. Other factors were the influence of solvents in the clearcoat, and the mechanical impact of spray application onto the pre-coat layer. Physical drying had therefore to be as efficient as possible, but the orientation of aluminium pigments, the most important precondition for the metallic effect, was not to be impaired. Aluminium pigments had to be perfectly distributed in the effect coat layer and to be homogeneously immobilised in that layer after physical drying. Finally, the two layers of metallic basecoat - the new name for the effect pre-coat - and clearcoat had to confer sufficient weathering resistance.

Metallic basecoats that met the described requirements contained a resin combination of a polymer offering rapid physical drying and resins that not only acted as plasticizers for the physical drying polymer but could also be crosslinked during stoving. The basecoats also contained rheological additives for immobilising the effect pigments during the drying process and protecting against the impact of clearcoat application. The first type of physical drying resin to be chosen was cellulose nitrate. Cellulose nitrate was inexpensive and distinguished by excellent application behaviour (easily applied by spraying and not sensitive to variations in application conditions). However, basecoats containing cellulose nitrate are not sufficiently weathering resistant. A great many other polymers were tested (e.g. polyvinyl chloride copolymers), before cellulose acetobutyrate was ultimately chosen as the physical drying resin. The combination resins chosen to act as plasticizer and to be crosslinkable were saturated polyesters together with amino resins: the first urea resins. Such metallic basecoats were introduced into the market at the end of the 1960s. In the 1970s, the urea resins were replaced by special melamine resins, which offered better weathering resistance and better intercoat adhesion.

Polymers for efficient physical drying coatings, e.g. cellulose acetobutyrate, contain rather large molecules (with high molecular weights). The solution viscosities of such polymers are high. In addition, the rheological additives increase the viscosity. Basecoats containing such polymers and rheological additives therefore only attain application viscosity at low solids content. For example, a silver metallic basecoat achieves an application viscosity of 15 s (cup 4 mm, 20 °C, DIN 3511 equivalent to ISO cup 58 s) at a solids content of 13 % by weight. Such low solids are conducive to appealing effects, but of course they fail to meet the goal of avoiding volatile organic compounds. Already in the late 1970s, tests were underway to develop solvent-borne basecoats with higher application solids. Since a high viscosity or low application solids is mainly attributable to the content of physical drying polymer, the biggest step is to completely eradicate such polymer types. Mainly in the USA, high-solid basecoats were developed that contained acrylic resins and low-molecular melamine resins. They yielded application solids of about 35 % per weight (silver) (see Chapter 3.7.4). However, the effects achievable with such basecoats are less brilliant than those of low-solids basecoats. Furthermore, the application process is not as robust as for low-solids basecoats. Preparing appealing and reproducible basecoat-clearcoat systems is more difficult and the application conditions must be chosen carefully. In Europe, a compromise was reached with regard to achieving high-solids. The quantity of physical drying polymer in basecoats was reduced, which caused the viscosity to decrease significantly. Some of the drying properties were transferred to the combination resins, but without any increase in viscosity. Such medium-solids basecoats achieved application solids (for silver) at 23 % by weight. The effects and application robustness were comparable to those of low-solids basecoats. Although the described development steps slashed solvent emissions in two (and by more in the case of high-solids), the goal of limiting emissions of volatile organic compounds was not met. This led in the early 1980s to the development of water-borne basecoats (see Chapter 3.7.5). Today, water-borne basecoats are well distributed across European OEM coatings applications. They are also widespread in Asia and Latin America (mainly in the "transplants" of European car producers). High-solid basecoats continue to predominate in the USA.

Initially, all effect basecoats contained aluminium pigments for creating the metallic effect. Later, effect pigments based on mica (potassium-aluminium-alumino silicate) were introduced that created a pearlescent effect. Mica pigments are coloured by doping with thin layers of heavy metal oxides. In recent years, there have been trials aimed at introducing a number of other effect substances into the coatings market. These developments are following the market demand for striking and individual colours and effects. Meanwhile, the advantages of basecoat-clearcoat systems (better weathering resistance, and mechanical and chemical resistance) are also being exploited in topcoat systems which do not contain effect substances. Solid-colour topcoats containing only conventional coloured pigments are being replaced by two-layer systems that contain a solid-colour basecoat and a clearcoat applied "wet-on-wet". If such solid-colour basecoats are solvent-borne, then application in conjunction with a solvent-borne basecoat will cause solvent emissions to increase. To meet the demand of eliminating volatile organic solvents, therefore, the solid-colour basecoats had to be water-borne.

Various trials have sought to develop and prepare powder coatings containing effect substances. However, only a few systems are suitable and available [106]. Of course, effect powder coatings are one-coat systems. Such powder coatings are inferior in performance to liquid effect basecoat-clearcoat systems.

3.7.2 Demands on conventional basecoats

3.7.2.1 Conferring effects

Of course, the first demand on basecoats is to confer an effect. The first effect substances were aluminium pigments. Aluminium pigments are flake-like, 3 to 25 µm in diameter and 0.1 to 2 µm thick. The effects stem from orientation parallel with the surface and homogeneous distribution of the flakes in the basecoat layer. The perceived lightness depends on the angle at which the surface is viewed. From above - that is, looking at the coating surface at 90° to the incident light - the lightness is high because light reflection from the surfaces of the flakes is a maximum. However, at an oblique angle - that is, looking at the coating surface at a narrow angle relative to the incident light - the lightness is low, because the light is scattered on reflection. The principle behind the different reflections and lightness is presented in Figure 3.7.1.

Since the impression of lightness varies significantly with change in viewing angle, it is called the "flip-flop" effect. The "flop" is the difference in lightness. The effect can be measured by determining the remission under different angles as a function of the incident light. Figure 3.7.2 (page 139) shows the remissions of different topcoat systems as measured under different reflection angles.

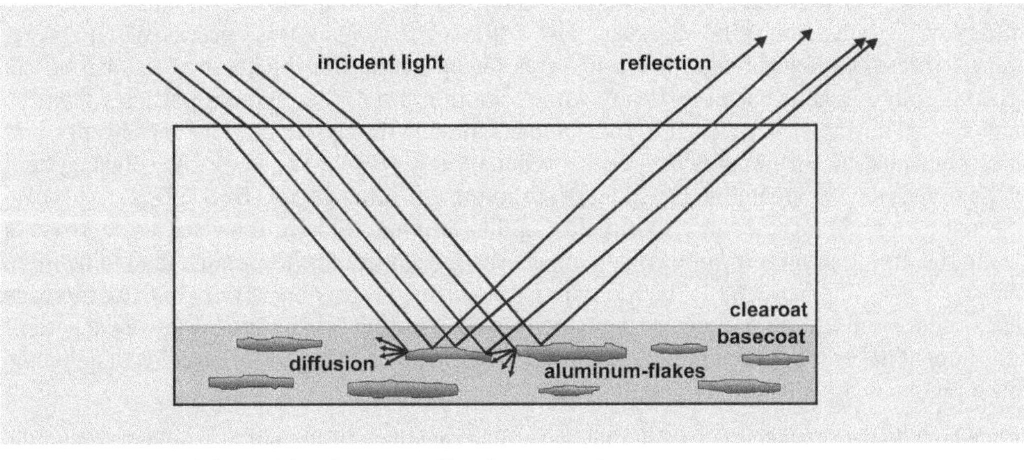

Figure 3.7.1: Principle behind the effect created by aluminium pigments

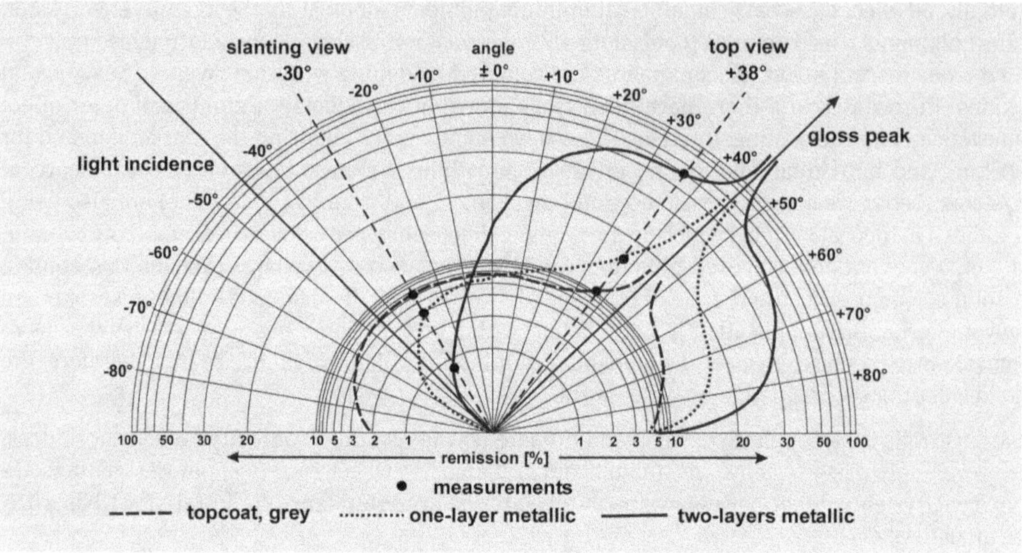

Figure 3.7.2: Remission of different topcoat systems as a function of viewing angle

Measurements are made on a grey solid colour topcoat, a one-layer metallic topcoat, and a metallic basecoat with clearcoat. The grey solid colour topcoat exhibits nearly the same amount of reflection irrespective of the reflection angle if the light is scattered on reflection. Reflection from the one-layer metallic topcoat varies as a function of the viewing angle. However, the differences in remission values as a function of viewing angle are much greater for the two-layer basecoat-clearcoat system. Reflection at 90° is ignored for the purposes of the evaluation as that represents maximum gloss due to surface reflection. Comparisons are made at remission values for angles close to 90°, e.g. 88° (37° to the coating surface) as a function of values at low reflection angles, e.g. 15° (or 125°). Comparison of the three value pairs for the different coating systems reveals particularly clear flip-flop effects, and explains why the two-layer systems are preferred to the one-layer metallic topcoats.

The parameters governing optimum effect are the basecoat formulation and the application conditions. It is essential throughout film-forming for the aluminium flakes to be well oriented, parallel with the surface of the substrate, and homogeneously spread horizontally. Experience shows that as soon as the spray particles hit the substrate, the more-or-less spherical droplets are flattened and that the aluminium flakes contained therein are oriented parallel with the surface of the substrate. If the basecoats have a low-solids content, it is necessary that several spray particles be deposited on the same area for optimum film thickness. This also boosts orientation of pigment flakes vertically within the film layer. The pigment particles must be immobilised during physical drying. This happens initially through evaporation of solvents, and is accompanied by shrinking, which boosts orientation very well. The lower the solids content of the basecoats, the greater is the extent of shrinking. Immobilising of effect substances is also effected by a rapid increase in viscosity during film forming. Initially, the viscosity increase is due to the drying properties of the polymers chosen for physical drying, which tend to be high-molecular (see Chapter 3.7.3.1). In addition, rheological additives boost the viscosity increase during film forming, preventing circulation within the film layer that might otherwise disorient the pigment flakes (see Chapter 3.7.3.5). The application process and the orientation of flake-like effect pigments are illustrated in Figure 3.7.3.

It is widely known that there are not only silver-like metallic effects, but also coloured metallic topcoat systems. The latter systems are prepared by blending aluminium flake pigments with

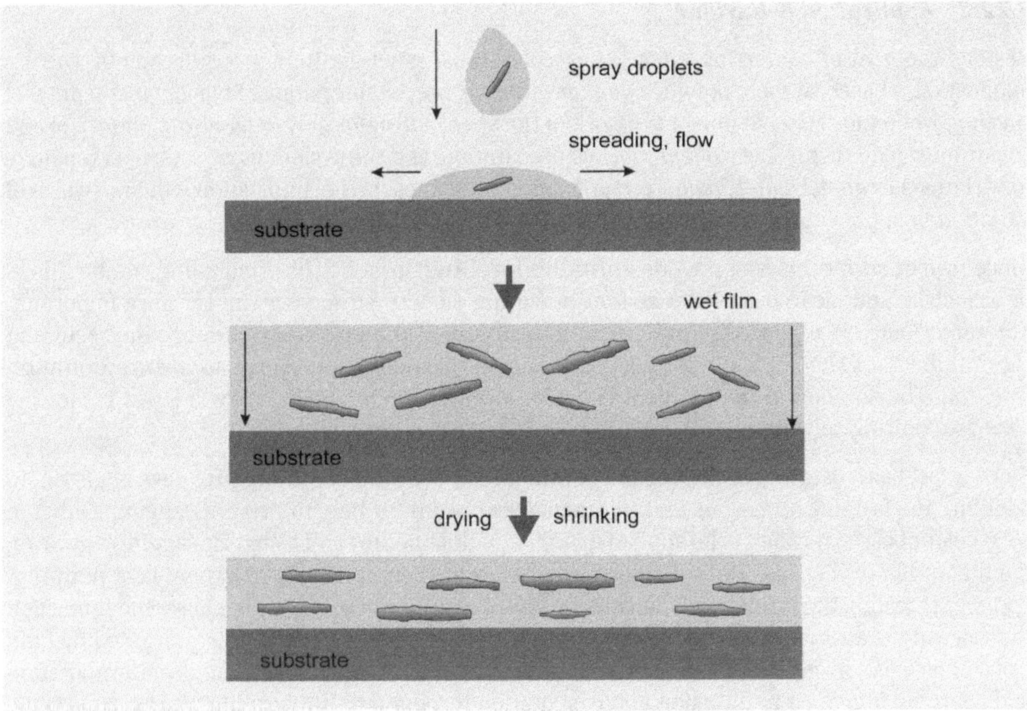

Figure 3.7.3: Model showing the orientation of flake-like pigments during application

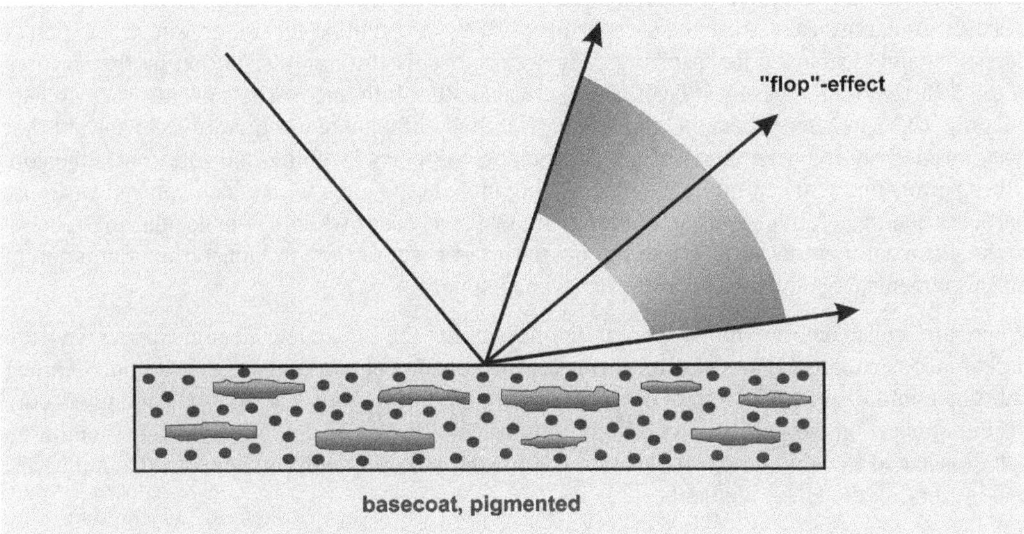

Figure 3.7.4: Effect of a coloured metallic basecoat

coloured pigments. For optimum effect, the pigments must have very small particle sizes and be readily dispersed, and generate transparent colour impressions without hiding reflection from the aluminium pigments. In that event, the flip-flop effect, as seen in plan view, is mostly reflection from the aluminium pigment (light and silver) but at low viewing angles is mainly the colour of the pigment (dark and coloured). The principle is shown in Figure 3.7.4.

A number of effect pigments have since been developed that confer a coloured flip-flop effect directly (see Chapter 3.7.3.6).

3.7.2.2 Application behaviour

Besides the goal of conferring effects, basecoats must meet additional requirements. During application, a basecoat must provide *optimum levelling*, prevent formation of blisters (*popping*) and *sagging*, while the clearcoat must not disorient the effect substances by *redissolving*, the effect and colour must lend themselves to *reproducible* preparation, and the system must not be too sensitive to variations in *application conditions* (application robustness). The application window (width of variation in application conditions) must be as wide as possible.

Since primer surfacers must provide **optimum levelling**, for topcoats - including combinations of basecoats and clearcoats - the goal of providing smooth surfaces is much more important. The rapid increase in viscosity due to physical drying of the polymer content of the resin and the influence of rheological additives necessitates particular measures to ensure optimum levelling. The solution to the problem is to use levelling agents and employ a lower content of slow-evaporating solvents.

Since effect basecoats tend to be applied in thin layers, the problem of **sagging** does not arise. In addition, the fast-dry properties and the rheological additives help to prevent sagging. Only the solid colour basecoats need to be applied in high film thicknesses and tested for sagging. However, the higher the drying rate and the thicker the basecoat layer, the more risk there is of **popping**. Popping is mainly the formation of blisters by high quantities of solvents that leave the film when the viscosity is already high and optimum diffusion is already restricted. It is prevented by judicious choice of solvent combination. The solvent combination must lead to uniform evaporation and keep the film open to diffusion and evaporation throughout film-forming. At the same time, the requisite solvent content of low evaporation rate must not facilitate redissolving.

Redissolving is the variation in effect that occurs when a clearcoat is applied to a basecoat which has only dried physically. Absence of redissolving is a pre-condition for wet-on-wet application of basecoats and clearcoats. Redissolving does not occur only through dissolution by the solvents of the clearcoat on the basecoat layer. There are also other influences. The solvents in clearcoats not only have a negative effect on the basecoat. Limited diffusion of low quantities of solvent that does not destroy the orientation of effect pigments supports levelling and intercoat adhesion. Other parameters that can influence redissolving include the force with which sprayed clearcoat particles impinge on the basecoat. Redissolving may also occur when low-molecular ingredients of the clearcoat material diffuse into the basecoat layer where they influence the orientation of effect pigments (e.g. low-molecular polyisocyanates).

Where areas of an object exhibit different lightness under the same viewing angle (plan view), the darker areas are called "clouds". **Cloudiness** is the outcome of non-uniform distribution of effect substance in the spray mist and of different orientation during physical drying. It mainly occurs during application of basecoats with high lightness and large flip-flop effect values. Cloudiness can be avoided by varying the application conditions, selecting optimum rheological additives, and by varying the effect pigments.

As the application conditions in automotive production lines may vary within specific ranges (application window), the basecoat-clearcoat system must provide a **reproducible effect and colour** (application robustness), although some variation may be permissible. This requirement is also crucial for coating of mounted parts (parts attached to car bodies). Since it is very common for customers to request that such parts match the rest of the body in terms of colour, reproducibility of colour and effect are also important in such parts. It is not so easy to meet this requirement, as such parts are usually made of a material different from that of the body, e.g. they may be of plastic, and therefore present a different surface for wetting and have a different heat capacity for drying and film-forming. The various stages of quality control play an essential role in the reproducibility of colours and effects, e.g. control over quality and availability of raw materials (pigments and effect substances), measurements and monitoring of characteristic values. The production process must be invariable and every batch must be tested (application test) before leaving the production unit.

3.7.2.3 Resistance

Although basecoats are ultimately over-coated with clearcoats, they must contribute to the resistance of the coating system as a whole. For optimum **weathering resistance**, the resins and pigments need to be weatherable. This is why cellulose acetobutyrate came to replace cellulose nitrate, which was less expensive but offered inferior weather resistance (see Chapter 3.7.1.). In addition, the pigments chosen for basecoats are developed to confer outstanding weathering resistance; some of them are also rather expensive.

Since clearcoats are not totally impervious to diffusion, basecoats must be **chemically resistant**. This requirement led to the development of combination resins, which provide not only plasticizing properties for the physical drying polymer, but can also be crosslinked. Crosslinking in resins confers far better chemical and solvent resistance than in resins which are unable to crosslink. It also contributes to better weathering resistance.

A combination of high-molecular polymer and crosslinked combination resins with plasticizing properties is also essential for boosting the mechanical resistance of the entire coating system. More particularly, the basecoat layer helps to improve **stonechip resistance**.

Another important property of basecoats is their **adhesion** (intercoat adhesion). Like the primer surfacer layer, the basecoat layer has to provide adhesion to both sides, namely to the primer surfacer below and the clearcoat above. It is not as easy as it seems to meet this demand. The film matrix of a basecoat that contains a physically drying polymer and effect substance with large particles is totally different from film matrixes of primer surfacer or clearcoat. Basecoat films have a different molecular network. The resins chosen must not only offer the possibility of optimum effect, colour, and resistance to weathering, chemicals and mechanical impact, but provide optimum intercoat adhesion. The resin types chosen contain building blocks that effect intercoat adhesion through physical interaction with the other layers. Crosslinking density greatly influences the intercoat adhesion as well.

Adhesion properties are commonly measured by the cross-hatch test. Basecoats are also subjected to the **vapour jet test**. The test panel is scored in the shape of a St. Andrew's cross with a knife, and the high-pressure vapour jet is played on that part of the panel for a specified length of time. The coating is rated according to the degree of peeling.

3.7.3 Composition of solvent-borne basecoats

Nowadays, nearly all solvent-borne basecoats from the different suppliers have the same resin composition, namely cellulose acetobutyrate as physically drying polymer, saturated polyesters, and melamine resin for plasticising and crosslinking. They contain different types of rheological additives. Originally, aluminium pigments were the effect substances. Later, mica pigments became popular. A large number of different new, effect pigments are now available. Coloured pigments for basecoats could be used equally well for topcoats, but the preferred types have very small particle sizes and are transparent in basecoat layers. The solvents for basecoats are chosen on the basis of the solubility of the basecoat resins. Other additives may be levelling agents, light protection additives, and reaction catalysts (for crosslinking).

3.7.3.1 Cellulose acetobutyrate

Cellulose is the most important building block of all plants. The cellulose molecule is a linear polysaccharide made up of glucose units connected by glycoside bridges. It has a molecular weight of 500,000 to 1,000,000 g/mol, i.e. 3,000 to 5,000 glucose units per molecule. The molecules are stretched and form crystalline micelles. This explains the outstanding strength of cellulose. However, polymers with crystalline micelles are insoluble in solvents and therefore make unsuitable coating raw materials. The most important use of cellulose is in the preparation of textile fibres. Modifications of cellulose (esters, ethers, nitrate) afford a means of using cellulose for paints. For solvent-borne basecoats, the most important cellulose derivative is cellulose acetobutyrate.

Production of cellulose butyrate requires very pure cellulose. In the past, this was prepared from cotton, but now is obtained from wood. The cellulose fibres are dispersed in acetic acid at 50 to 80 °C and then treated with a mixture of acetic anhydride and butyric anhydride, and some sulphuric acid. Nearly all the hydroxyl groups are transformed into ester groups. The virtually totally esterified cellulose is soluble in acetic acid. To an extent depending on the temperature and duration, the large molecules of cellulose are cleaved into smaller molecules. Acidic cleavage of glycoside bridges affords molecules containing only 100 to 350 glucose units. For optimum solubility and compatibility, some of the ester groups are saponified by adding water to the reaction mixture. Some free hydroxyl groups are generated thereby. This is accompanied by decomposition of polymer molecules. Figure 3.7.5 shows the structure of a typical molecular segment of cellulose acetobutyrate.

The available types of cellulose butyrate differ in molecular size (as measured and defined by the viscosity of a test solution), the number of acetate and butyrate groups, and the number of free hydroxyl groups. A typical cellulose acetobutyrate for metallic basecoats [124] consists of 13.0 to 16.0 % by weight acetate groups, 33.0 to 38.0 % butyrate groups, and has an average hydroxyl content of 1.8 % by weight. The dynamic viscosity is 790 to 1,900 mPa s (test solution 15 % by weight in butyl acetate). This type of cellulose acetobutyrate is soluble in esters, ketones, glycol ethers, and glycol ether esters; it is thinnable with aromatic hydrocarbons or alcohols, or both. Although there are free hydroxyl groups, their participation in crosslinking reactions is implausible. While the crystallinity of cellulose is widely degraded by esterification, the acetobutyrate molecules - mainly the high-molecular types - still have a tendency to associate. This tendency to form molecular associations is the reason that domains form in the film structure. Provided that such domains do not influence the effect, they are advantageous as they boost the resistance of basecoats to redissolving. The domains are visible under an interference microscope. The higher the molecular mass of cellulose acetobutyrate, the better is its resistance to redissolving, the better the resistance of basecoat films to chemicals, solvents and mechanical

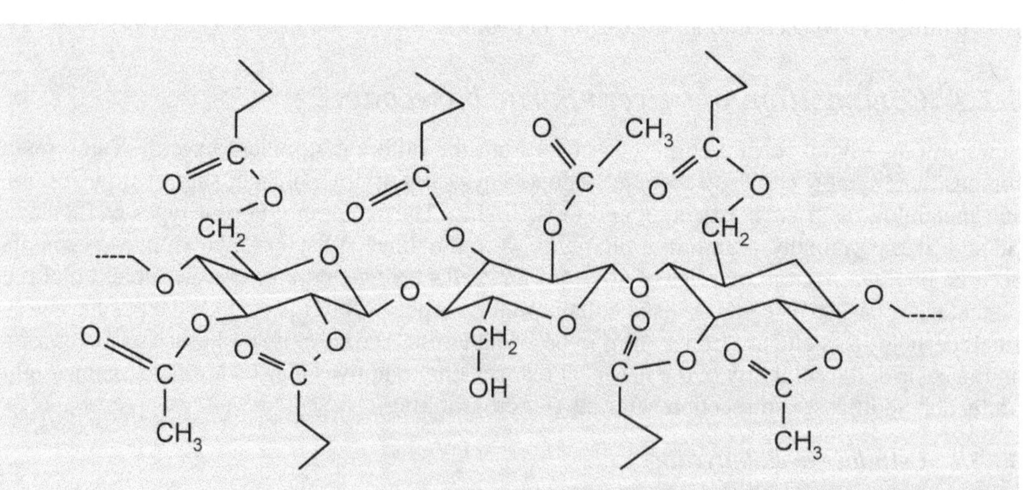

Figure 3.7.5: Structure of a molecular segment of cellulose acetobutyrate

stress, the lower is the compatibility with other resins and the lower is the solids content at application viscosity of the basecoat thereof. The influence of low-molecular cellulose acetobutyrates on film properties is quite the opposite. High quantities of butyrate promote solubility and compatibility, while high quantities of acetate support resistance to redissolving. High quantities of hydroxyl groups promote compatibility with combination resins [125].

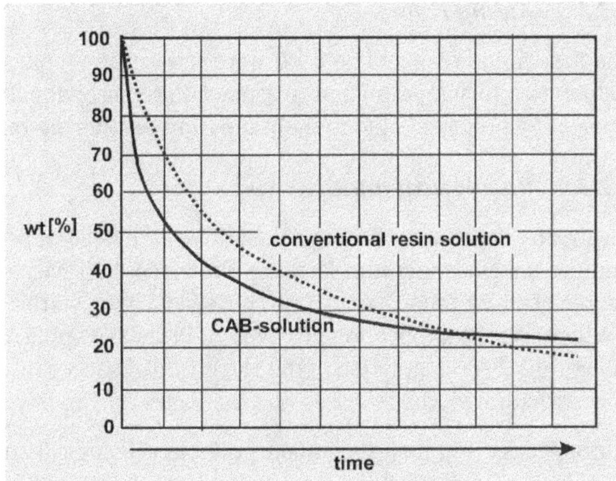

Figure 3.7.6: Comparison of evaporation of solvents from cellulose acetobutyrate and other resins

On account of the molecular structure of cellulose acetobutyrate and the residual tendency to form molecular associations, the drying properties of solutions of cellulose acetobutyrate are quite different from those of conventional resins. Initially, the solvents evaporate much faster than other resins. However, at the end of physical film formation, there is greater retention of residual solvents in the coiled molecular structure of the cellulose acetobutyrate film. Such drying properties are advantageous for resistance to redissolving and also promote levelling. The different evaporation properties of cellulose acetobutyrate solutions and other resins are compared in Figure 3.7.6.

3.7.3.2 Polyesters

Theoretically, the polyesters used for primer surfacers (see Chapter 3.5.3.1) are also suitable for basecoats. Optimum compatibility with cellulose acetobutyrate is achieved by basecoats composed of polyesters that have rather low molecular weights, are not highly branched, and contain some less-polar building blocks. Table 3.7.1 shows the composition and physical characteristics of polyester that is typically suitable for basecoats [126].

Such polyesters also make ideal resins for preparing pigment dispersions.

Table 3.7.1: Composition and characteristics of a polyester for basecoats

n = m/M	Components	M	m (example)	wt-‰
1.829	ethylene glycol	62	113.4	94.3
1.371	neopentyl glycol	104	142.6	118.6
1.372	HPN	206	279.8	232.7
0.686	trimethylol propane	134	91.9	76.5
1.826	isophthalic acid	166	303.5	252.5
1.354	HHPA	154	208.5	173.4
1.371	adipic acid	146	200.2	166.5
	sum		1339.9	1114.4
7.646	water	18	137.6	114.4
	yield (AN = 5.0)		1202.3	1000.0

average molecular weight [g/mol]	1482	acid number [mg KOH/g]	5.0
degree of branching [mol/kg]	0.57	OH-number [mg KOH/g]	102.5

3.7.3.3 Amino resins

The first amino resins to be used were rather high-molecular urea resins that supported solvent evaporation (drying rate) and resistance to redissolving. However, specific melamine resins later came to be preferred, due to their superior weathering resistance and better intercoat adhesion.

3.7.3.4 Rheological additives

As already described, rheological additives are necessary for immobilising the orientation of aluminium flakes during film forming. Preferred additives are those that do not generate a certain degree of structural viscosity until physical drying starts. The advantage is that such additives do not increase the application viscosity or lower the application solids. The viscosity curves shown in Figure 3.7.7 as a function of the shear rate of basecoats containing optimum additives illustrate how structural viscosity arises and increases during physical drying.

Various classes of materials make suitable rheological additives: modified bentonites (swellable by solvents), polyethylene wax dispersions (prepared by precipitation in solvents of hot wax solutions), crystalline ureas (low-molecular weights, precipitated *in situ* from primary amines and polyisocyanates during formation in a resin solution), polyamides and polyurethanes which are partly crystalline.

3.7.3.5 Effect substances

The first effect substances to be used for coatings and which are still the most important are **aluminium flakes**. These pigments contain highly pure aluminium.

Physical data of pure aluminium for pigments [127]:

Melting temperature	660.2 °C
Boiling temperature	2495 °C
Density (20 °C)	2.699 g/cm³
Density (liquid, at 660.2 °C)	2.357 g/cm³
Crystal structure	Cubic, lattice length 4.0496 * 10^{-10} m at 298 K
Electrical resistance	2.5 μΩ cm

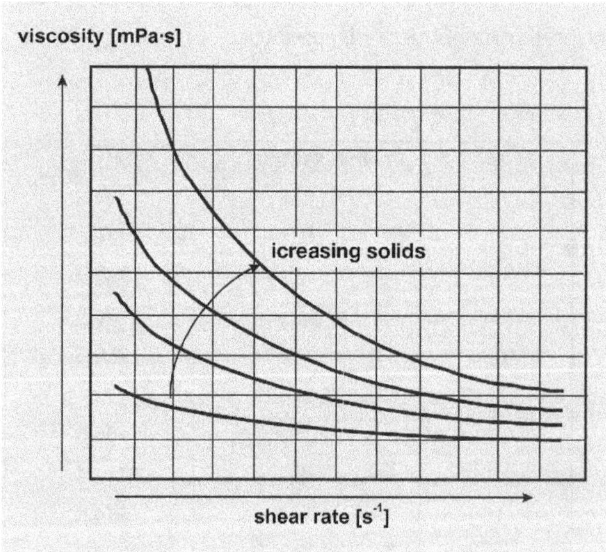

Figure 3.7.7: Change in viscosity of basecoats containing rheological additives during physical drying

Aluminium is essentially obtained from pure aluminium oxide (Al_2O_3). Pure aluminium oxide is rare in Nature but is found in the minerals corundum and natural emery. Corundum is also the essential ingredient of rubies and sapphires. Relatively high quantities are contained in bauxite (50 to 60 % by weight Al_2O_3, plus silica). However, various bentonites lend themselves to the preparation of aluminium metal as well. A number of dressing procedures are employed to separate the aluminium oxide from by-products such as silica and iron oxides.

The aluminium metal is prepared from aluminium oxide by electrolysis in a fused state. Since pure aluminium oxide has a melting temperature of 2,050 °C, a flux has to be used. A suitable flux is cryolite (Greenland spar, sodium hexafluoroaluminate, Na_3AlF_6) which in the past was mined in Greenland, but now is made synthetically by making aluminium oxide react with hydrofluoric acid and sodium carbonate. The melting temperature is 1,010 °C. The eutectic mixture of cryolite (81.5 % by weight) and aluminium oxide (18.5 % by weight) melts at 935 °C. The electrolysis is initiated by adding aluminium metal to increase the conductivity and is carried out at 940 to 985 °C. The cryolite dissociates into sodium and hexafluoroaluminate ions. Hexafluoroaluminate reacts with aluminium oxide to form fluoroxoaluminates. All aluminate ions are decomposed to aluminium ions and fluoride ions. The theoretical decomposition voltage is 2.2 Volts, but to overcome the electrical resistance a higher voltage of 6 to 7 Volts is chosen. The excess current power is converted into heat to boost the melting process. The discharged aluminium is deposited on the cathode, which is the floor of the electrolysis cell. The sodium is not discharged. When the voltage rises, more aluminium oxide is added. The molten aluminium is covered and protected from oxidation by the overlying melt, and can be removed from the cell. The electrodes are made of carbon. Since oxygen is formed at the anode, the carbon is oxidized and must be replaced. Aluminium prepared by this process has a purity of more than 99.9 % by weight.

Aluminium of such high purity is suitable for the preparation of aluminium pigment flakes. The pure metal is easy to shape (it is ductile). In addition, there is the advantage that pure aluminium is much more resistant to chemicals (acids) and more weather resistant than aluminium that might contain small quantities of silica or iron oxide. The pigments are produced by atomising melted metal in a cyclone. The resultant powders are passed through screens. Wetting materials, e.g. fatty acids and aromatic hydrocarbons to act as solvent, are added to batches of defined particle size which are then shaped in ball mills. The resultant slurries are screened again and the excess solvent is removed. In a kneader, other solvents and additives are added. The delivery form is a paste containing mainly 65 % by weight aluminium and 35 % aromatic hydrocarbons, along with small quantities of additives.

Aluminium pigments are either leafing or non-leafing, the difference arising from the degree of surface treatment. Leafing types are covered with wetting agents which are solid, very non-polar, and not easy to remove from the flake surface (e.g. stearic acid). After application of paints containing leafing pigments, the aluminium flakes float on the surface to form a mirror-like layer (like leaves on a lake). Non-leafing effect pigments are covered with liquid, more polar additives that are removable (e.g. oleic acid or fatty amines). Such pigments are distributed within the film layer, and do not float on the surface.

Leafing types are suitable for heat-resistant paints and corrosion-protection paints. They can also be used to create coating layers with a so-called chrome effect. Such effect coatings are very sensitive and cannot be recoated. Non-leafing types are suitable for metallic basecoats because distribution of the particles is one of the preconditions for the flip-flop effect. Different influences govern the effect created by aluminium flake pigments. Most important - of course - are optimum distribution in the basecoat layer, efficient wetting of the particle surfaces, and good compatibility with the resin matrix. Also important are the viscosity, the application solids, the application process and its conditions, and the type of clearcoat and its method of application. Shrinking and the effect of rheological additives during film forming by physical drying are important for immobilising the aluminium particles to achieve the requisite metallic effect. The effect also depends on the particle size of the aluminium pigments, the particle size distribution, and the shape of the particles. The conventional process of atomising and milling aluminium yields irregularly shaped flakes. These are called "corn-flake" aluminium [128] because of their shape under a microscope. Figure 3.7.8 (page 148) shows scanning electron microscope (SEM) micrographs of conventionally prepared aluminium pigments after atomization and milling.

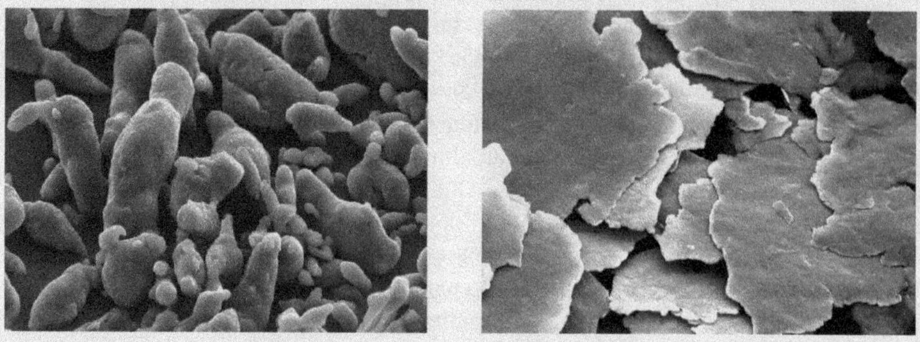

Figure 3.7.8: SEM micrographs of particles and flakes of corn-flake aluminium pigments

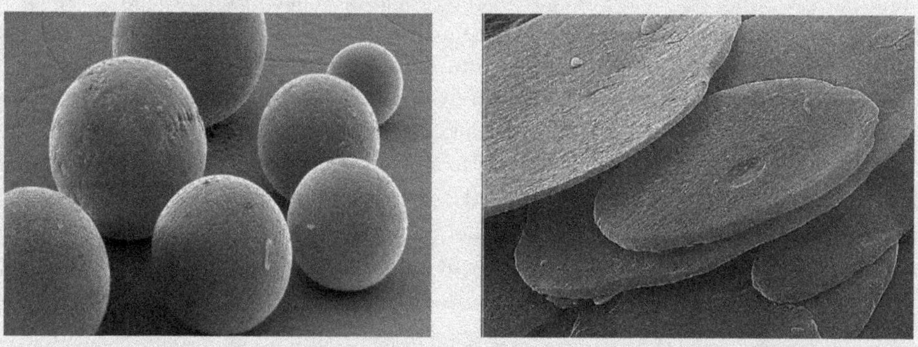

Figure 3.7.9: SEM micrographs of particles and flakes of silver-dollar aluminium pigments

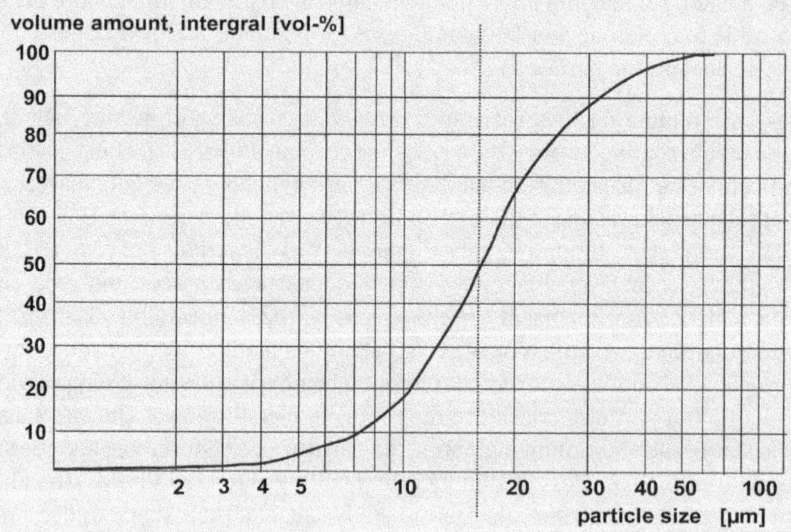

Figure 3.7.10: Particle size distribution (flake diameters) of a typical aluminium pigment

The process for preparing aluminium pigments is by now well optimized. It is possible by special means to prepare spherical particles which have nearly the same diameter. Such particles are shaped very carefully to form flakes that have very smooth edges and look like very small coins. These are therefore called “silver dollar” aluminium pigments [128] (see SEM micrographs of both types in Figure 3.7.9). Such aluminium pigments have been available since the late 1980s.

The more uniform the aluminium pigment particles are and the smaller the particle size distribution is, the more brilliant is the effect of the basecoat made there from. Aluminium pigments are classified and traded in specific particle sizes. Aluminium flakes for automotive paints have average diameters between 3 and 25 µm. The flake thicknesses are 0.2 to 1.0 µm. Figure 3.7.10 shows the particle size distribution (flake diameters) of a typical aluminium pigment for OEM coatings.

Increasing average flake diameters of aluminium pigments in basecoats of the same pigment volume concentration lead to increasing brilliance, increasing lightness in plan view, and a more pronounced flip-flop effect. However, hiding power and distinctness of image (DOI) are diminished. Customers in the USA prefer sparkling effects within the coating layers. These are obtained by using aluminium pigment with rather large flake particles. If the flakes have large diameters but low thicknesses, the pigments are sensitive to mechanical stress. Where shearing stress occurs in the pumps or reducing valves in the circulation lines during application, pigment particles can become deformed, lose their effect and change colour. Where application equipment is fitted with centrifugal pumps instead of piston pumps (mainly in the USA), the aluminium pigments employed must be of adequate flake thickness if deformation is to be avoided. One such example is ETR pigments.

Development of pearlescent pigments

Since ancient times, mankind has been fascinated by the gloss and effect of natural pearls and mother of pearl. These consist of thin layers of calcium carbonate in the modification of the mineral aragonite, along with small quantities of calcite. The thin layers are connected by layers of proteins. The specific effect - a particular gloss and slight colour flop - stems from differential light refraction on the very thin layers, leading to interference. Interference is the superposition of reflected coherent light beams. If the reflected beams are in phase with each other, the light is intensified (constructive interference); if they are out of phase, they cancel each other out (destructive interference). Figure 3.7.11 shows the principle behind constructive and destructive interference of light reflected from very thin layers.

In the past, attempts were made to reproduce pearlescence on decorations. The first reported product to be used for this was **fish-silver** [129], which was obtained by extraction of fish scales. These, in turn, were obtained from bleak (a type of carp) and other fish (herring and mackerel). Extraction from aqueous slurry yielded a paste that was treated with solvents and then cleaning. The delivery form is a paste of crystals of guanine and small quantities of hypoxanthine (two types of purine bases).

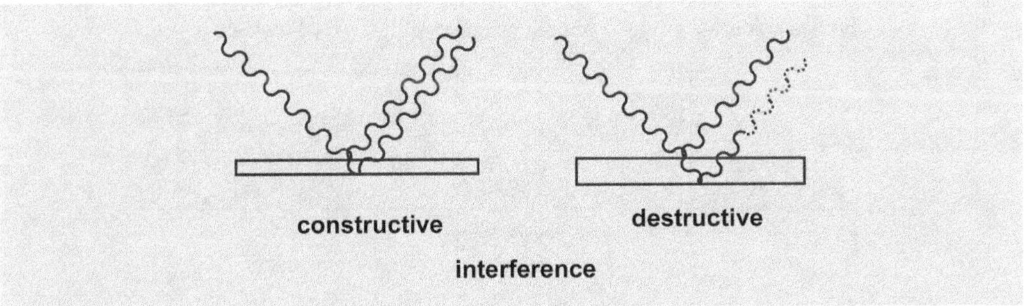

Figure 3.7.11: Principle behind constructive and destructive interference of light reflected from very thin layers

The refractive index is relatively high (parallel n_D = 1.79, vertical n_D = 1.91), which is essential for reflection and interference. The production of about 500 g fish silver requires up to 20,000 fish. The product is therefore very expensive and unsuitable for industrial use; hence its employment in specific applications, e.g. high-class nail varnishes. Annual global production is less than 50 t [130].

The first synthetic material to confer pearlescence was **basic lead carbonate** ($Pb[OH]_2 \cdot 2\ PbCO_3$). This is precipitated from dilute aqueous solutions of lead acetate and carbon dioxide [130]. Given the right precipitation conditions, hexagonal flakes are formed (diameter 20 µm, thickness 50 nm) that have a refractive index of 2.10. The particles are very sensitive to mechanical impact. As a result, and since lead compounds are classified as toxic, the product no longer finds industrial use. Global consumption is estimated to be less than 1,000 t [130].

Bismuth oxychloride (BiOCl), which is prepared by hydrolysis of acidic bismuth chloride solutions, forms flake-like pearlescent particles under specific reaction conditions. The product has a refractive index of 2.15. It has low resistance to light and mechanical impact. Its sole use is in cosmetics and the global market is estimated at about 400 t [130].

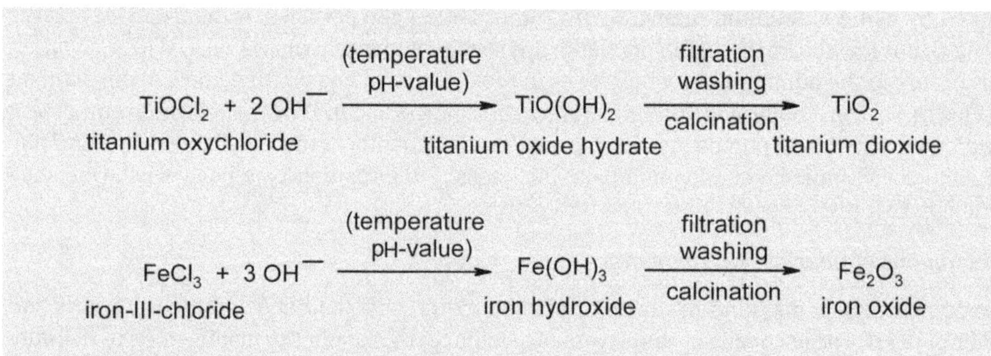

Figure 3.7.12: Reactions which occur in the doping of mica surfaces

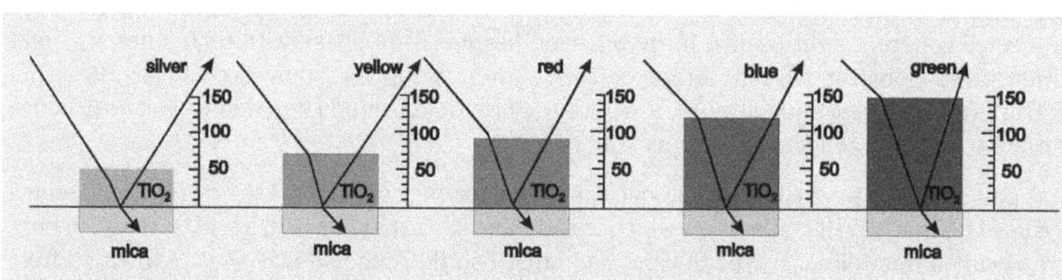

Figure 3.7.13: Colour effects of mica pigments doped with different layers of titanium dioxide

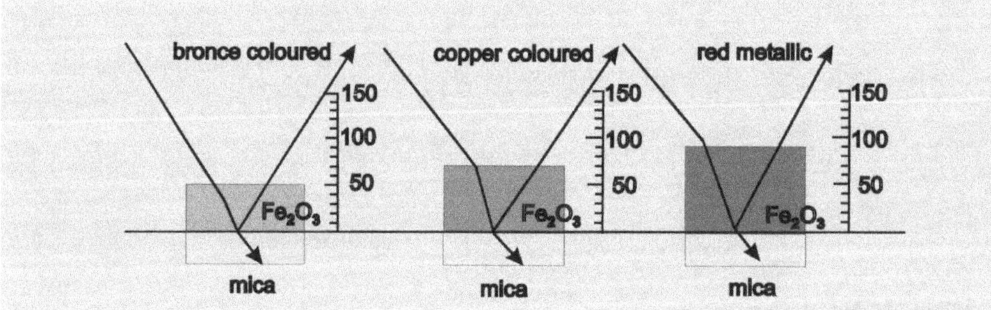

Figure 3.7.14: Colour effects of mica pigments doped with different layers of iron oxide

Mica pigments

Eventually, pearlescent pigments were prepared from mica. Mica is potassium aluminium alumino silicate, also known as the mineral muscovite ($KAl_2[OH]_2[AlSi_3O_{10}]$). It is often contaminated with iron. Deposits of mica are numerous and so the raw material for pigments is relatively inexpensive. The mineral is conditioned by milling, cleaning, and classification. Mica is composed of a thin-layer crystal lattice; it is easy to fragment into thin flakes that have thicknesses of 200 to 500 nm and mean flake diameters of 5 to 200 µm. Although the refractive index is relatively small at 1.60, the flakes are pearlescent.

Mica pigments were described as long ago as 1942, but they were not introduced into the pigment market until the late 1960s [131]. The first mica pigments conferred translucent pearlescence. On account of the low hiding power, such pearlescent basecoats had to be combined with coloured primer surfacers, or more particularly with white primer surfacers. The first formulations suffered from problems with levelling and topcoat hold-out. Not until it proved possible to prepare particles with a reproducible narrow particle size distribution did the market for mica pearlescent pigments grow. Since then, the mica pigments have usually been combined with aluminium pigments and coloured pigments. However, the most important step in the development of new effect pigments was the development and introduction of mica pigments which were doped with thin layers of metal oxides [132]. Besides pearlescence, such pigments confer colour effects (colour flop). The first colours developed were very “gaudy”. However, since the mid-1980s, muted colours have also become available which better meet the tastes of car buyers. At the moment, more than 50 % of effect basecoats contain pearlescent pigments - mostly based on mica. Surface doping of the mica flakes is carried out in an aqueous dispersion. The most popular doping layers consist of oxides whose higher refractive indices are much greater than that of mica. These are usually titanium dioxide and iron oxide. The process converts the dispersion of mica flakes into a dilute aqueous solution of metal chlorides. Under carefully controlled conditions (temperature, concentration, time), alkali hydroxide solution is added to precipitate a thin layer of metal hydroxides on the mica particles. The particles are filtered, washed and calcined at temperatures of 800 to 900 °C to yield the corresponding metal oxides. The reaction is illustrated in Figure 3.7.12.

On account of the different refractive indices and the ability of metal oxides to absorb some wavelengths of visible light, colour effects (colour flop) are also obtained in addition to interference effects. Since interference depends on the thickness of the layers and the absorption of light, the colour effect also depends on the layer thicknesses. For example, doping with titanium dioxide generates silver greys at 40 to 50 nm, yellows at 60 to 70 nm, reds at 80 to 90 nm, blues at 110 to 120 nm, and greens at 140 to 150 nm. Interference effects produced by titanium dioxide layers of different thickness are shown in Figure 3.7.13.

Doping mica surfaces with iron oxide also yields colour effects that vary with the thickness of the oxide layers. The light absorption properties of iron oxide cause all the colours to have a metallic sheen. At layer thicknesses of 40 to 50 nm, the effect is that of bronze; at 60 to 70 nm, copper; and at 80 to 90 nm, metallic red to gold. The dependence of colour effects on the layer thickness of iron oxides is illustrated in Figure 3.7.14.

Combinations of different metal oxide layers (titanium dioxide together with iron oxide) leads to other colour effects, mainly various shades of gold. If mica particles are treated with titanium oxychloride and ammonia, the surface layer after calcination at 800 to 900 °C consists of some titanium compounds in a low oxidation state and probably of compounds containing nitrogen (TiO, TiO_xN_y, TiN) in addition to titanium dioxide. This type of mica pigment confers bluish-silver effects.

As mica pigments have a tendency to settle, it is necessary to use wetting agents and suitable rheological additives. The low hiding power of mica pigments is compensated by combining them with aluminium pigments and coloured pigments. The degree of pearlescence increases with increase in particle size of the mica pigments. Large mica pigments give rise to a sparkle effect comparable to that of aluminium pigments.

Other inorganic flake pigments

In line with the trend in the automotive industry towards appealing, custom effects, many new effect pigments have been developed in the last few years. New effects have been developed, too. Besides micas, flake-like particles have been doped with ultra-thin layers of metal oxides and launched onto the coatings market.

First, there are the **aluminium flakes**, which are doped with thin layers of iron oxide. Doping proceeds by means of **chemical vapour deposition** (CVD) in the gas phase and in a fluid bed. Aluminium particles are fluidized by nitrogen. Then, at 450 °C, iron pentacarbonyl and oxygen are injected into the fluid bed. This leads to the formation of iron oxide layers on the aluminium flakes, and carbon dioxide. The thickness of the layers depends heavily on the reaction duration. These pigments range from gold metallic at 20 nm Fe_2O_3 to orange metallic at 30 nm Fe_2O_3 [133]. An extension of the process consists in doping the aluminium flakes first with silica (in a wet process) and then with iron oxide by means of CVD. The resultant pigments exhibit greater colour flop than the pigment types described first. To an extent depending on the layer thickness, the colour changes from gold to grey metallic or from red metallic to golden [134].

Another pigment class is prepared by doping aluminium flakes by means of **physical vapour deposition** (PVD) to form ultra-thin layers of magnesium fluoride and chromium [135]. The resultant products exhibit extensive colour flop.

It is possible to prepare very thin, uniform layers of silica on metal sheets. **Silica flakes** prepared in this manner are doped with titanium dioxide or iron oxide in a CVD process. The resultant pigments exhibit extreme colour flop, from violet to green, from red to gold, from green to red, and from gold to blue [136].

Aluminium oxide flakes (α-Al_2O_3, corundum) are prepared by controlled crystallisation in molten sodium sulphate. The flakes are also doped with iron oxide or titanium dioxide. These effect pigments exhibit significant colour flop. Unlike the other pigments already described, these pigments are also suitable for creating sparkle effects.

It is also possible to crystallise **iron oxide** (α-Fe_2O_3, hematite) as flakes. Such flakes are also available as a natural mineral, micaceous iron oxide. The natural modification serves as corrosion protection pigment and is dark grey. The flakes are prepared synthetically by a hydrothermal process. When iron is combined with specific quantities of other components, such as aluminium or manganese salts, the resultant flakes have much larger diameters than those prepared from pure iron oxide. The resultant effect pigments have colours in different cuprous shades [134].

PVDA pigments

A new class of effect pigments are the PVDA pigments (physical vapour deposited aluminium). The process consists in vapour deposition of aluminium onto a foil covered with a strippable coating. The ultra-thin layer of aluminium is removed, crushed, and screened. The resultant flakes are much thinner than those prepared conventionally. The ratio of mean diameter to thickness is very high. The plan view is very light and bright. Remission at an oblique angle is much lower than that of conventional aluminium pigments. This produces a particular depth of metallic effect (liquid metal effects). Basecoats containing such PVDA pigments are not easy to apply as the effect is greatly influenced by the application conditions. In addition, the ultra-thin aluminium foil can be texturised in a PVD process to produce holographic effects or interferences, which yield different degrees of colour flop.

LC pigments

Liquid crystal structures can generate interference and colour flop, too. Special organic compounds that form nematic molecular structures are combined with molecules which have chiral structures. Both are combined by polymerisation (UV crosslinking). The resultant films are crushed and screened to form LC pigments. LC pigments are available that are based on aromatic polyesters [137] or polysiloxanes [138]. Currently available pigments have rather large particle sizes (up to 4 μm), which give rise to problems with application behaviour and inferior mechanical stability. Comprehensive information is available in the literature [139].

3.7.3.6 Coloured pigments

Effect pigments may be combined with coloured pigments to produce colour flop. The plan view is dominated by reflection from the flake-like effect pigments. Remission at an oblique angle is dominated by the coloured pigments. The optimum impression is obtained when organic pigments are chosen which are transparent. Most organic pigments and carbon blacks are transparent. In other words, they absorb fractions of visible light, while the remaining coloured light is reflected from the substrate and from the surfaces of effect substances. Transparency is gained from pigments whose particles are smaller than half the wavelength of visible light. It is therefore important to disperse such pigments very well. Optimum hiding power of pigments is achieved when their particle size is half the wavelength of visible light or even longer; for transparency, the particle sizes must be lower. Dispersion of one and the same pigment for solid colour topcoats (optimum hiding power) is different from dispersion for effect basecoats (optimum transparency). Figure 3.7.15 shows how colour strength, hiding power and corresponding particle size depend on the dispersion effort (length of dispersion in a batch stirrer mill).

For effect basecoats, the pigment groups employed are the same as those for topcoats (see Chapters 3.6.3.4 to 3.6.3.6). Transparency is ensured through optimal dispersion of small particle sizes.

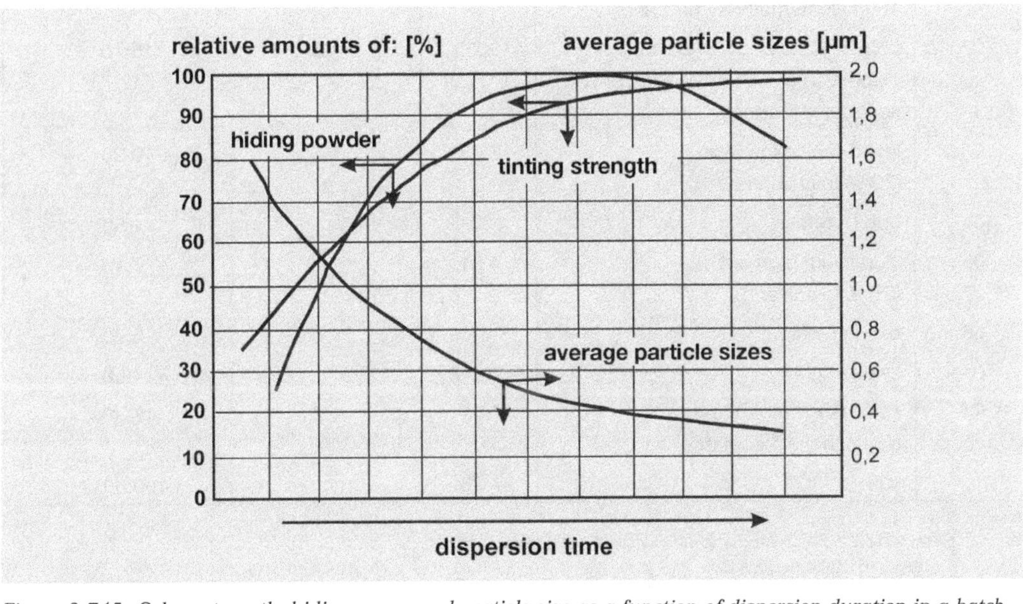

Figure 3.7.15: Colour strength, hiding power and particle size as a function of dispersion duration in a batch stirrer mill

Pigments with high hiding power, for example titanium dioxide, greatly reduce the flip-flop effect. However, there are titanium dioxides available that have very small particles (transparent titanium dioxide). Such types, in combination with aluminium pigments, are responsible for the ice effect. Of course, optimum dispersion requires that the chosen pigments are easy to wet and resistant to weathering, solvents, and migration. An important aspect is reproducibility of effect and colour. This is much more difficult to achieve than reproduction of a solid colour. The requirements on guaranteeing the same, pure shade, same particle size and particle size distribution, optimum dispersion properties of pigments are much more critical for basecoats than for solid colour topcoats.

As is the case with the production of solid colour topcoats, basecoats can be made in two different ways, namely with tinting pastes and tinting colours. In both cases, the dispersion process is tailored to optimum dispersion of just one pigment. Tinting pastes are usually prepared by making the polyester resin solution serve as grinding medium. The tinting colours contain all the resins of the basecoat formulation. The basecoat is completed by adding a dispersion of effect substances, carefully prepared to prevent destruction of the flake-like particles, which may also contain the rheological additives and some resins.

3.7.3.7 Solvents

On account of the solubility of cellulose acetobutyrate and the specific application process (wet-on-wet application), the solvents in basecoats are different from those in solid colour topcoats. The solvent composition contains solvents which have a more polar character and shorter evaporation times. The most important solvents here are esters (e.g. butyl acetate), aromatic hydrocarbons (e.g. xylene) and alcohols (e.g. butanol). However, to ensure optimum flow and levelling during film forming by physical drying, some slow- evaporation solvents, e.g. glycol ethers (e.g. methoxy propanol) or glycol ether esters (e.g. methoxy propyl acetate), must be added.

Table 3.7.2: Formulation and characteristics for a metallic basecoat (silver coloured)

Pos.	Components	Solid m	Density g/ml	Formulation wt-%
01	saturated polyester. modified by crystalline urea (50 % in aromatic 100/xylene/methoxy propanol)	19.65	1.2	39.30
02	polyethylene copolymer wax dispersion (4,7 % in xylene, butyl acetate, n-butanol)	0.74	1.0	15.70
03	imino-melamine resin (70 % in n-butanol)	7.63	1.5	10.90
04	butyl acetate			1.50
05	aluminium pigment (65 % in solvent)	4.81		7.40
06	butyl acetate			5.90
07	xylene			1.50
08	cellulose acetobutyrate (25 % in butyl acetate)	4.45	1.2	17.80
	sum	37.28		100.00

pos.

01 Setal 90173, 50 % in aromatic 100/xylene/PM (Nuplex)
02 Ceratix 8461 (Altana Cera BV)
03 Setamin US 138 70 % in n-butanol (Nuplex)
05 Stapa Metallux R 9755 (Altana Eckardt)
08 CAB 381-0,5, 25 % in butyl acetate (Eastman Chemicals)

3.7.3.8 Other additives

Other additives may be added to the basecoats in addition to the aforementioned rheological additives. Optimum transparency can be promoted by adding a wetting agent when pigment dispersions are being prepared. Only a few levelling agents are suitable for basecoats. Since they act in the surface, they must not interfere with the wet-on-wet process, where it is much more important than in the case of primer surfacer for the clearcoat perfectly wet the surface of the basecoat. After stoving, both layers must afford excellent intercoat adhesion. Levelling agents that float on the surface and are incompatible with the next layer (clearcoat) would impair the requisite properties. The optimum solvent composition therefore makes the most important contribution to levelling.

3.7.3.9 Formulation example

Table 3.7.2 contains the formulation and characteristics of a sample solvent-borne silver basecoat for an automotive OEM coating [140]. It consists of cellulose acetobutyrate, saturated polyester, melamine resin, and aluminium pigment. The rheological additive is a dispersion of polyethylene wax in solvents and a crystalline urea (dispersion in the polyester). The added solvents are butyl acetate and xylene.

Application solids are 23 % by weight, while the application viscosity is 15 s (cup 4, 23 °C, DIN 3511 equivalent to ISO cup 58 s). The film thickness (dry) is 12 µm. The flash-off conditions of the basecoat are 5 minutes at 80 °C. The pre-dried basecoat is then over-coated with clearcoat and finally – after a second brief flash-off – stoved for 20 minutes at 140 °C.

3.7.4 Basecoats with increased application solids

In the past, conventional effect basecoats had a very low solids content at application viscosity, e.g. for a silver basecoat: 13 % by weight at 15 s (cup 4, 23 °C, DIN 3511 equivalent to ISO cup 58 s). Low-solids, solvent-borne basecoats were the first systems to be formulated with higher solids to reduce emissions of volatile organic compounds (VOCs). The first step was to reduce the viscosity by lowering the molecular weights of the resins for the basecoats. However, when low-molecular polyesters and melamine resins are used for basecoats, there is little effect on the viscosity. A decisive influence on viscosity is exerted firstly by the high-molecular cellulose acetobutyrate and secondly by the rheological additives. The rheological additives are chosen such that they do not influence the viscosity at application state, but rather act efficiently during the drying process by rapidly increasing the viscosity.

Besides the commonly employed medium-and-high molecular cellulose acetobutyrate types, there are low-molecular types available commercially. Such low-molecular cellulose acetobutyrates lower the solution viscosities and hence the basecoat viscosities significantly. Unfortunately, however, effect development and application robustness are also greatly reduced. The effect pigments are no longer optimally immobilised and the basecoat layer exhibits redissolving in wet-on-wet spray application of the clearcoat. Also, the combination of high-molecular types with low-molecular types of cellulose acetobutyrate did not lead to a practicable solution.

It finally proved necessary to search for other ways to lower the application viscosity or increase the application solids. Two solutions were found:

1. Reduce the quantity of cellulose acetobutyrate and transfer its role and properties to the other resin components.
2. Replace the cellulose acetobutyrate with resins that play the same role and have the same properties but have a lower solution viscosity.

The first measure led to the development of medium-solids basecoats, which are currently widespread throughout Europe as substitutes for all low-solids basecoats. Low-solids basecoats achieve solids content in the application state of 23 to 25 % by weight (silver). This represents cuts the solvent content relative to low-solids basecoats by nearly half. The basecoats contain less cellulose acetobutyrate (medium molecular weights). Combination resins are chosen which are able to contribute to physical drying properties. These are either saturated polyesters with building blocks that promote physical drying, or acrylic resins that contain hydroxyl groups and monomers that also support physical drying. Both resin groups are combined with melamine resins acting as crosslinker. These melamine resins, too, are chosen for their ability to boost physical drying.

Two classes of resin have replaced cellulose acetobutyrate:

- polyurethane elastomers
- special acrylic resins

Polyurethane elastomers

High-molecular polyurethane elastomers contain linear polyesters by way of soft segments, diisocyanates as hard segments, along with polyols or diamines for molecular chain extension. The first step is to polyurethane prepolymers by making polyesters react with an excess of diisocyanates. Chain extension is then effected by reaction with polyols or amines. The resultant products have a relatively high molecular weight and contain urethane and urea groups, which readily form hydrogen bonds. The type of polyester building blocks, the type of diisocyanate, the molecular weight, and the hydrogen bonds influence the physical drying properties of these polyurethane elastomers. The polyurethanes have lower solution viscosities than cellulose acetobutyrate. Combinations of such polyurethane elastomers with polyesters and melamine resins serve as the basis for the formulation of basecoats which achieve application solids of 25% by weight and more [141]. Said basecoats are distinguished by particularly excellent application behaviour. They exhibit no redissolving in wet-on-wet processes, not even in automotive coatings applied at high conveyor speeds. In addition, due to the excellent flexibility conferred by polyurethane elastomers, they are particularly suitable for application to plastic parts.

Special acrylic resins

Acrylic resins, prepared by solution polymerisation (see Chapter 3.8.3.1), can emit solvents much faster than, for example, polyesters. The reason is that the acrylic molecules are linear but coiled, much more so than other resin molecules, and the colloidal particles in solution are relatively dense, so that there is less interaction with the solvent molecules. It is possible to boost this property by using monomers and modifications which serve to increase the glass-transition temperature. Acrylic resins prepared in this manner may have higher solution viscosities than others, but their viscosity is much lower than that of cellulose acetobutyrate solutions. Such acrylic resins are not ideal for wetting effect substances and yield relatively brittle films. To circumvent these problems, acrylic resins modified with ε-caprolactone have been developed. They have low solution viscosities while the ε-caprolactone modification confers excellent flexibility. Nevertheless, due to the segmented structure of uniform polyester side chains generated by addition of several molecules of ε-caprolactone, the drying properties are fairly good. Combinations of such acrylic resins with melamine resins and suitable rheological additives in effect basecoats lead to application solids of 35 to 40 % by weight. The rheological additives are preferably non-aqueous dispersions (NADs) of crosslinked acrylic resins, also known as microgels [142]. Such microgels are able to swell under the influence of solvents to bestow structural viscosity above specific solids content, which immobilises the effect substances very well. Microgels also contribute to optimum physical drying.

The described basecoats are called high-solid basecoats because the solids content in the application state exceed 30 % by weight. Theoretically, it is more difficult to achieve reproducible bright and light effects with such basecoats. The reason is that the shrinking effect - which boosts orientation of the pigment flakes - is much less pronounced than in lower solids versions. Very light and brilliant metallic effects, which are preferred mainly by customers in Europe, are difficult to achieve with high-solid basecoats. However, in the USA, where basecoats with a sparkling effect are preferred, such basecoats perform very well. To further reduce VOC emissions during effect basecoat application and to overcome the problems of achieving all effects with high-solid basecoats, water-borne basecoats were developed.

3.7.5 Water-borne basecoats

Although high-solid basecoats contain 30 to 40 % by weight application solids and save a lot of solvent in comparison to low-solids basecoats, solvent emissions still amount to 60 to 70 % by weight. In addition, high-solid basecoats have poor application robustness, and some colours and effects are not available. Both these reasons led to the development of water-borne basecoats. The components for water-borne basecoats were chosen in accordance with the experiences gained with solvent-borne basecoats. For water-borne basecoats, too, the resin combination chosen contained physically drying resins and crosslinkable resins with plasticizing properties. There was also the need for very efficient rheological additives. In transferring the principles from solvent phase into water phase, it proved necessary in some cases to change the class of product, but not the function and properties.

3.7.5.1 Resins for water-borne basecoats

Resins for water-borne basecoats that boost physical drying are acrylic dispersions and polyurethane dispersions. Resins with plasticizing properties are water-thinnable, saturated polyester or acrylic resins that contain hydroxyl groups which are crosslinked by water-thinnable melamine resins.

Primary acrylic dispersions

There are different types of aqueous acrylic dispersions, namely primary and secondary. Primary acrylic dispersions are prepared from acrylic monomers directly in the water phase, following the emulsion polymerisation process. The process utilise water-soluble polymerisation initiators (e.g. ammonium peroxydisulphate) and emulsifiers (e.g. alkyl sulphonates, ethoxylated alkyl sulphonates, and alkyl ethoxylates). Emulsifier and initiator are dissolved in water. According to the most widespread, classical theory of emulsion polymerisation [143], the emulsifier forms micro-micelles that contain initiator molecules. The aqueous phase is heated to about 80 °C, the reaction temperature for polymerisation. The mixture of acrylic monomers is admixed to yield emulsions with relatively large particles. Monomers which are mainly insoluble in water diffuse into the micelles, starting a free-radical polymerisation process. The free-radical chain polymerisation process is comparable to solution polymerisation (described in detail in Chapter 3.8.3.1) and consists of the reaction steps: initiator decomposition, initiation, chain extension (propagation), and chain termination through recombination of free-radicals or chain transfer. This process can yield high-molecular polymers. The reaction enthalpy (exothermic reaction) is transferred into water and removed by cooling the aqueous reaction mixture. The particle sizes are determined by the type and quantity of emulsifier. The more emulsifier and the greater it efficiency is, the smaller are the particles and the more stable is the emulsion. Attainable average particle sizes range from 150 to 300 nm (diameter). Figure 3.7.16 (page 158) shows the principle behind emulsion polymerisation.

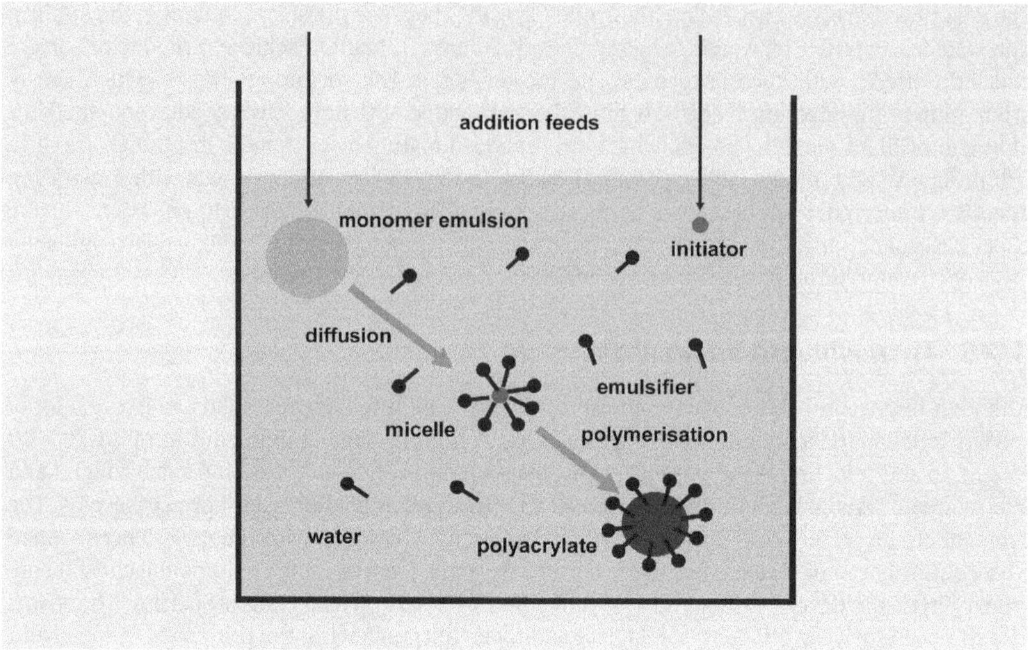

Figure 3.7.16: Principle behind emulsion polymerisation

The physical properties of the polymer (hardness, flexibility, glass-transition temperature) depend on the composition of the monomers. Theoretically, the same conditions are employed as for solution polymerisation products (see Chapter 3.8.3.1). The film-forming behaviour of such primary acrylic dispersions is influenced by the interaction of the small particles. As already mentioned under the general description of dispersions, the viscosity at low solids content is relatively low. However, from a certain concentration on, the viscosity increases rapidly with increase in solids content. The reason for this is that fine particles interact efficiently due to the large surface energy possessed by small particles. This property provides strong support for film formation and the immobilisation of effect substances. This behaviour is intensified when polymer dispersions are used that have very large molecules or, even better, crosslinked molecules. Such dispersions are able to form gels in the aqueous phase by interacting with the cosolvent. Such polymer gels act like rheological additives. Partly crosslinked acrylic polymers bearing free carboxyl groups can form gels if they are partly neutralised by amines. Specific quantities of such products in a water-borne basecoat can have benefits for the application behaviour: immobilisation of the effect substance, no sagging of the film on vertical parts of the coated object, support for physical drying, and no redissolving.

The particles of primary acrylic dispersion are able to form films only above a specific temperature after water has evaporated. This temperature is called the minimum film-forming temperature (MFT) and is a little bit higher than the glass-transition temperature (T_G). Film forming of dispersions takes place by fusion of particles by their outer shells. As the film-forming temperatures of basecoats are initially relatively low, acrylic polymers with low minimum film-forming temperatures should be used. Unfortunately such polymers are relatively soft and not resistant to mechanical impact, solvents or chemicals. To compensate, the dispersions chosen have polymers with higher glass-transition temperature that confer hardness and resistance. The particles of such polymers are embedded into the film matrix by the combination resins, which are not dispersions, but rather are water-soluble and able to fill the hollow spaces between the said particles. So, at latest during stoving with the clearcoat, the basecoat layer is fused very well, although it contains polymers with a high MFT. However, it must be assumed that full interdiffusion does not occur. Domains of relatively high-molecular acrylic polymers still remain. Such areas cannot contribute to the crosslinking process, even though they may contain functional groups for crosslinking reactions. Normally, therefore, the polymers of dispersion do not contain monomers with additional functional groups. The useful properties possessed by acrylic polymer dispersions are their drying properties, influence on the rheology, and physical resistance. They are less suitable for wetting substrates and pigments, promoting flow and levelling, or supporting flexibility and crosslinking.

Polyurethane dispersions

Polyurethanes are prepared by addition reactions in which the chemical equilibrium lies virtually completely on the product side. It is therefore possible - unlike the case for making polyesters - to achieve polymer molecules with segmented structures. The addition reaction also explains the ability to achieve products with relatively high molecular weights. Such large molecules that have a segmented structure and contain hard and soft segments are a precondition for forming elastomers. The elastomeric character mainly stems from the molecular association of hard segments between different molecules. In the case of polyurethanes, the hard segments consist of urethane and urea groups, which associate mainly by forming hydrogen bonds. Since urethane bonds are much more resistant to saponification, at high pH values as well, polyurethanes offer the possibility of preparing water-borne systems with optimum storage stability. Polyurethanes are thus ideal for formulating aqueous coatings materials, especially for water-borne basecoats. Aqueous polyurethane dispersions are usually not prepared with the aid of emulsifiers, but rather by means of anionic stabilisation. The anions may consist of carboxylates, sulphates, sulphonates, or phosphates.

Polyurethane dispersions for water-borne basecoats contain soft segments (linear polyester diols, polyether diols, and polycarbonate diols) and hard segments. The building blocks for hard segments are diisocyanates, and polyols or diamines as chain extender. The diisocyanates employed are aliphatic (e.g. hexamethylene diisocyanate) or cycloaliphatic (e.g. isophorone diisocyanate, or 4,4'-diisocyanatodicyclohexylmethane [HMDI]), which confer optimum weathering resistance on the basecoats. At least parts of the building blocks for hard segments have to generate the ionic group as a carrier group for solubility in water. The most important compound for this is dimethylol propionic acid (DMPA).

The production process for polyurethane dispersions starts with the formation of a polyurethane prepolymer by an addition reaction of the polymeric diol (soft segment) and an ionic carrier compound with excess polyisocyanate at elevated temperatures between 60 and 120 °C. The reaction is carried out in a process solvent (ketones or N-methylpyrrolidone), as well as in the melt. Chain extension is performed to produce the molecular weight by making the polyurethane prepolymer react with the chain extender, i.e. polyols or diamines. The ions are then formed by neutralisation with amines. Neutralisation may also take place at an earlier stage of the process. Chain extension with amines and neutralisation can also be performed in the water phase. Chain extension with polyols, and neutralisation, must be carried out in organic solutions. After preparation of the aqueous dispersion, the process solvents are stripped by distillation (mainly under vacuum) to afford solvent-free dispersions. The N-methylpyrrolidone remains in the dispersion to act as cosolvent. However, if actually N-methylpyrrolidone is defined as hazardous, there are trials and efforts to eliminate that solvent, producing the polyurethane dispersions without solvent or replacing the N-methylpyrrolidone by other solvents which are harmless and environmentally friendly. Solvents to replace the N-methylpyrrolidone make suitable process solvents of course only if they do not react with isocyanates (they must contain reactive hydrogen atoms).

The physical state of aqueous polyurethane dispersions is totally different from that of aqueous primary acrylic dispersions. The particle structure of polyurethane dispersions is comparable to that of the molecular coils of colloidal aqueous solutions of polymers with much lower molecular weights. Unlike the emulsifier-stabilized dispersion, the particles of ionically-stabilised polyurethane dispersion do not have sharply defined phase borders. The aqueous dispersion forms, without the need for shear, from the interaction of ionic carrier groups and the neutralisation agents. Neutralisation agents are described along with water-borne polyesters for primer surfacers in Chapter 3.5.6.1. The most important examples of neutralisation agents are triethylamine (TEA) and N,N-dimethyl ethanolamine (DMEA). The degree of neutralisation is 40 to 80 % of the potential anionic carrier groups. The particle size is difficult to measure on account of the lack of defined phase border. It varies with the molecular weight of the polyurethane, the mobility of the molecule chains (glass-transition temperature), the quantity of potential carrier groups, the type of neutralisation agent, and the degree of neutralisation. By virtue of their particle structures, polyurethane dispersions can form dense films easily, even at lower temperatures. Physical drying properties are influenced by the molecular weight, the hydrophobic nature of the building blocks (mainly of the soft segment), and the ability to form hydrogen bonds. Such polyurethane dispersions have excellent wetting properties for substrate surfaces, pigments, and effect substances. If they contain other functional groups (besides the ionic carrier groups), they may participate in the crosslinking reactions during stoving, albeit to a lesser extent due to the relatively high molecular weight. They have less influence over the rheology of the basecoats than acrylic dispersions. Table 3.7.3 describes the composition and characteristic values of a polyurethane dispersion suitable for water-borne basecoats [144].

Table 3.7.3: Composition and characteristic values of a polyurethane dispersion suitable for water-borne basecoats

n = m/M	Components	M	m (example)
0.625	polyester of hexandiol-1.6. dimer fatty acid	784	490.0
0.490	poly tetrahydro furane	1000	490.0
0.658	dimethylol propionic acid	134	88.2
2.100	HMDI	262	550.2
	catalyst: dibutyl tin dilaurate		1.0
0.293	trimethylol propane	134	39.3
	sum (solids)		1657.7
0.565	dimethyl ethanol amine	89	50.3
	water. deionised		3817.7
	sum. delivery form		5525.7

solid (60' 130 °C): 25 wt-% neutralisation degree 0.86
acid number 25 mg KOH/g pH-value 8.2

Table 3.7.4: Polyester for water-borne basecoats

n = m/M	Components	M	m example 3	wt-‰
7.362	hexane diol-1.6	118	868.7	331.0
2.374	dimer fatty acid	567	1346.2	512.9
2.325	isophthalic acid	166	386.0	147.1
1.076	trimellitic anhydride	192	206.6	78.7
	sum		2807.5	1069.6
10.148	water	18	182.66	69.6
	yield (AN = 30)		2624.84	1000.0

average molecular weight [g/mol] **M_n** 1372
acid number [mg KOH/g] **SZ** 30.0
degree of branching [mol/kg] **v'** 0.41
OH-number [mg KOH/g] **OHN** 74.8

Polyurethane dispersions and primary acrylic dispersions have different properties, advantages and disadvantages. It is therefore not surprising that some water-borne basecoat contain both resin types. In addition to that, there are examples where both resins are prepared in combination directly [145].

Water-borne polyesters

Water-borne polyesters are preferred partners for combination resins for acrylic dispersions and polyurethane dispersions in basecoats. Theoretically, the same resin types can be used as for water-borne primer surfacers (see Chapter 3.5.6). The polyesters contain excess hydroxyl groups for crosslinking, and carboxyl groups for neutralisation with amines to achieve colloidal solubility in water. Table 3.7.4 describes polyester for water-borne basecoats [146] that contain hydrophobic building blocks for conferring saponification resistance in the aqueous phase, even at high pH values.

Secondary acrylic dispersions

Unlike primary acrylic dispersions, secondary acrylic dispersions are prepared indirectly in two steps. As described in more detail in Chapter 3.8.3.1, acrylic resins are prepared by the solution polymerisation process with the aid of free-radical initiators. The preparation of secondary acrylic dispersions requires process solvents which are at least partly water-soluble. After that, polymers are prepared at elevated temperatures by the following reaction stages: initiator decomposition (peroxides, azo compounds), chain initiation, chain propagation, and chain termination by recombination of free-radicals, or chain transfer. The molecular weight and molecular weight distribution of the resultant acrylic polymers depend on the polymerisation temperature, the type and quantity of initiator, polymerisation rate of the monomers, the type and quantity of solvent, and the type and quantity of optional regulation agent. Suitable acrylic resins for secondary dispersions contain hydroxyl groups for crosslinking and carboxyl groups. The carboxyl groups are neutralised at least partly by adding amines to form anionic carrier groups as carrier groups for water-solubility. Water is added to the neutralised solution to yield dispersions containing colloidally dispersed particles. It is possible to strip off the solvents by distillation to produce solvent-free dispersions. It is also possible to leave the solvent in or to add special solvents, which ultimately serve as cosolvents for the preparation of water-borne basecoats. The advantage of secondary acrylic dispersions over water-soluble polyesters is their better saponification resistance. They also contribute more to physical drying. By contrast, polyesters confer better wetting, levelling, and flexibility on basecoats.

Water-thinnable melamine resins

The crosslinkers for water-borne basecoats are mostly melamine resins (see Chapter 3.5.3.2). Melamine resins which contain methanol by way of etherification alcohol are water-soluble, or at least water-thinnable. There are types which still contain free NH groups, and HMMM resin (hexymethoxymethyl melamine resins). When HMMM resin is used, acid catalysts must be added for effective crosslinking at the stoving temperatures commonly employed for automotive OEM coatings. The carboxyl groups of polyesters, acrylic resins, and polyurethanes can have a catalytic effect. Amine salts of sulphonic acids also serve as additive catalysts.

3.7.5.2 Rheological additives

Rheological additives for water-borne basecoats are: bentonites, colloidal silica, associative thickeners prepared from hydrophobically modified ethylene oxide urethanes (HEUR), natural carbohydrate thickeners, and the aforementioned high-molecular or crosslinked acrylic dispersions.

The choice of rheological additives (thickeners) is much more important for water-borne basecoat than for conventional solvent-borne basecoats. On account of its high evaporation enthalpy, water evaporates from film layers more slowly than do most solvents. Therefore, the solids content of

$$Al + 3\,H^+ \longrightarrow Al^{3+} + 1{,}5\,H_2$$

$$Al + 3\,H_2O + OH^- \longrightarrow [Al(OH)_4]^- + 1{,}5\,H_2$$

Figure 3.7.17: Reactions of aluminium in acids and bases

basecoat films after the application process is much lower than in the case of solvent-borne systems. This explains the cause of lower viscosity, including the risk of sagging on vertical parts of the substrates and of less immobilising of effect substances. This can only be compensated by influencing the rheology - i.e. generating structural viscosity. Of course, the addition of such effective additives must not impair flow and levelling of the basecoat.

3.7.5.3 Pigments and effect substances

Coloured pigments suitable for water-borne basecoats are the same as those described for topcoats (see Chapter 3.6.3.6) or for solvent-borne basecoats (see Chapter 3.7.3.6). Of course, it is very important for the pigments to be resistant in aqueous phase at high pH values. They must also be dispersed easily in the aforementioned resins for water-borne basecoats. The preferred dispersion resins are water-thinnable polyesters and polyurethane dispersions. The chosen pigments must not flocculate in aqueous phase.

Unfortunately, some of the effect substances used in solvent-borne basecoats are not stable enough in the aqueous phase. Mica pigments must be chosen with regard to the stability of their surface doping. More problematic are the aluminium pigments. On account of its amphoteric nature, aluminium can become soluble in acids or in bases. The reactions of aluminium metal are shown in Figure 3.7.17.

The relatively large surface area of the aluminium particles causes its reactions to proceed relatively quickly. The destruction of aluminium particles leads to a loss of effect and formation of a grey colour. The aluminium or aluminate ions formed by the reactions may cause coagulation of the resins that make up the water-borne basecoat. In addition, hydrogen is generated. Therefore, when water-borne basecoats with aluminium pigments are being prepared, it is essential to protect the surface of the pigments against reaction with water phase. Specific resins stabilise aluminium by wetting the surface of particles, e.g. epoxy resins and some acrylic resins bearing carboxyl groups. However, the main additives or agents chosen react with the surface of aluminium to form very thin layers of compounds (complex compounds). Such agents are phosphoric esters, alkyl phosphonic acid, special silicates, chromates, and chromic acid [147].

3.7.5.4 Cosolvents and additives

Most water-borne basecoats contain certain amounts of cosolvents. Cosolvents may originate from the delivery form of resins, but may also be added separately during basecoat production. Cosolvents are at least partly water-soluble. The standard cosolvents are glycol ethers (e.g. butyl glycol = butyl cellosolve = ethylene glycol monobutyl ether) or alcohols (e.g. butanols). Cosolvents improve the storage stability of water-borne basecoats, while boosting wetting and levelling. They promote more uniform evaporation along with water. Of course, to fulfil the requirement of preventing emissions of volatile organic compounds, the content of cosolvents in water-borne basecoats is restricted.

Additives for all coating materials aim to optimise application behaviour. For water-borne coating materials, it is important to prevent foaming and to improve deaeration. Wetting of substrate surfaces and pigment surfaces is already supported by resins, which are ionically stabilised, and may act like surfactants. Typical levelling agents are acrylic polymers which contain hydrophobic monomers and carboxyl groups which are neutralized by amines. Silicone oils, modified by polyethers, are used in only low quantities, even though such products are water-soluble, because they can adversely affect recoatability.

3.7.5.5 Formulation example and application

Table 3.7.5 contains the formulation and characteristic values of a water-borne basecoat consisting of a polyurethane dispersion, a polyester, and a melamine resin. The effect substance is a stabilised aluminium pigment. The thickener is a sodium magnesium silicate with layer-like crystal lattice (hectorite) [148]. The basecoat has an application viscosity of 14 s (cup 4, 23 °C, DIN 3511, equivalent to ISO cup 53 s); the solids content is 15.4 % by weight. The quantity of cosolvent is 6 % by weight. The basecoat is applied by pneumatic spray gun onto a panel coated with primer and primer surfacer. The flash-off conditions are 5 min at 80 °C. After cooling, a two-component clearcoat (see Chapter 3.8.3.4) is applied to the pre-dried basecoat. After a second short flash-off, the layers are stoved for 20 min at 140 °C.

More effort is needed for pre-drying water-borne basecoats than solvent-borne basecoats, because water has a higher evaporation enthalpy than the organic solvents. Not only is the flash-off temperature important, but the efficiency of air circulation must also be taken into consideration. Elevated temperatures and short flash-off times may be the reasons for popping (forming of blisters) and poor levelling.

3.7.6 Solid colour water-borne basecoats

Following the successful implementation of effect basecoats, the wet-on-wet process was extended to solid colour topcoats as well. This led to significant improvements in weathering resistance, resistance to chemicals and, most important, gloss retention. However, there were application problems.

Table 3.7.5: Formulation and characteristic values of a water-borne basecoat

Pos.	Components	Solid wt-%	Formulation wt-%
01	thickener dispersion (3 % in water)	0.75	25.0
02	polyurethane dispersion (25 % in water)	6.00	24.0
03	polyester (80 % in n-butanol)	4.00	5.0
04	DMEA		0.5
05	melamine resin (70 % in water)	1.40	2.0
06	aluminium paste (65 % in water)	3.25	5.0
07	butyl glycol		5.0
12	deionised water		33.5
	sum	15.4	100.0

application solids: 15,4 wt-%
application viscosity 14" (DIN 4/23 °C), equivalent to ISO cup 53 s
solvent amount 6 wt-%

pos.
01 dispersion of precipitated Hectorite (sodium-magnesium-silicate)
02 polyurethane dispersion, example in DE 3825278
05 polyester of neopentyl glycol, isophthalic acid and trimellitic anhydride (acid number 39 mg KOH/g, average molecular weight 1439 g/mol)
06 stabilised aluminium pigment paste 65 % in water

Perfect execution of the wet-on-wet process requires that the resultant film thickness of the basecoats must be relatively low. Increasing the film thickness of basecoats can lead to blistering due to evaporating components such as water and cosolvents (popping). There is also the risk of sagging on vertical parts of substrate surfaces. Furthermore, there will also be a tendency to redissolve. However, if solid colours are to provide sufficient hiding power, they must be applied in optimum film thicknesses. For conferring optimum hiding power, mainly in the case of yellows and reds, the film thicknesses must be greater than for black, blue or green colours or for effect basecoats. Hiding power depends not only on the film thickness but also on the pigment volume concentration in a coating layer. If basecoats do not have to be glossy, it is possible to further increase the pigment volume concentration as in the case of solid colour topcoats applied in one layer. However, there are limits to this in that the hiding power does not always increase with increase in pigment quantities as these have a tendency to flocculate. These problems are circumvented in various ways. Agents for preventing popping can be added, and also judicious quantities of cosolvent can serve to prevent formation of blisters during the film-forming process. Modifying the application conditions can also boost optimum application behaviour for thicker layers.

The choice of suitable pigments and, again, modifying application conditions can help to yield higher film thicknesses and to improve hiding power. Finally, there is the possibility of applying solid colour basecoats over primer surfacers with modified colours (see Chapter 3.5.3.4).

3.7.7 Alternatives

Several studies have been performed on the use of alternative application processes for preparing effect coatings. For special cases, powder coatings are available that contain effect substances, such as aluminium pigments. Since film-forming in powder coatings is totally different from film-forming in wet coating materials, the effect substances in powder coating also behave totally differently. It is therefore impossible to generate comparable effects with powder coatings.

Effect substances can also be incorporated into plastic films. Again, the achievable effects are totally different from those generated by wet application of effect basecoats. Plastic films offer the possibility of creating holographic effects.

Wet application of effect basecoats is unlikely to be replaced by other application methods in the near future, since it is the best way of creating appealing, individual colours and effects.

3.8 Clearcoats

3.8.1 Development of clearcoats

The introduction of clearcoats into automotive coatings is, of course, closely connected to the introduction of effect basecoats. The experience gained with clearcoats in other application fields was applied here. In the 1960s, stoving clearcoats for general industrial application consisted of alkyd resins and amino resins (melamine resins). To obtain clearcoats with the least possible yellowing tendency, alkyds were chosen which had been modified with saturated fatty acids. The sources were coconut fat or coconut fatty acids. At first, only melamine resins with low reactivity were available for crosslinking. Therefore, crosslinking during film forming was supported with dehydrated castor oil fatty acids (DCO fatty acids) in addition to the coconut fatty acids. At that time, DCO fatty acids were regarded as relatively resistant to yellowing. Initially, such clearcoats were applied together with ready-stoved metallic pre-coats. Later, they were used for wet-on-wet application in conjunction with metallic basecoats in automotive OEM coating processes. Optimum hardness and resistance were provided by alkyd resins that had been modified by coconut precursor fatty acids (C_8- and C_{10}-acids). Nowadays, such fatty acids are very expensive raw materials for the production of high-class lubricants and biodegradable liquid surfactants. Later, clearcoats contained both alkyds that had been modified by synthetic fatty acids from the oxo-synthesis and relatively highly reactive melamine resins.

Meanwhile, in the USA, stoving enamels were being developed on the basis of hydroxy acrylic resins crosslinked by melamine resins. In Europe, and predominantly in Germany, clearcoats containing alkyds were retained because of the application advantages (improved wetting properties, better levelling, prevention of popping) and appearance (optimum filling power, excellent gloss and clearcoat holdout). However, complaints arose about the lack of weathering resistance and cracking on exposure to weathering (Florida tests). Clearcoats based on alkyds were subsequently displaced from the automotive coatings market. However, even the clearcoats based on acrylic resins often provided inadequate weathering resistance. This prompted extensive studies of weathering resistance. Accelerated weathering tests involving the use of UV lamps were introduced into clearcoat development [149]. The results showed that, to be sure, the clearcoats showed no cracking after exposure, but they became matt and the film layers degraded (loss of film thickness). There were also symptoms of damage in the basecoat, such as greying and blistering. The breakthrough in the development of clearcoats with optimum weathering resistance came with the introduction of UV absorbers and free-radical quenchers. Nowadays, light stabiliser packages are included in all clearcoats.

Subsequent stages in the development of clearcoats included improvements in chemical resistance and compatibility with high-solid basecoats and, later, water-borne basecoats. The outcome was the introduction of two-component clearcoats containing hydroxy acrylic resins and also polyisocyanate adducts as crosslinker for automotive OEM coating. Many car producers in Europe now employ this technology, although some have stuck by one-component clearcoats. Two-component clearcoats are notable for their chemical resistance. With a view to improving the chemical resistance of one-component clearcoats as well, several crosslinking methods were investigated. The result was the introduction of clearcoats containing acrylic resins and blocked polyisocyanate adducts. It has been found that a combination of melamine resins and blocked polyisocyanate adducts yields the best crosslinking properties (hybrid crosslinked clearcoats).

Even today, most clearcoats for automotive OEM application contain solventborne systems. Of course, there have also been attempts to cut emissions of VOCs from clearcoats, too. There are now clearcoats available with application solids of about 60 % by weight. Applications for water-borne clearcoats are very few. Powder clearcoats are available which have virtually zero VOC emissions.

Since the late 1990s, car producers have been attempting to improve the scratch resistance of clearcoats. This can be accomplished in different ways: optimising the resin composition (acrylic resins) and crosslinking, the use of special additives, and the selection of totally new application methods.

Attempts are currently underway to combine the improvements in clearcoat properties with reductions in solvent emissions. To this end, a large number of studies are pursuing alternative crosslinking methods. However, so far, these studies have yielded only a few products for the automotive coatings market. There are clearcoats containing carbamate groups instead of hydroxyl groups for crosslinking with melamine resins. There are clearcoats which crosslink by reaction of unsaturated molecules in UV light, and clearcoats which crosslink both in UV light and by other crosslinking methods, for example those involving polyisocyanate adducts (these are called dual-cure systems).

3.8.2 Requirements on automotive OEM clearcoats

Clearcoats are used in automotive OEM coating to protect the sensitive metallic effect basecoats against the influence of weathering, chemicals, solvents, and mechanical impact. The various influencing factors and test methods for characterising the properties of clearcoats are described below. As already discussed in Chapter 3.7.1, coating layers with effect substances require protective surfacers. The first effect basecoats contained aluminium pigments. Appealing metallic effects are only achieved if the effect coatings contain sufficient quantities of aluminium pigments. The resin matrix of such coating layers is unable to provide sufficient protection for the effect substances. A protective clearcoat is therefore applied as overcoat.

3.8.2.1 Application behaviour and surface of clearcoats

Since most car producers were not prepared to apply four layers in separate sections of their application lines, the two layers of effect basecoat and clearcoat had to be applied wet-on-wet. The term wet-on-wet is not entirely accurate if the basecoat layer is allowed to dry physically and exposed to special flash-off conditions before the clearcoat is applied. That is the basic process currently employed in all car assembly lines, and also in the application of solid colour systems.

The first requirement imposed on a clearcoat is therefore the establishment of a wet-on-wet process that is compatible with a basecoat layer that has dried by physical means only. In other words, the clearcoat has to wet the basecoat perfectly and spread over the basecoat surface efficiently. The clearcoat must not destroy the orientation of the effect substances in the basecoat layers as that would lead to loss of both the basecoat lightness in plan view and the flip-flop effect. A negative influence of this kind on a basecoat layer upon overcoating with a clearcoat is called redissolving. Redissolving is not only the dissolution effect exerted by the clearcoat and its solvent. There is also the mechanical impact of the spray mist falling on the surface of the basecoat. In addition, there may be influences exerted by the low-molecular resin content of a clearcoat, e.g. low-molecular polyisocyanate adducts may disorient the effect substance of the basecoat layer in the first phase of the stoving process by diffusing into the basecoat before the products are incorporated into the crosslinked molecular network of the clearcoat layer.

After clearcoat application (see Chapter 3.8.4), stoving and cooling, the following properties are visually assessed: levelling, gloss, distinctness of image, and clearcoat hold-out.

Optimum levelling is exhibited by a clearcoat surface that has only a few structures. Structures of different sizes are observed and can be determined physically. Measurements of the quantity of scattered light and by means of Fourier analysis enables the quantity of structures with different wavelengths to be determined. Car makers specify limits on short-waves and long-waves (measured by short-wave and long-wave scanning, [150]). Factors affecting levelling are surface wetting, cohesion, flow (viscosity of the clearcoat during film forming, including application, flash-off, and the first stage of stoving).

The gloss of clearcoats is dictated by the transparency, refractive index of the film, and the surface structure. Certain amounts of very fine surface structures (short-waves) yield a matt clearcoat. The gloss value is defined by measuring surface reflection at different angles of reflection (20° and 60°). The human eye barely notices differences in readings caused by different refractive indices whereas it sees surface structures very well.

Clearcoat hold-out is a measure of the filling properties of a clearcoat. The human eye evaluates the totality of or interaction between clearcoat transparency, gloss, and surface structure. Short-wave surface structures are perceived to be negative, and so are smooth and plain (waveless) film surfaces. Surprisingly, small quantities of long wave structures are most pleasing to the human eye.

The totality of levelling, i.e. surface structure, of gloss and of clearcoat hold-out is called the **appearance**. Here, too, the distinctness of image (DOI or sharpness of mirror-like pictures on the clearcoat surface) is crucial for any evaluation of appearance.

A further very important property is the behaviour of the clearcoat film on vertical parts of the coated object. These parts tend to suffer from **sagging**. Sagging occurs when some of the film material runs down vertical parts, leaving tracks and droplets behind. If large areas of the coating film start to run, the defect is called sliding. The damage not only impairs the appearance of the clearcoat. The basecoat may also be affected. This can lead to disorientation of the effect substances and thus to a loss of effect and colour impression. Sagging resistance is tested by a special application method. A large panel with a horizontal series of holes in the middle is coated with clearcoat (after basecoat application) in different layer thicknesses (giving rise to a wedge). The application conditions are well defined; the quantity of clearcoat applied per unit of time is relatively large. The panel is then placed vertically in the oven and stoved. After stoving, the sagging resistance is evaluated by measuring the extent of sagging at the holes. The sagging limit is defined as the layer thickness at which the sagging track is less than 1 cm [107]. A balance between levelling and sagging resistance is obtained by choosing the optimum solvent combination for the current application conditions of the application line. The crosslinking rate influences levelling and sagging as well. Solvent evaporation and crosslinking affect the viscosity during flash-off and the first stage of stoving as a function of temperature and time. The viscosity is a measure of the possibility of optimum flow and levelling, and prevention of sagging. The film viscosity can be influenced via rheological additives that must be chosen by type and quantity very carefully if the balance between optimum levelling and sagging prevention is to be obtained (see Chapter 3.8.3.9).

3.8.2.2 *Weathering resistance*

Weathering resistance firstly refers to the ability of coating layers to resist the effects of sunlight. Sunlight contains significant quantities of UV light, starting at wavelengths of 290 nm. The UV content of sunlight is shown in Figure 3.8.1.

Clearcoats are naturally transparent to all wavelengths of visible light, but can absorb UV light. The energy-rich UV light is capable of destroying molecular bonds of organic compounds, such as polymers. UV-induced destruction of the film matrix of clearcoats leads to matting, degradation of the film layer, cracking, embrittlement, and loss of adhesion. Clearcoats are much more sensitive to UV light than pigmented topcoats because pigments absorb fractions of the UV light and prevent the light from affecting the lower chapters of the coating layer. In addition, the interfaces of the pigment particles and the resin matrix prevent propagation of stresses.

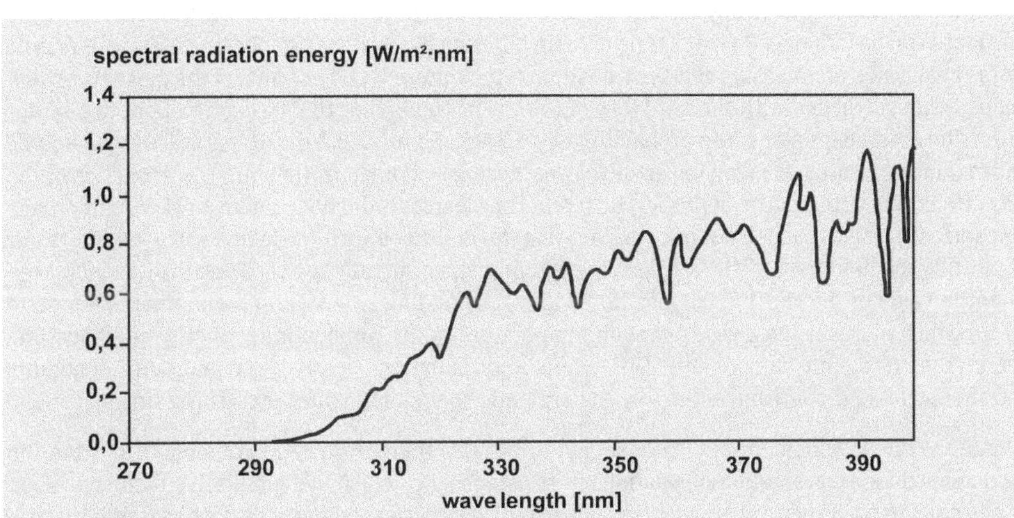

Figure 3.8.1: Spectrum of sunlight between 270 and 400 nm

Natural weathering involves more than the effects of UV light. Other factors are the effect of oxygen, humidity, and temperature fluctuations that can damage the coating layers. Humidity results not only from rainfall; air always contains some degree of humidity. Sunlight can heat coating layers to surprisingly high temperatures. For example, temperatures in excess of 70 °C have been measured on coating layers on car roofs. Not only that, the damage will be exacerbated if electrolytes are present on the coating layers in addition to humidity. Such situations can arise in the event of environmental pollution, e.g. acid rain following industrial emissions.

Molecular decompositions in coating films are mainly oxidation reactions initiated by oxygen free-radicals and hydroxyl free-radicals (formed from water). Such oxidation reactions lead to decomposition of the molecular networks in clearcoat film matrix. Decomposition and the formation of oxidation products containing carbonyl and carboxyl groups cause the film to become brittle. In the past, it was believed that UV light forms free-radicals which trigger post-crosslinking that could lead to embrittlement. Nowadays, it is known for certain that post-crosslinking increases the elastic component of flexibility. The best example of this is the flexibility of UV coatings, which is high, irrespective of the parallel high crosslinking density. Low-molecular decomposition products, by contrast, are never flexible. The molecular decomposition generates holes in the film matrix. Ultimately, such holes, frequently under the influence of temperature fluctuations, can lead to macroscopic (visible) damage. Either the clearcoat layer loses thickness or cracks are formed. Excessive cracks look like crackleware. The reason clearcoat layers behave in different ways on exposure to weather is as follows: aromatic compounds in the film matrix absorb UV light via their double bond system. Consequently, lower sections of the clearcoat layer are protected against the UV light. However, absorption of UV light by aromatic double bonds leads to decomposition reactions, e.g. the ligands on aromatic rings are cleaved. The decomposition of such molecules is much less pronounced than in molecules that have no aromatic building blocks. Molecular networks without aromatic compounds are degraded faster, this resulting in matting and loss of layer thickness, even after only brief exposure. However, as far as the value of passenger cars is concerned, this type of physically measurable degradation (matting and loss of film thickness) is perceived totally differently. Clearcoats are washed and polished, and so matting and loss of film material is compensated or accepted. Although clearcoat layers whose film matrix contains aromatic compounds are much more resistant and have relatively good gloss retention, damage in the form of cracking, which may occur after a relatively long time, is not accepted by customers.

An additional factor governing damage by cracking is the physical hardness of the clearcoat film. Hardness can be influenced via the type of resin building block. For example, monomers for acrylic resins are chosen on the basis of glass-transition temperature (T_G). The value of the glass-transition temperature correlates with the hardness of the resin. The higher the glass transition, the harder is the film and the greater is the probability of cracking. Figure 3.8.2 illustrates cracking in acrylic and melamine clearcoats following artificial weathering by UV lamps (in hours) as a function of the glass-transition temperature of the acrylic resin. The clearcoats did not contain any UV light protection and the test was carried out in a weather-o-meter equipped with UV lamps and edge-filters (for separating the light fractions). The glass-transition temperature of the acrylic resins was increased from the value of the starting composition (-15 °C) by adding two different monomers (from 0 to 40 %), which increase the glass-transition temperature in the same manner: methyl methacrylate and styrene (aromatic component). The results after exposure clearly show that both monomers greatly accelerated cracking, especially the aromatic component (shorter cracking times).

Ever since the development of clearcoats got underway, there have been attempts to shorten the time needed for an evaluation of weathering resistance, i.e. to achieve qualitative results as soon as possible. Initially, natural areas were chosen which experienced particularly high UV light irradiation and possibly high humidity. In these areas, the coatings producers and also the assembly

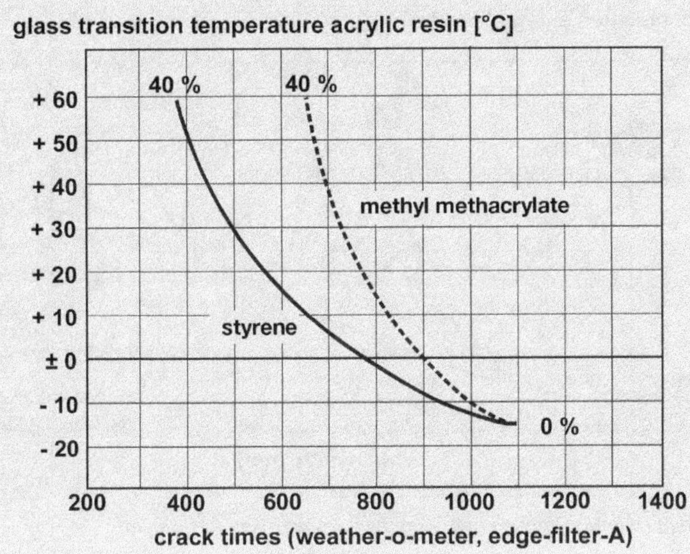

Figure 3.8.2: Cracking times of clearcoats as a function of the glass-transition temperature and the content of aromatic monomer

plants installed test stations for exposing panels to sunlight and humidity. Test stations of this kind were installed in Florida (USA; the best known test area), Arizona (USA; high UV fraction, but low humidity), Okinawa (Japan), Allunga (northern Australia), and other less important places. To intensify the impact of the sunlight and humidity, special conditions were created for exposing the panels. The panels have a special angle of inclination (e.g. 5°) and face south. Alternatively, they are surrounded with a frame whose inside is coated with a matt black topcoat that raises surface temperature of the test panels (heated box).

Car makers specify guarantee times for the weathering resistance of clearcoats. For example, the most common is two years' outdoor exposure in Florida, with negligible decrease in gloss and resistance. The car makers extrapolate these results to offer their customers a guarantee of 10 years (normal European use conditions), and the promise of no cracking, no loss of adhesion, and no corrosion.

Two years is a long time for the introduction of a new clearcoat system into the market. Not only that, but in contrast to the widely held view, the results from Florida exposure can vary from year to year in accordance with the different weather periods there. Consequently, in the early days of development, short-term, artificial weathering tests using UV lamps were introduced. Ever since such weather-o-meters have been used, the question has arisen as to how the times produced by Florida exposure correlate with the results of artificial weathering. UV lamps emit higher quantities of light of very short wavelength, and in some cases waves that have a much shorter wavelength and are more energetic than those emitted by sunlight. Figure 3.8.3 (page 170) compares the radiation emitted by UV lamps and by sunlight as a function of wavelength.

Very short UV light radiation is highly energetic and can trigger decomposition reactions different from those caused by sunlight. This means that natural weathering is not being properly simulated. For this reason, edge filters for certain fractions of UV light were fitted to the UV lamps. This simulated the sunlight very well, but at a higher intensity that created a time-lapse effect. The test machines were also equipped with sprinkling equipment to simulate rain and temperature fluctuations [149]. All sorts of weathering test equipment are available nowadays, and

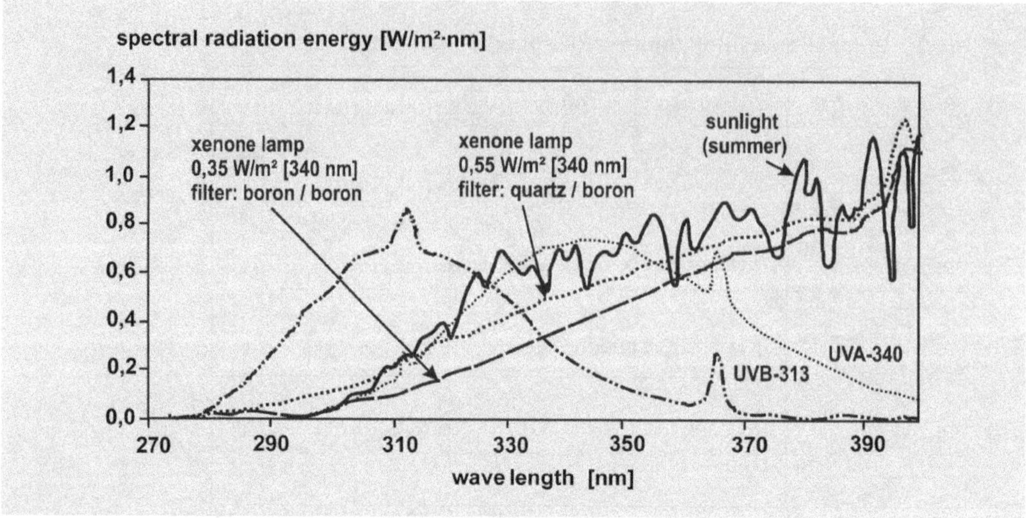

Figure 3.8.3: Spectra of radiation emitted by UV lamps in comparison to sunlight

their results can vary. The type of equipment employed must therefore always be stated with the results. The suppliers of the various test stations provide correlation data for Florida exposure (or mostly natural exposure), but those have no general validity as they depend on the type of clearcoat tested [151].

3.8.2.3 Chemical resistance

Coatings, especially, clearcoats must be resistant to chemicals. Here, the definition of chemicals is rather broad. Acid from the car battery could be spilled onto the coating, although the probability of this is relatively low. Tests of chemical resistance are principally an extension of the weatherability tests since the test substances are chosen for their ability to simulate the various environmental influences. Therefore, besides resistance to dilute **acids** and alkalis (caustic solutions), coatings are tested for resistance to more complex compounds or substances that simulate environmental influences, such as **oil soot** (mixture of tar, sulphuric acid and iron sulphate); **bird droppings** (represented by the chemical compound pancreatine) **tree rosin** (greenfly excrement as a solution of colophony). In particular, bird droppings are highly corrosive to clearcoat layers. Studies have shown the main reason to be the solvent power of the ingredients, which leads to swelling of the clearcoat layers. Furthermore, the ingredient does not evaporate fast enough and become covered by other ingredients, this giving rise to lasting impact; significant chemical reactions do not occur. It was also thought at one time that the ingredients of bird droppings could act as enzymes, but this has proved to be incorrect. Tests are also performed on **solvents** (for example premium-grade fuel whose ethanol content is important for the dissolution effect) and finally on **deionised water**.

The chemicals and solvents work initially through diffusion, i.e. they penetrate into the films. This causes swelling of the film or accumulation of the substances at the interfaces of the coating system. Swollen molecular networks are highly sensitive to mechanical impact; the film can be destroyed easily. Accumulations at interfaces can cause delamination. Finally, the substances can react with the polymer network of the film matrix to trigger chemical decomposition. Most noticeable in this regard is the decomposition of methylol ethers of melamine resins that act as crosslinkers in the clearcoat film. Urethane and ester linkages are relatively stable to acids. If acids manage to diffuse through the entire clearcoat layer, they may damage the aluminium pigments, a fact which leads to loss of effect and greying.

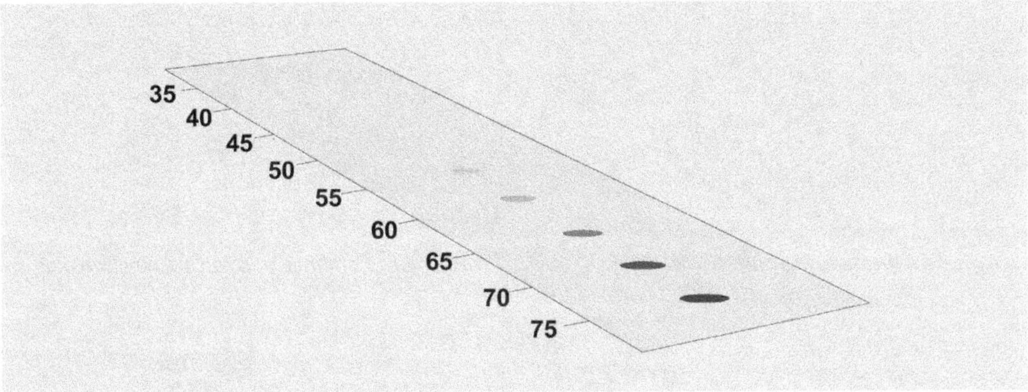

Figure 3.8.4: Measuring chemical resistance in a gradient bar oven

Diffusion processes as the cause of damage to clearcoats are dictated by temperature and time. Chemical resistance is therefore tested at defined temperatures until such time as damage is observed. The other method is to determine the impact time at different temperatures. The temperature at damage is not observed is recorded [152]. This test utilises a ***gradient bar oven***. The oven consists of a metal bar in which a defined gradient temperature is generated. A long test panel which has been coated with the full coating system and bears several series of droplets of the various chemicals is placed on the bar for a defined time. After the time has elapsed, the temperature at which damage to the coating (matting, swelling or other destruction) just fails to be observed is recorded for each chemical. Figure 3.8.4 shows an example of such tests.

3.8.2.4 Mechanical resistance

The primary meaning of the mechanical resistance of a coating layer is resistance to shock and blows. As already described for primer surfacers and topcoats, stonechipping is the primary source of impact on automotive coatings. A stone impinging on a coating at high velocity will definitely cause damage. As drivers know, stones can also destroy car windscreens. The goal of conferring adequate stonechip resistance is to minimise the damage. The damage must cover as small an area as possible and must not penetrate as far as the substrate surface (metal of car body). All layers of the full coating system are required to offer resistance to stone chipping, but especially the primer surfacer. The clearcoat must also contribute to stonechip resistance. To improve stonechip resistance, clearcoats must mainly provide optimum intercoat adhesion via the basecoat layer to the entire coating system.

3.8.2.5 Scratch resistance

Scratch resistance is the resistance of film surfaces to avoid damage by small, relatively hard particles or by fibres or paper. This type of damage includes not only the erosion caused by particles acting on a moving car, but also the formation of various scratches that mainly occur during washing. Nowadays, many car washes use textiles, instead of the polypropylene brushes previously used, to prevent too much impact on the coating surfaces. Nevertheless scratches still occur (washing scratches). These stem from dust particles from the road surface and from dirt that are rubbed onto the coating surface during the washing process. These types of scratches are observed more easily at an oblique angle on dark cars standing in sunlight just after having been washed. The scratches are circular and merge into each other. Now that clearcoats have been optimised as regards weathering resistance and chemical resistance, there are no longer complaints about these issues. Customers and car producers are now focusing on scratch resistance.

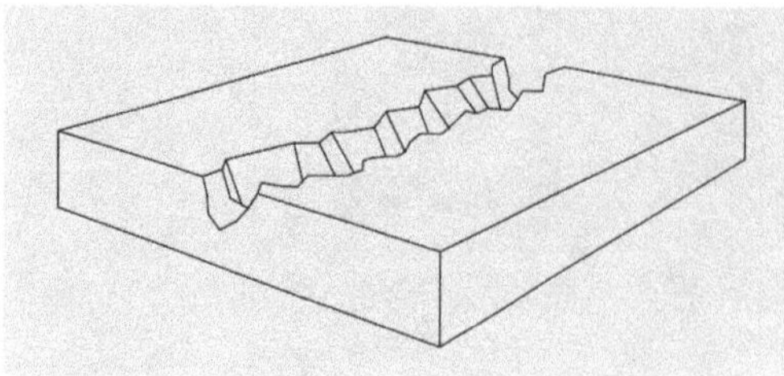

Figure 3.8.5: Abrasion scratches on clearcoats

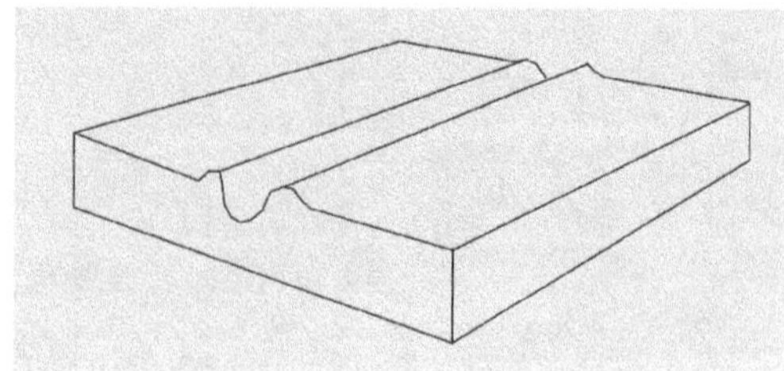

Figure 3.8.6: Deformation scratches on clearcoats

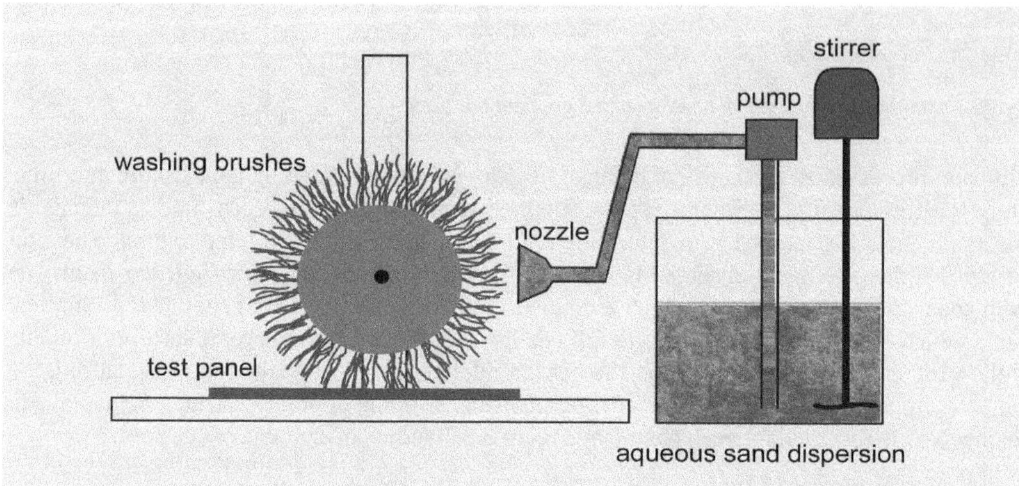

Figure 3.8.7: Test apparatus featuring rotating wash brushes and aqueous sand dispersion as test medium

When small particles act on a clearcoat surface, two types of scoring occur. Abrasion scratches, where material from the clearcoat layer is totally removed in the direction of particle rubbing. Such scratches have sharp and irregular edges at the microscopic level (see Figure 3.8.5).

Abrasion scratches are observed on clearcoat layers which are very hard and brittle. In contrast, deformation scratches occur when the particles move the material sideways in the direction of particle rubbing. A microscopic wall is created, but little or no material is removed from the clearcoat (see Figure 3.8.6).

Such scratches are formed on flexible clearcoats. They can heal themselves after some delay (cold flow) or, much more quickly, at high temperatures. Theoretically, clearcoats which yield small quantities of deformation scratches should be preferred to the others. Such clearcoats have greater plasticity (in addition to elasticity; both parts contribute to flexibility). However, they are less stable to diffusion and dissolution, which means they have less chemical and solvent resistance. They are more easily swelled by solvents, and swollen paint layers are sensitive to mechanical impact and relatively easy to damage. The demand for improved scratch resistance still exists and has been the most important demand for several years. However, in line with what has been said above, what really is called for is optimisation of the totality of chemical resistance and scratch resistance. Gradual improvements have come about through modifications to the resin composition, with mainly acrylic resins developed that offered improved scratch resistance. However, the demands of the automotive industry for improve scratch resistance remain. Additional studies and solutions have been proposed: clearcoats with optimised scratch resistance may contain nanoparticles as additives (see Chapter 3.8.3.11). Scratch resistance is improved by using alternative crosslinking methods (see Chapter 3.8.8). In addition, special sealers can be applied that offer better scratch resistance (see Chapter 3.8.3.11).

Several items of test equipment for determining scratch resistance. Most of them employ specific, well defined media (e.g. sand or textile tissues) which impinge on the coating surface under well defined conditions, e.g. rotating brushes or plungers operating at defined temperatures and times. The impact is assessed visually. However, it can also be measured by means of physical data, e.g. derived from measurements of light scattering generated by scratches on the clearcoat surface. Figure 3.8.7 shows a model of a test apparatus equipped with rotating washing brushes employing an aqueous sand dispersion for simulating the car wash process.

3.8.2.6 Windscreen bonding

Windscreens and other window parts are installed in car bodies by being bonded in the grooves of window frames with special sealants [153]. On one hand, the windows must be firmly connected to the frame while, on the other, the bond must offer some play to allow for differential thermal expansion of the window and frame. The installed window must also contribute to the stiffness of the entire car body.

The adhesive material consists of plastic beading. The binders are moisture-curing polyurethane prepolymers. Such prepolymers are prepared by making polyether polyols react with excess aromatic diisocyanates to afford polymers with terminated isocyanate groups. The polyether polyols contain mixtures of difunctional compounds with smaller quantities of polyethers of high functionality for creating crosslinked elastomer molecules after moisture curing. The curing starts by addition of atmospheric moisture to the isocyanate group to yield a carbaminic acid. The carbaminic acid is unstable and decomposes into an amine and carbon dioxide. The amine reacts very quickly with remaining isocyanate groups to form urea linkages (crosslinked polyureas). These reactions are accelerated by catalysts, e.g. dibutyltin dilaurate and bismuth octoate. The adhesives also contain pigments, mainly carbon black, up to 20 % by weight. The critical property here is adhesion of the adhesive material - during application but more important after crosslinking - to the clearcoat layer on the window frame. The adhesion properties must be tested carefully. Crucial factors governing adhesion are the building blocks in the clearcoat resin composition, the type of crosslinker used for the clearcoat, the crosslinking density, atmospheric humidity and temperature. Furthermore, different adhesive materials react differently with different clearcoats. It is very important to match the adhesive type and its constituent materials to the clearcoat. Car producers stipulate close cooperation between windscreen-adhesive producers and coating-materials producers as they have to guarantee that there will be no loss of adhesion of windows in the car-body frames during use. This is tested by applying a bead of adhesive from a cartridge to panels treated with the full coating system. The adhesive is allowed to crosslink under well defined conditions. The end of the bead on the panel is cut and the bead pulled off under high force. The force needed to remove the bead is determined. The best result is one in which very little of the clearcoat surface is exposed and cohesion failure occurs in the adhesive itself.

3.8.2.7 Protection in transport

The coatings on cars must withstand all kinds of impact when the cars are transported from the assembly plants to the dealerships. Such transport often involves long distances by truck, train, and ship. For top-of-the-range cars, especially, it is necessary to protect the coatings against impact during transport. The protection is composed of films or aqueous polymer wax dispersions. The protective layer has to prevent the coating surface from impact by dust or other fall-out and scratching and erosion. For presentation purposes or for selling the cars, these additional layers must easily and completely removable. The various products employed therefore contain release agents. The release agents must not impair the properties of the clearcoat layers. It is important that the protective materials do not adversely affect the clearcoat during application, transport, storage, or removal. The reason is the potential for swelling. If the protective films are not laid smoothly over the clearcoat surface, condensation may collect under the folds and cause swelling of the clearcoat. That said, however, clearcoat layers must be resistant to such swelling.

3.8.3 Composition of clearcoats

The resins for solventborne clearcoats are mainly acrylic resins that contain hydroxyl groups by way of functional (reactive) groups for crosslinking. The crosslinkers are either melamine resins (one-component clearcoats) or polyisocyanates (two-component clearcoats). The preferred solvents for these conventional clearcoats are aromatic hydrocarbons, combined with esters, alcohols, glycol ethers, and glycol ether esters. The most important additives are light-protection agents, levelling agents, and rheological additives.

3.8.3.1 Acrylic resins

Acrylic resins rank among those resins prepared by polymerisation (in the narrower sense of that definition). Acrylic resins for solventborne clearcoats are made by solution polymerisation of acrylic esters, methacrylic esters (monomers), and other unsaturated compounds that have nearly the same polymerisation parameters as the acrylic or methacrylic esters (co-monomers). The polymerisation initiators are compounds which decompose easily into free-radicals, namely peroxy and azo compounds.

Free-radical chain polymerisation

Initiators, such as peroxides, generate free-radicals spontaneously by decomposition at elevated temperatures (initiator reaction). The free-radicals open the double bonds of monomers, creating new free-radicals (initiation). These monomer free-radicals can undergo stepwise addition to other monomers, with a free-radicals always forming at the end of the growing chain (propagation). Chain growth terminates when two free-radicals of growing chains react with each other (recombination) or when the free-radical of a growing chain is transferred to a molecule (regulator,

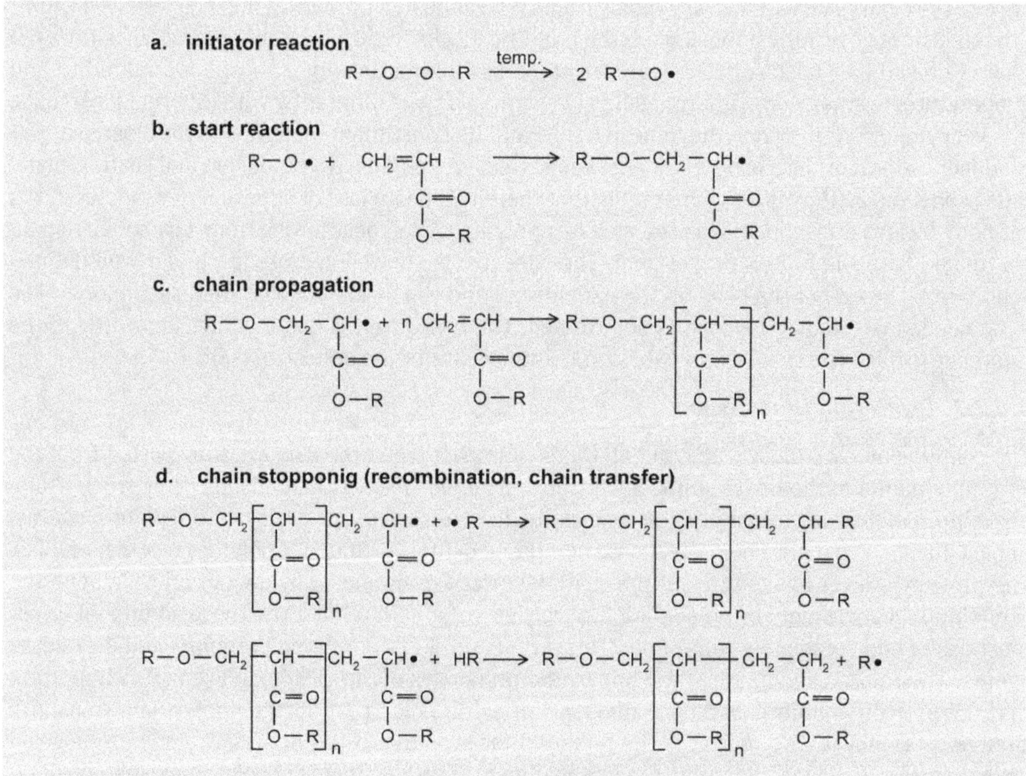

Figure 3.8.8: Free-radical chain polymerisation

solvent) which is not added to the chain, but causes chain termination and the formation of a new initiator free-radical by itself (see Figure 3.8.8). These reactions take place in fractions of seconds and are extremely exothermic.

Production process

Unlike the production of polyester resins (see Chapter 3.5.3.1) which form continuously growing molecules throughout the reaction, the polymer molecules are formed in fractions of seconds, as already described. Since the reaction is extremely exothermic, removing the thermal energy is a prime consideration as otherwise the process would get out of control. The exothermic reaction is managed by spreading the entire reaction over a period of time. This is best done by a feed process in which the monomers and initiator are fed separately but concurrently into the reactor. Both feeding and the reaction are carried out in a solvent that is added to the reactor before monomer and initiator feed commence. Furthermore, before the reaction is commenced, the solvent is heated to the temperature most conducive to initiator decomposition. The solvent chosen is the one which is contained in the delivery form of the acrylic resin and which fits the coating formulation. The solvent may also act as regulator.

Figure 3.8.9 illustrates the process diagram for the production of acrylic resins by solution polymerisation. In the first step, the process solvent is heated to the reaction temperature. Then, continuous addition of monomer mixture and initiator commences. There is an advantage in commencing addition of the monomer mixture somewhat later (and then concurrently) than addition of the initiator. Judicious choice of feed time ensures that the content of free monomer in the reaction mixture remains more or less the same over the addition time. This is essential for forming molecules of nearly the same size (comparable molecular weight and molecular weight

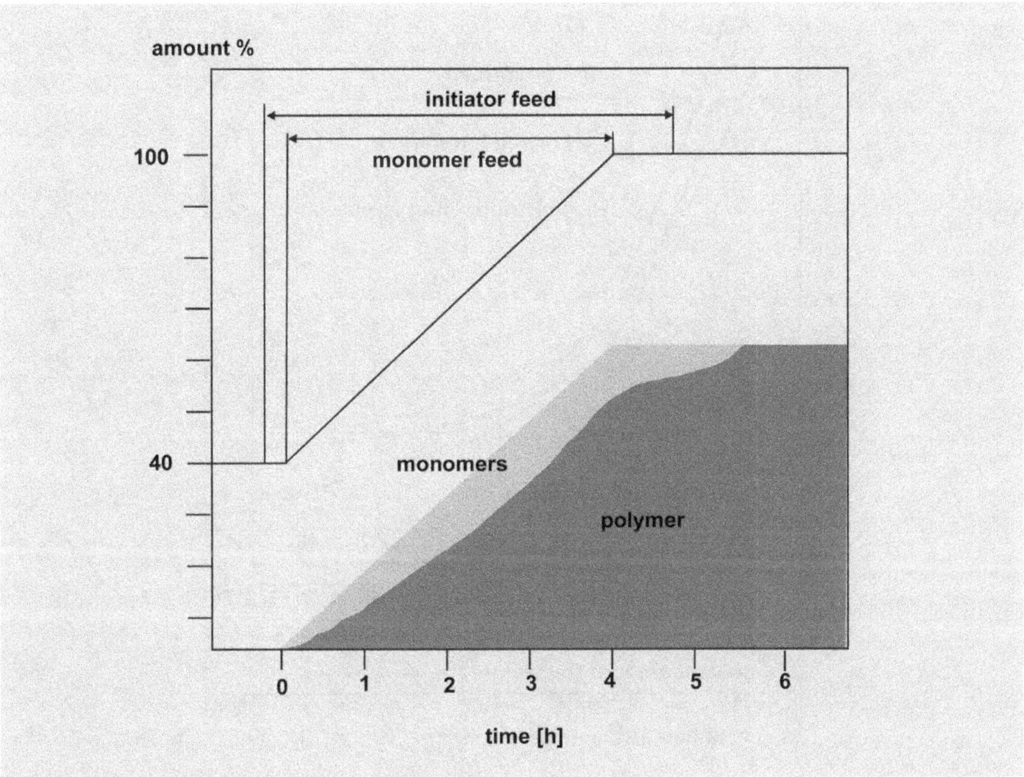

Figure 3.8.9: The solution polymerisation process.

distribution). Under these conditions, the reaction is virtually in a steady state. The temperature is kept constant for the entire reaction by cooling. When addition of monomer is complete, addition of initiator is continued to make sure that all free monomer is fully polymerised. To complete the process, the characteristic values of the acrylic resin solution, namely solids content (nonvolatiles fraction) and viscosity, are measured. Older processes employed either evaporative cooling by the solvent or separate addition of initiator after the monomer. However, both methods do not provide the best conditions for the preparation of reproducible acrylic polymers.

Factors governing the polymerisation process

Factors governing the molecular masses of polyesters are molecular excesses of polyols, the reactivity of the functional groups, the degree of condensation, and the degree of branching. These are mainly material factors. By contrast, the critical factors governing the production of acrylic resins are the process conditions. It is possible to analyse the process-determining factors independently of the different compounds that are used for the preparation of acrylic resins. The most important are the initiator concentration, the polymerisation temperature, the ratio of solvent and quantity of monomers (i.e. the concentration of free monomer in the reaction mixture), and the dosing time of the monomer mixture, which also generates a specific free monomer concentration during the process. Figure 3.8.10 shows that the values of attainable molecular weight, together with the viscosity of the resultant resin solution, will decrease significantly with increase in initiator concentration, with increase in polymerisation temperature, increase in solvent content, and increase in dosing time.

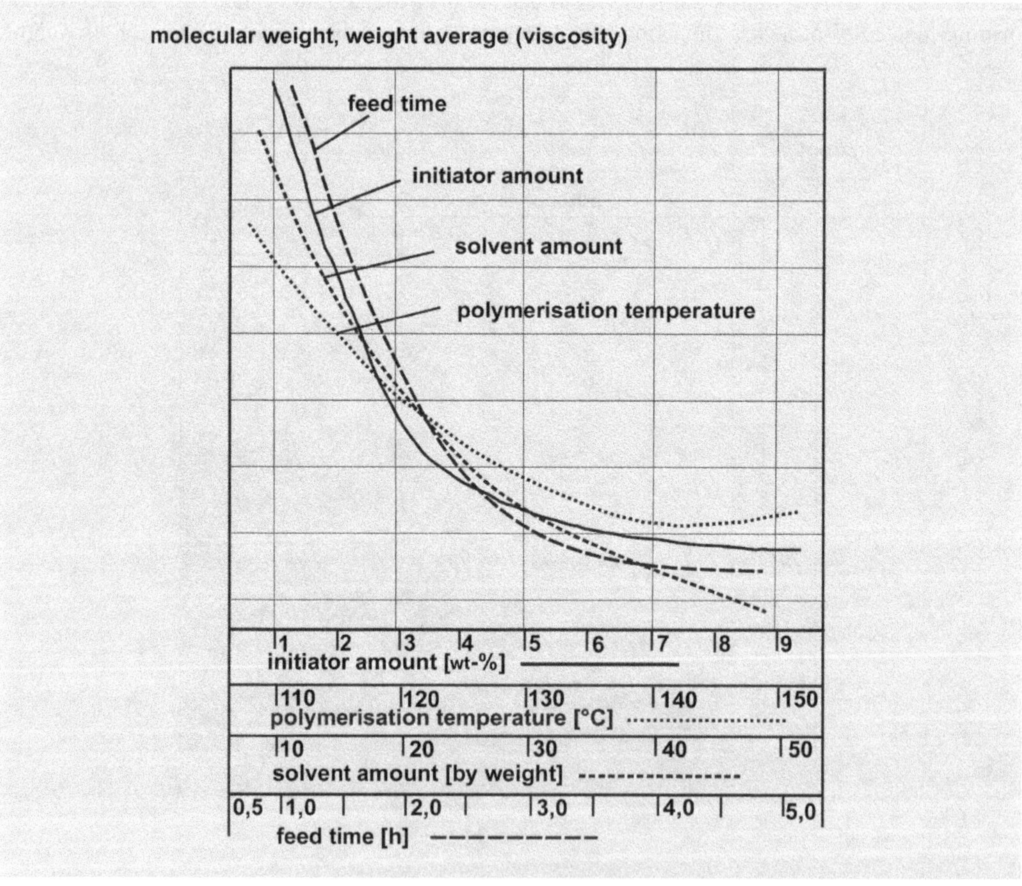

Figure 3.8.10: Process factors governing the molecular weight or viscosity

Most of the curves decline, but change very little change at high values of the governing factors. However, with increase in solvent content, the molecular weight or viscosity continues to decrease. The reason is that the solvents react by free-radical transfer which terminates chain propagation and are thus competing against the monomers which extend the chains. Thus, the more solvent molecules in the vicinity of growing chains and the lower the monomer concentration, the shorter the chains become. At elevated temperatures, initiator free-radicals can recombine, with the result that no initiation reaction occurs, or side reactions occur (transesterification, disproportionation), both of which increase the molecular weight and viscosity. All these factors work in concert and, of course, there are interactions between them. To prepare high-solid acrylic resins (see Chapter 3.8.4), it is necessary to follow the rules of the governing parameters.

Material factors

Besides the process parameters, there are also material factors. The raw materials for the production of acrylic resins are monomers, solvents, initiators, and regulators. The monomers for acrylic resins are esters formed by acrylic acid or methacrylic acid and aliphatic monoalcohols that contain different alkyl chains. Examples are methyl, ethyl, n-butyl, isobutyl, tert.-butyl, ethylhexyl, and isodecyl esters. There are other monomers that have nearly the same polymerisation parameters as the acrylic or methacrylic esters and that form copolymers easily; they are called comonomers. The most important comonomer is styrene. Due to their different polymerisation reactivity, methacrylate monomers always form lower molecular weights under comparable conditions. However, as a result of the higher stiffness of polymer chains containing methacrylates, the solution viscosity of such resins is always higher than that of polymers containing mainly acrylic esters. Polymers containing mainly methacrylate monomers also confer greater hardness and resistance to solvents and chemicals. Monomers containing esters of alcohols with long aliphatic chains promote solubility, compatibility, wetting, levelling, and flexibility of the resins and coating films prepared there from. Esters of alcohols with short aliphatic chains boost hardness and durability (due to better diffusion resistance). Esters of acrylic or methacrylic acid with cycloaliphatic alcohols (cyclohexyl, tert.-butylcyclohexyl, isobornyl, and tricylodecanemethyl [TCD]) form an optimum compromise of the different properties, mentioned above. Contents of styrene lead to higher hardness and gloss values, but lower flexibility. The quantity of styrene in acrylic resins for clearcoats is restricted as high levels lead to lower weathering resistance (cracks after exposure to UV light). For crosslinking, the acrylic resins contain hydroxyl groups. Different monomers are available for introducing hydroxyl groups into the acrylic resin molecule. These are monoesters of diols, namely 2-hydroxyethyl acrylate, 2-hydroxypropyl acrylate, 2-hydroxyethyl methacrylate, and 2-hydroxypropyl methacrylate. Acrylic resins also always contain a small quantity of free acid (acrylic acid or methacrylic acid), which are responsible for the acid number of the resin. In contrast to polyesters, the quantity of functional groups (hydroxyl number, acid number) does not depend on the production process; the values are governed by the monomer composition. Table 3.8.1 (page 178) describes an acrylic resin which is typical of the formulation of clearcoats [154].

There is a special class of acrylic resins developed mainly for clearcoats whose hydroxyl groups for crosslinking have been replaced by carbamate groups. The carbamates in this case are esters of carbaminic acid that bear free amide groups (primary amides). As in the preparation of amino resins, the carbamate groups react with methylol and methylol ether groups of amino resins by cleaving water or alcohols to form urethane linkages. These reactions afford a way of forming urethanes without the use of isocyanates. Urethane groups are widely known for their superior resistance to chemicals (mainly at high pH values). Thus, films prepared by crosslinking carbamate acrylates with etherified melamine resins are more resistant to chemicals than are films prepared by the reaction of hydroxy acrylic resins and such melamine resins. They also feature better weathering resistance and are more resistant to mechanical impact. Studies here show that not only the urethane group is responsible for the described improvements. Since carbamate

Table 3.8.1: Formulation of acrylic resin for clearcoats

Pos.	m example	Components	wt-% monomers
01	1140	aromatic hydro carbons	
02	562	tert.-butyl acrylate	46.3
03	182	n-butyl methacrylate	15.0
04	364	2-hydroxypropyl methacrylate	30.0
05	73	4-hydroxybutyl acrylate	6.0
06	33	acrylic acid	2.7
07	73	tert.-butyl perbenzoate	
08	73	aromatic hydro carbons	
	- 526	aromatic hydro carbons	
09	101	1-methoxy propyl acetate	
	2075	sum	100.0

Pos. 01, 08 aromatic 100 (boiling range 158 – 172°C)

characteristic values:
OH-number (NV) 139 mg KOH/g
acid number (NV) 23.6 mg KOH/g
NV 60.0 % in aromatic 100/methoxy propyl acetate 32/8
viscosity 390 mPa·s (plate-and-cone viscosimeter, 23 °C)

groups react more readily with the functional groups of melamine resins than hydroxyl groups, the extent of co-crosslinking is greater and the networks are more extensive and more homogeneous. Such networks are essential for properties such as flexibility, high weathering resistance, high chemical resistance, improved adhesion, and greater resistance to mechanical impact. The greater crosslinking efficiency is also beneficial to the formulation of one-component clearcoats with high-solid content and low VOC levels under application conditions (see Chapter 3.8.4).

3.8.3.2 Polyesters

Polyesters are characterised - mainly in the US literature - as being insufficiently weathering-resistant. That literature cites the ester linkages as being the weak points of polyester molecules. However, most ester groups are sensitive to high pH values, yet fairly stable in acid conditions (which are related to weathering: "acid rain"). As already mentioned, aromatic resin building blocks and in the film matrix absorb UV light and can transform the light energy into chemical energy followed by decomposition reactions. In clearcoats, those aromatic compounds can cause film damage, such as cracking. Since saturated polyesters and also alkyds usually contain aromatic building blocks (isophthalic acid, phthalic anhydride), clearcoats prepared with such resins suffer this damage. The same symptoms of damage are exhibited by clearcoats based on acrylic resins that have high levels of styrene. Polyesters without aromatic compounds, which contain only aliphatic and cycloaliphatic building blocks, behave comparably to acrylic resins without styrene. They show the same cracking resistance, and sometimes even better properties. Due to the differences in the structures of their colloidal particles in organic solutions, polyesters offer some advantages over acrylics. The diffusion processes between resin and crosslinker are more effective, and lead to more efficient crosslinking, creating an optimum balance of hardness and flexibility and boosting resistance to weathering, chemicals and solvents. In addition, they can provide better wetting and levelling. Apart from the aforementioned misconception, there are two objective reasons why polyesters are not often used for clearcoat applications. To obtain films of adequate hardness, the polyesters must contain not only aliphatic building blocks but also sufficient quantities of cycloaliphatic compounds. Unfortunately, cycloaliphatic compounds (e.g. cyclohexane dimethanol, hexahydrophthalic anhydride) are more expensive than acrylic monomers. What is more, polyesters containing these compounds generate rather steep viscosity curves as a function of solids content and of temperature during heating in the stoving oven. The first of these is an advantage as it allows the application solids content to be increased and the VOC values to be lowered. The second also seems to be an advantage as a low viscosity during the heating phase in the stoving oven is essential for optimum flow and levelling. However, it could also prove to be disadvantageous to the wet-on-wet process (clearcoat over basecoat). If viscosity of the clearcoat film during the stoving process reaches a minimum, sagging may occur. More seriously, the basecoat may show redissolving, followed by disorientation of effect matter and loss of effect and brightness. Therefore, there are genuinely objective reasons which militate against the wider use of polyesters in automotive clearcoats - but poorer weathering resistance is not the main one.

These problems can be circumvented mainly by using water-borne basecoats that do not show redissolving relative to solventborne basecoats of increased solids content, and by using use clearcoats with optimum rheological additives (sag control additives) and polyisocyanate adducts as crosslinker. Such clearcoats are available commercially; they are distinguished by better appearance, flow and levelling, and high gloss in comparison to clearcoats based on acrylic resins. They also offer superior weathering resistance.

3.8.3.3 Amino resins

Theoretically, the melamine resins that are used for topcoats and basecoats are suitable for clearcoats. However, regional preferences exist with regard to the melamine preferred. For medium-solid, one-component clearcoats that are used in Europe, the chosen melamine resins are relatively low-molecular, highly etherified and contain free imino-groups; they exhibit adequate reactivity and do not require catalysts under the right automotive stoving conditions. The application solids content achieved with suitable acrylic resins is about 50 % by weight. During crosslinking, some methylene groups form by self-crosslinking. Methylene groups impart relatively good weathering resistance and better resistance to chemicals (acids) than methylene ether groups.

One-component, high-solid clearcoats (solids content up to 60 % by weight) contain mainly HMMM resins (hexamethoxymethylmelamine). These are particularly important in the USA (see Chapter 3.8.4).

3.8.3.4 Polyisocyanates

Automotive clearcoats that feature optimum weathering require the use of aliphatic and cycloaliphatic polyisocyanates as crosslinker. Since isocyanate groups react at ambient temperatures with hydroxyl groups, the clearcoat with resins containing hydroxyl groups and additives (base lacquer) has to be delivered separately from the isocyanate crosslinker (hardener). Such clearcoats are therefore classified as two-component coatings. Both components are mixed just before application. Ready-mixed clearcoats have to be use within a certain period of time known as the pot life. During this time, the viscosity undergoes exponential growth that ultimately leads to application problems, e.g. loss of flow and levelling properties and forming of specks in the clearcoat layer. For the application of two-component clearcoats in automotive coating processes, the problem of pot life is resolved by using two-component spray guns. This application system has separate feed lines for the base lacquer and hardener. The two are mixed just prior by a static mixer just prior to reaching the gun and the mixture is sprayed immediately. Consequently, the pot life has no impact on the application process. It is only necessary to ensure that the spray gun is cleaned, mainly during breaks.

Diisocyanates are toxic (by inhalation) because they have a measurable vapour pressure at ambient temperature. They therefore make unsuitable hardeners. To render them suitable, they are transformed into oligomers by various oligomerisation methods. Such oligomers still present physiological concerns, due to the quantity of free reactive isocyanate groups, but they are not toxic by inhalation on account of their resin-like properties, with practically no vapour pressure at ambient temperatures. Nonetheless, users must ensure to avoid direct contact between such products and skin or mucous membranes. The oligomerisation reactions are nearly exclusively based on hexamethylene diisocyanate (HDI) and isophorone diisocyanate (IPDI). They involve the formation of different adducts, such as urethanes, allophanates, biurets, uretdiones and, most commonly, isocyanurates (see Figure 3.8.11).

These polyisocyanate adducts react much more clearly with hydroxyl groups from partner resins than with the functional groups of melamine resins. Consequently, the mixing ratios for resins containing hydroxyl groups and polyisocyanate adducts are calculated stoichiometrically. The optimum is an equivalent ratio of both components ($n_{NCO} = n_{OH}$), but it is possible to vary the ratios within certain limits, without loss of properties. Only the reaction of isocyanates with atmospheric moisture need be taken into consideration. This side-reaction leads to the formation of urea groups. For stoving coatings, such as automotive OEM clearcoats, this side-reaction is not very extensive.

As already mentioned, isocyanates can react at ambient temperatures with resins containing hydroxyl groups. However, the gradient of the reaction rate as a function of temperature (Arrhenius plot) is much smaller than that for the reaction of functional groups of melamine resins. It makes sense, then, to crosslink two-component clearcoats containing polyisocyanate adducts at the same temperatures as the one-component clearcoats containing melamine resins with a view to obtaining efficient crosslinking efficiently and perfect film properties.

Compared with films resulting from melamine resin crosslinking, films containing polyisocyanate adducts are much more homogeneous and the molecular network is more extended. This explains the better chemical resistance, better resistance to chemicals and better flexibility. However, as a rule, the network arc length is greater than for the networks in films crosslinked by melamine resins. That leads to lower hardness and opens the possibility of diffusion effects. However, both groups of polyisocyanate adducts behave in totally different ways. Crosslinkers based on hexamethylene diisocyanate are distinguished by high reactivity and crosslink optimally to afford relatively flexible and weathering-resistant clearcoat films that can be swollen by strong chemicals and specific solvents. IPDI adducts are less reactive and lead to less efficient crosslinking than

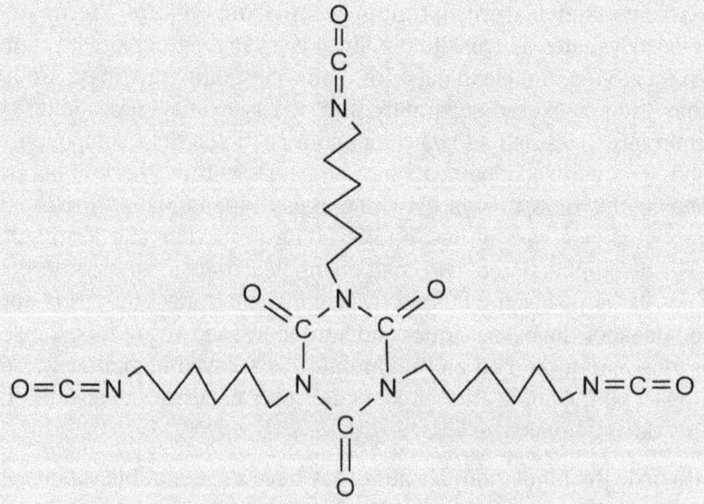

Figure 3.8.11: Isocyanurate adduct of hexamethylene diisocyanate

HDI adducts. However, due to their cycloaliphatic structure, the resultant films are much harder and more resistant to diffusion processes, which can cause swelling. Therefore, the films offer better chemical and solvent resistance, but are also less flexible.

3.8.3.5 Blocked polyisocyanates

Blocked polyisocyanates offer the possibility of exploiting the crosslinking properties of isocyanates without having to pay attention to pot life. They are prepared by making isocyanates react with compounds containing active hydrogen atoms to yield adducts which can be cleaved at elevated temperatures (reversible addition reaction). The reaction of isocyanates with hydroxyl groups at room temperature is blocked, so the resins containing hydroxyl groups can be mixed with the blocked polyisocyanates without reaction (one-component paints). In the past, it was believed that, at elevated temperatures, the blocked polyisocyanate become de-blocked, leading to the formation of free isocyanate groups that react with the free hydroxyl groups from the combination resins. It is now known that the reaction of blocked polyisocyanates is a substitution reaction which leads to a chemical equilibrium. Since, as in most cases, the blocking agent is removed from the reaction media, the balance is shifted to the products which are expected. The principle behind the reaction is comparable to that of the transesterification reaction (see Figure 3.8.12).

There are blocked polyisocyanates which do not lead to cleavage of the blocking agent. For example, polyisocyanates blocked by diethyl malonate react at elevated temperatures with hydroxyl groups in a transesterification reaction and releasing ethanol; the amide structure from the blocking agent remains in the film. Suitable commercial blocking agents can be classified according to the temperatures at which they react efficiently under industrial application conditions (stoving time: 15 to 30 minutes) with partner resins containing hydroxyl groups (polyesters). The effective reaction temperatures of suitable blocking agents are given in Figure 3.8.13, page 182.

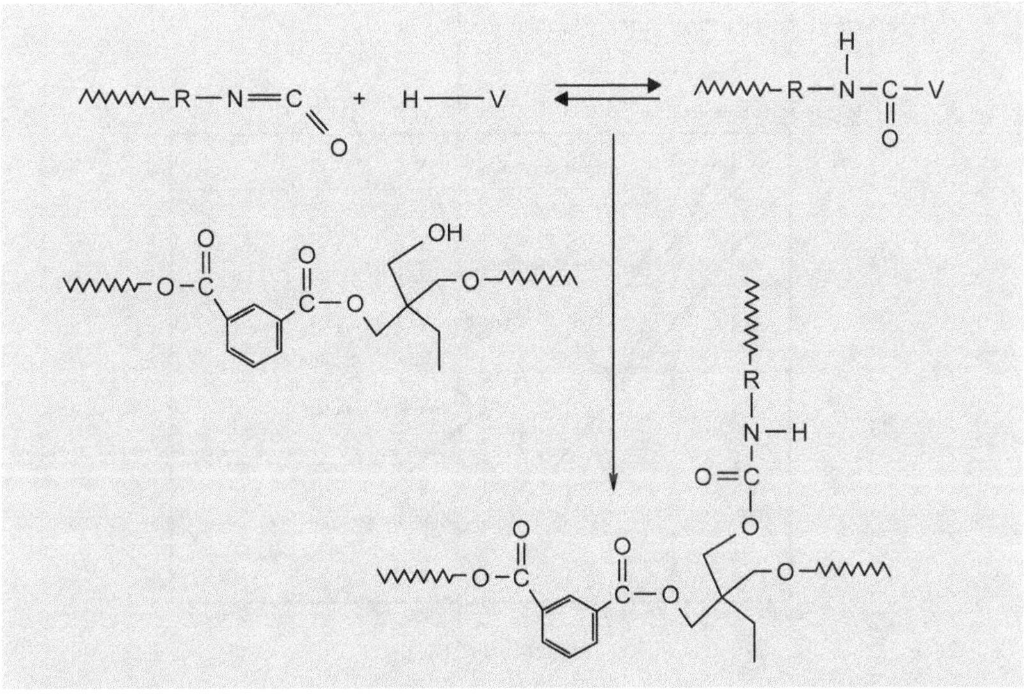

Figure 3.8.12: Principle behind the reaction of blocked polyisocyanates

Since automotive OEM clearcoats are usually stoved at temperatures below 150 °C, blocking agents are chosen which react up to that temperature, namely diethyl malonate, ethyl acetoacetate, dimethyl pyrazole. When using blocked polyisocyanates, it is necessary bear the specific properties of the compounds in mind: problems with solubility for diethyl malonate, tendency to yellowing for ethyl acetoacetate, remaining in films for dimethyl pyrazole. As blocked polyisocyanates cannot meet all requirements on crosslinkers for clearcoats, they are mixed with melamine resins (hybrid crosslinking). In addition, they are more expensive than melamine resins. Surprisingly, the mixture does not lead to the expected compromise on properties. Instead, the overall properties are greater than the sum of the individual ones. The explanation is that the crosslink-

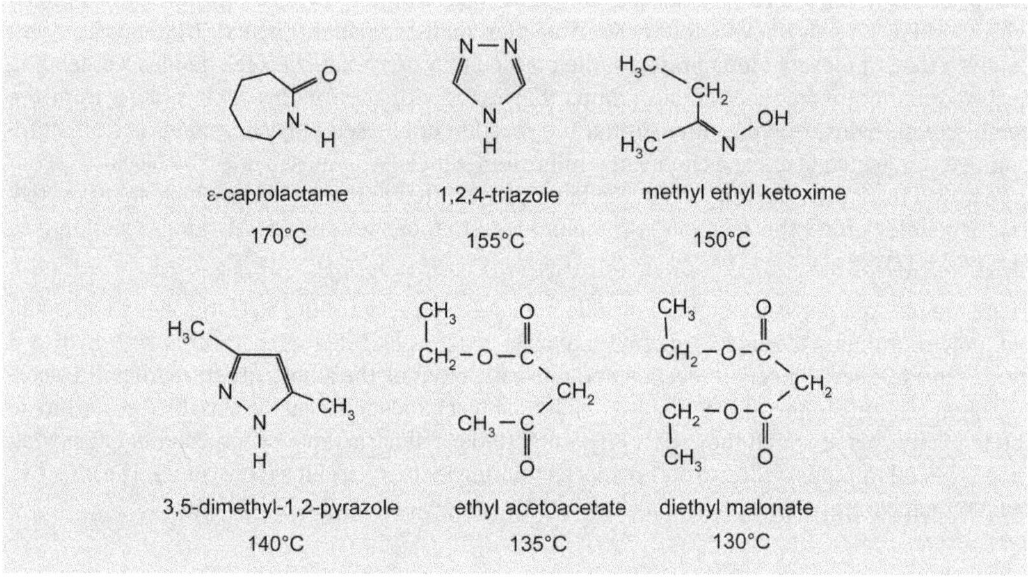

Figure 3.8.13: Effective reaction temperatures of suitable blocking agents

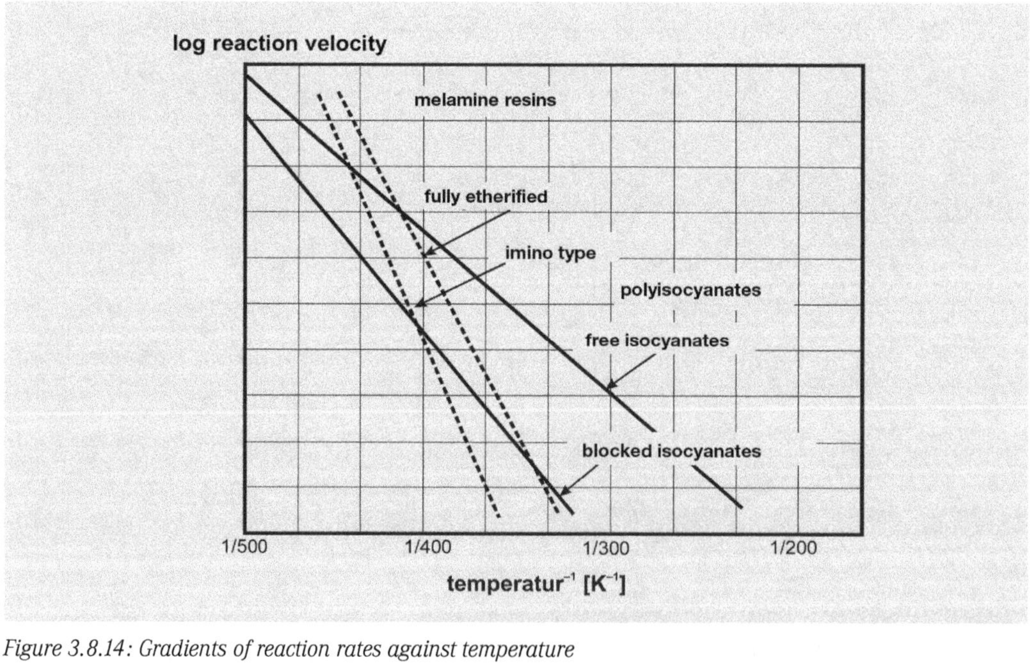

Figure 3.8.14: Gradients of reaction rates against temperature

ing reactions take place concurrently but at different reaction rates. Such step-wise reaction leads to the formation of very effective molecular networks (large extension). Figure 3.8.14 shows different reaction rates (logarithmic) against temperature (reciprocal plot) that might provide an explanation for this.

Hybrid-crosslinked clearcoats exhibit optimum values for resistance to weathering, chemicals and solvents, and a balance of hardness and flexibility. A special blocked polyisocyanate is tris-alkyl carbamato triazine (TACT) [155], whose molecular structure is given in Figure 3.8.15.

The product combines the positive properties of triazines, e.g. resistance to heat and chemicals and a high refractive index that leads to high gloss values, with the positive crosslinking behaviour of blocked polyisocyanates. The crosslinker reacts with hydroxyl groups of partner resins, cleaving the monoalcohols of the carbamate group. Although monoalcohols as blocking agents for "normal" isocyanates do not react at temperatures below 200 °C, clearcoats containing TACT are crosslinked effectively at temperatures of 130 to 140 °C. The reason is the directing effect of the double bond system of the triazine ring. Nevertheless, the resultant networks are relatively stable.

3.8.3.6 Solvents

The most important solvents for conventional automotive OEM clearcoats are aromatic hydrocarbons with medium boiling rates, e.g. C_3-alkyl aromatic hydrocarbons with boiling temperatures between 155 and 175 °C. They are mixed with smaller quantities of more polar solvents. The polar solvents are alcohols (e.g. n-butanol, iso-butanol) that are mainly introduced into the clearcoat formulation via the delivery form of the melamine resins. Polar solvents that have somewhat higher boiling points are glycol ethers (e.g. butyl glycol, butyl diglycol, methoxypropanol) or glycol ether esters (e.g. butyl glycol acetate, butyl diglycol acetate). Ester solvents with very high boiling rates may be contained in small quantities in clearcoats (e.g. glycol acid butyl ester, dimethyl ester of a mixture of low-molecular dicarboxylic acids [dibasic ester, DBE]). The selection criteria for clearcoat solvent composition are mainly the application conditions (conveyor line speed, flow rate of clearcoat, type of spray guns, flash-off conditions, temperature and the length of the stoving oven). The solvents must not lead to redissolving of the physically dried basecoat. They must evaporate uniformly. It is an advantage if a smaller quantity of the more polar solvents remains longer in the film than the unpolar solvents during physical film forming the evaporation rate of solvents and residual quantities of solvent during the film forming phase, together with the increase in temperature, influence the viscosity curve, which in turn has an effect on flow and levelling.

3.8.3.7 Levelling agents

The most important precondition for optimum levelling of clearcoats is wetting of the basecoat surface. The impinging spray particles must spread very well by interacting with this surface. They must be able to absorb subsequent particles to form a smooth and homogeneous film layer (spray mist take-up). The resultant film must become smooth during the following phase of film forming.

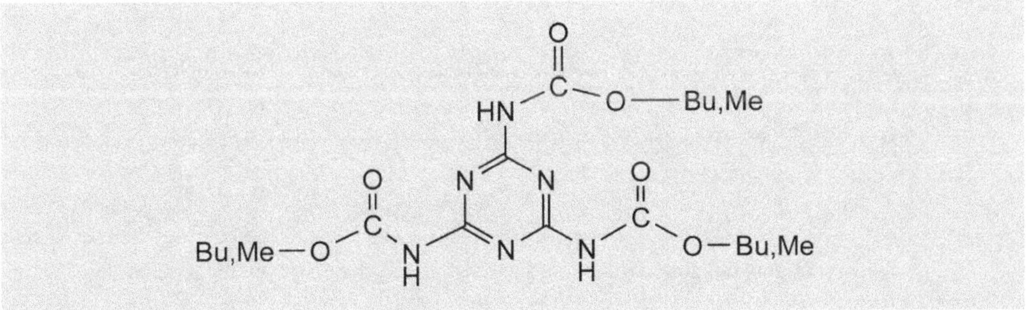

Figure 3.8.15: Molecular structure of trisalkyl carbamato triazine (TACT)

It must not sag on vertical parts of objects, nor sag during flash-off or at increasing temperatures in the stoving oven. However, the clearcoat film must perfectly cover the remaining unevenness (which is known as telegraphing) of the other film layers. The clearcoat must not contour this telegraphing. Therefore, the levelling agents for clearcoats are chosen for their ability to support wetting properties. The most important classes of material for levelling agents are polyacrylates (non-functional, high-molecular polyacrylates containing unpolar monomers) and modified silicone oils.

Silicone oils are polycondensation products of dimethyl silane diols. Polydimethyl siloxanes are incompatible with most resins for coatings. The incompatibility increases with increase in molecular weight. Very low-molecular silicone oils are compatible with resins, but only in organic solution. The special feature of silicone oils - their acting on surfaces - stems from their very low surface tension. In the past, it was believed that silicone oils reduce the surface tension of the entire coating system, thereby conducing to better wetting and levelling. However, in reality, the silicone oils work by floating to the surface of the film layer during the film-forming process. After floating, the silicone oil spreads over the surface of the layer efficiently due its the low surface tension. It is believed that such behaviour occurs on the spray particles, which are then capable of boosting wetting, spray mist take-up and levelling. Levelling stems from the formation of a secondary structure on the surface of the film. If the additive is able to influence the surface structure, it is possible to create a structure comprising a small quantity of long waves that covers all other remaining unevenness and creates the visual impression of high filling power (clearcoat hold-out).

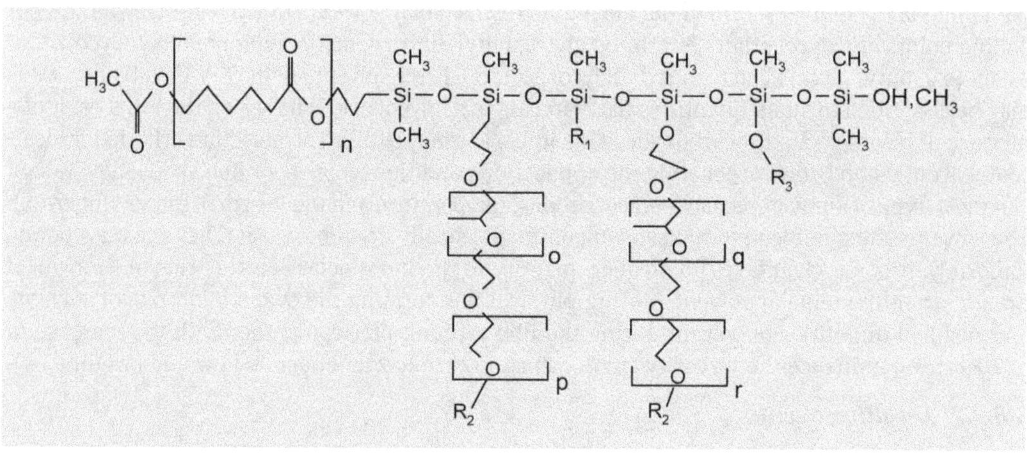

Figure 3.8.16: Possible modifications of silicone oils

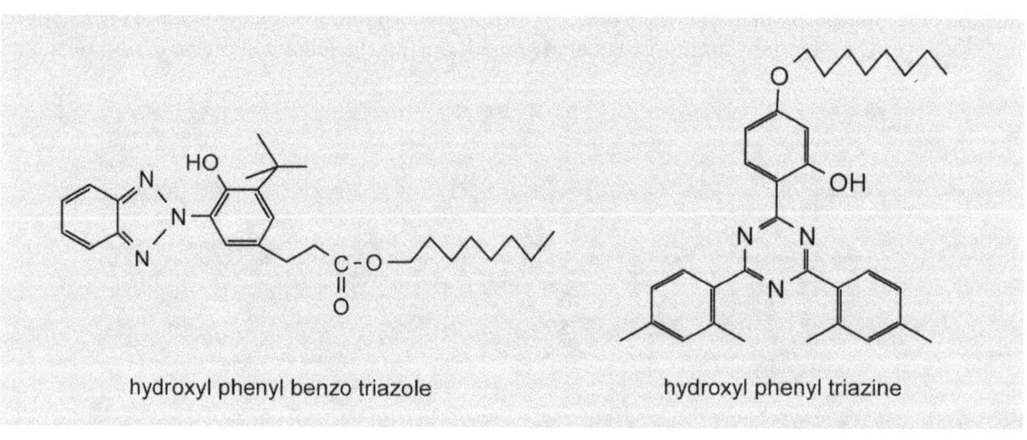

Figure 3.8.17: Chemical structure of typical UV absorbers

The sharp border between effectiveness and incompatibility in the case of unmodified silicone oils has been accentuated by modifying them. The modifications are more polar than the silicone oils themselves and increase the compatibility of the products with resins, without diminishing the surface effects. Modification leads to easier handling of silicone oils and prevention of film defects. It is also possible to choose the modified silicone oil to suit the type of resin in the paint formulation and to strike a balance between compatibility and floating effect. The most important modification of silicone oils is the addition of polyether chains (EO and PO polyesters). Furthermore, replacing some dimethyl siloxanes by siloxanes with longer aliphatic chains increase compatibility. There are also silicone oils that have been modified with polyester chains (e.g. polycaprolactones). The modifications are obtained by making alkyl silane triols react with the modification molecules, or adding allyl ethers to silanes. The modification may be a chain termination or a side chain. Figure 3.8.16 show all the ways in which silicone oils can be modified.

There are many different commercial products, which are chosen for their specific effects. Some of the modified silicone oils exhibit an additional influence on the viscosity behaviour of the paint, namely by conferring structural viscosity. It is necessary to ensure that polyether-modified silicone oil do not render the films sensitive to moisture.

Theoretically, all coating layers should lend themselves to re-coating. Although clearcoats are the last layer of the automotive coating system, it must be possible for them to be re-coated in the event of repairs. The surface additives chosen must not prevent recoatability.

Some silicone oils that spread over coating surfaces generate slip effects. Slip effects are conducive to scratch resistance. But they can also pose recoatability. Subsequent coatings might not wet or adhere to the surface of the last layer. This may be an advantage where the requirement is for an anti-graffiti effect. However, it is definitely a disadvantage for repair coatings if the surface has been sanded completely before recoating can occur. Therefore, a crucial requirement imposed on the use of silicone oils as levelling agents is their mobility. They must create excellent surfaces, but they must also be readily absorbed by the next coating. The mobility of silicone oils is of course also very important for use in primers, primer surfacers, and basecoats.

3.8.3.8 Light stabilisers

The various behaviours of resins which do and do not contain aromatic compounds are described in Chapter 3.8.2.2. The conclusion drawn there is that all resins are influenced by UV light, even if they react differently. Thus, clearcoats must be protected separately against UV light. Currently, all clearcoats contain a light stabiliser package. The light stabilisers consist of UV absorbers and free-radical quenchers. UV absorbers are quasi dyestuffs which absorb UV light instead of visible light. In this regard, it is important that the light absorption is concentrated only on the UV content of the light, as otherwise the UV absorber would influence colour and transparency of the clearcoat, e.g. cause yellowing. UV absorbers must offer optimum compatibility in the liquid phase and in the film matrix. The compounds chosen are therefore modified with particularly organophilic molecular building blocks. They must be resistant to evaporation and migration. Of course, they must not decompose under the influence of light. The first UV absorbers to prove suitable for automotive clearcoats were the oxalanilides. They were replaced by the 2-hydroxyphenyl benzotriazoles. 2-Hydroxyphenyl triazines are currently touted as the best UV absorbers on the market. Figure 3.8.17 shows the chemical structure of a typical member of both classes of compounds. Figure 3.8.18 (page 186) shows the absorption spectra of these types of UV absorbers.

UV absorbers absorb UV light in accordance with the Beer-Lambert Law (see Chapter 2.1.2.1), which describes the dependence of absorption on the layer thickness. The filter effect of UV absorbers in the upper parts of the layer is quite low; formation of free-radicals followed by decomposition reactions may take place. In addition, UV light may also generate free-radicals with other molecules besides the film matrix, e.g. with water to generate atmospheric humidity that may also impact the clearcoat layer. That is why UV absorbers are always combined with free-radical quenchers when intended for clearcoats. Free-radical quenchers are compounds that form low-energy free-radicals which have long life-times and which can absorb other energy-rich free-radicals by recombination reactions, converting their energy into chemical energy. The quencher free-radicals then re-form. The most important class of free-radical quenchers for clearcoats is that of sterically hindered amines, which are variously substituted 2,2,6,6-tetramethyl piperidines. Different substituents are attached to the nitrogen atom and the piperidine ring. These products

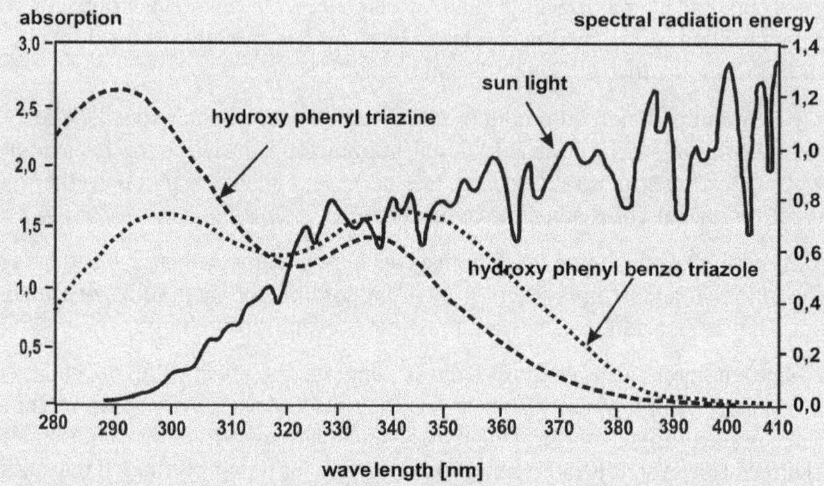

Figure 3.8.18: Absorption spectra of UV absorbers

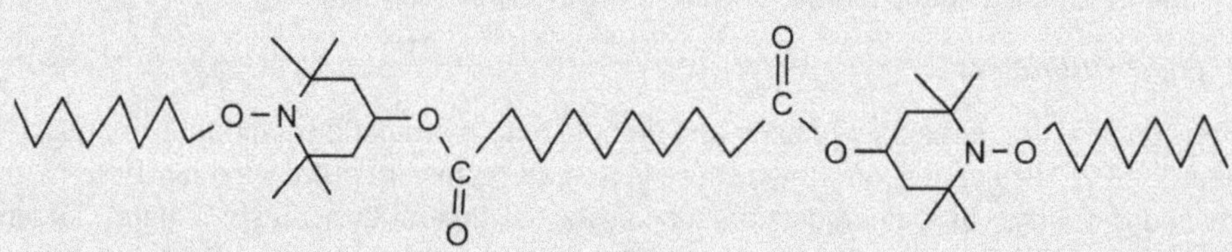

Figure 3.8.19: Molecular structure of a free-radical quencher

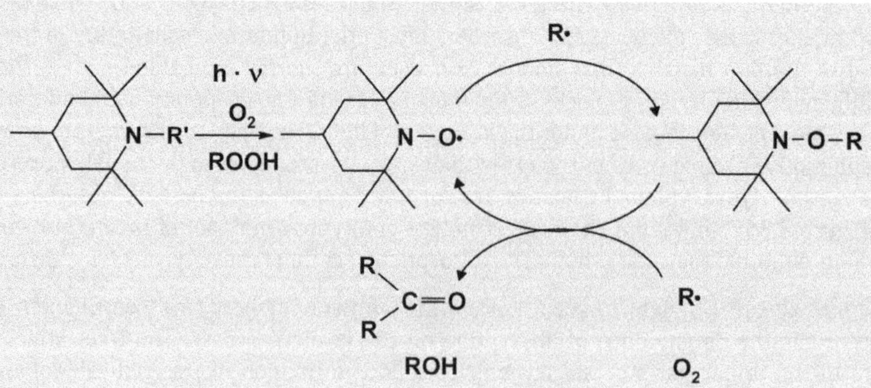

Figure 3.8.20: Proposed reaction mechanism of free-radical quenchers

are called hindered amine light stabilisers (HALS). The molecular structure of a typical HALS is given in Figure 3.8.19 [156]. The reaction mechanism of free-radical quenchers is thought to be that shown in Figure 3.8.20 [156].

The other requirements imposed on free-radical quenchers are the same as for UV absorbers: compatibility with the resin matrix, very low vapour pressure, no migration, and resistance to light.

Since piperidines with secondary nitrogen atoms and N-alkyl substituted piperidines are basic, they have a pro-catalytic effect on the isocyanate reaction and an anti-catalytic effect on melamine reactions. The N-acetyl compounds behave indifferently, but make less effective free-radical quenchers. The best in this regards are the N-alkoxy compounds of the piperidines. They are not basic and have the highest yield of free-radicals.

3.8.3.9 Rheological additives

The requirements on rheological additives for clearcoats are prevention of sagging by thick layers on vertical parts and the creation of a surface structure which generates the visual impression of filling power (clearcoat hold-out) and good appearance. Of course, they must not impair transparency or gloss of clearcoats either. The preferred chosen rheological additives are pyrogenic silica [121] and crystalline ureas [157]. Pyrogenic silica (already described in Chapter 3.5.3.4) is prepared by flame hydrolysis of silicon tetrachloride. The very fine primary particles of the products have average particle size diameters of 7 to 25 nm. Such pyrogenic silicas have a tendency to form aggregates and agglomerates. It is therefore necessary to disperse the products very well, preferably in parts of the clearcoat main resin. The quantities of additives contained in clearcoats are relatively small. High quantities of and poorly dispersed silica cause adversely affect levelling and lead to loss of gloss. There is some evidence that incorporating silica improves the scratch resistance of clearcoats (see Chapters 3.2.8.5 and 3.8.3.11).

Crystalline ureas, mentioned above along with solventborne basecoats, are also suitable for clearcoats. These ureas must be precipitated in the main resin (acrylic resin) of the clearcoat. When used in the right quantities, they have no negative influence on the viscosity of the clearcoat in the application state and nor do they lower the application solids. Only during physical drying of the clearcoat do they generate a certain degree of structural viscosity through particle-particle interaction, which increases the viscosity rapidly, preventing sagging of thick layers on vertical parts. The additives also generate a specific surface structure that creates the visual impression of filling power and good appearance, without decreasing the gloss. This product class also includes ureas that form gel-like dispersions [158].

Nano-particles, described in Chapter 3.8.11, can influence the viscosity of clearcoats, too.

3.8.3.10 Formulation example

Table 3.8.2 (page 188) shows a sample formulation for a solventborne automotive OEM clearcoat [159]. The clearcoat consists of an acrylic resin containing hydroxyl groups, part of which has been modified with crystalline urea; a melamine resin as crosslinker; an UV absorber and free-radical quencher: and levelling agents as additives. Apart from the solvents from the delivery form of the resins and xylene, it contains high-boiling aromatic hydrocarbons and glycol ether for boosting flow and levelling.

3.8.3.11 Nanoparticles for clearcoats

Hardly any other term has been mentioned as often in connection with the term “innovation” as “nanotechnology”. The term “nano” is applied to many different processes, products and applications, and in some cases even when it should not be. The term “nano” is derived from the Latin word “nanus” which means “dwarf”. In the CI measuring system, “nano” means the billionth part

Table 3.8.2: Sample clearcoat formulation

Pos.	Components	Solid m	Formulation wt-%
01	acrylic resin. containing hydroxyl groups (65 %ig aromatic 100/n-butanol 97/3)	27.87	42.88
02	acrylic resin, containing hydroxyl groups, modified by SCA (60 % in aromatic 100/n-butanol)	8.73 0.62	15.59
03	melamine resin (70 % in n-butanol)	15.61	22.30
04	silicone oil, polyether-modified (2 % in xylene)	0.05	2.60
05	radical quencher (N-Methyl-2,2,6,6-tetramethylpiperidine-derivate)	0.52	0.52
06	UV-absorber (95 % in MPA) (hydroxyphenyl benzotriazole, substituted)	0.49	0.52
07	butyl diglycol		0.52
08	aromatic solvent (bp.: 180 – 210 °C)		9.36
09	xylene		5.71
	sum	53.89	100.00

pos.

01 Setal 1761, 65 % in aromatic 100 /n-butanol 97/3 (Nuplex)
02 Setal 711761, 60 % in aromatic 100 /n-butanol 98/2 (Nuplex)
03 Setamin US 138, 70 % in n-butanol (Nuplex)
04 Baysilon OL-17, 2 % in xylene (Borchers)
05 Tinuvin 292 (Ciba Chemical Specialties)
06 Tinuvin 384 (Ciba Chemical Specialties)
08 Solvesso 150 (Exxon)

(10^{-9}) of a basic measuring unit. One metre (m) is thus one billion (10^9) nanometres (nm) and one micron (µm) is 1,000 nanometres (nm). The wavelengths of visible light are between 420 and 750 nm while the distance between carbon atoms in an aliphatic hydrocarbon (single bond) is 0.154 nm. There are nanofibres, nanostructures, and nanoparticles. Nanoparticles are defined as particles with particle diameters below of 100 nm.

Due to the enthusiasm surrounding the results of recent studies of nanoparticles, as evident from the number of publications and press releases on the subject, the expectations on such products are very high. It is time to get back to the facts and to describe the properties and efficiency of such particles more scientifically. These particles are remarkable for their special properties, even from physical aspects. Even though they are solid matter in the physical sense, they do not have a regular crystalline structure. Of course, this is also true if the particles are organic polymers. The most important property possessed by such particles is that they have a very large surface area relative to their mass.

To get an impression of the size of the surface area of such small particles, consider the following thought experiment. A sphere with a surface area of 1 cm^2 has a diameter of 0.5642 cm and a volume of 0.0940 cm^3, i.e. 9.4 mm^3. Spheres with diameters of 10 nm (10^{-7} cm) have volumes of $5.236 * 10^{-22}$ cm^3 and surface areas of $3.1416 * 10^{-14}$ cm^2. From the volume of the first sphere, it would be possible to create $1.7959 * 10^{20}$ (179,590,000,000,000,000,000) spheres with a diameter of 10 nm and a total surface area of 5,640,000 cm^2, i.e. 654 m^2. On the surface of such matter, the cohesion forces are free to generate the surface tension. In the case of such tiny particles, the surface tension is very large relative to their cohesion. This explains their outstanding particle-particle interactions, their tendency to agglomerate, and also their rheological action on dispersions of the particles.

Nano-particles have long been widely known in the coatings industry, but they were not called by that name. The aforementioned pyrogenic silicas (see Chapter 3.8.3.8) have been commercially available since 1942. Since they have particle size diameters down to 7 nm (primary particle size), they can definitely be counted as nanoparticles. Carbon black pigments, too, have long been commonplace in the paint industry (see Chapter 3.6.3.4) and can have particle diameters of about 20 nm.

Nanoparticles in a clearcoat matrix convey the impression of generating or supporting scratch resistance. Several concepts have been advanced to explain this. The assumption here [160] is that the particles dissipate the mechanical energy of scratching within the clearcoat film, transforming the energy into thermal energy and preventing damage. This requires that the particles be incorporated into the film matrix very well, preferably by chemical (covalent) bonds. The matrix itself should be very flexible. The linkage between the flexible film matrix and nanoparticles absorbs the energy of scratching. The nanoparticles move inside the film matrix under the mechanical influence but return to their stating position due to the flexible response of the matrix; energy is absorbed (dissipated) and damage does not occur. The reaction is comparable to that of elastomers that consist of hard and soft segments. But the molecular dimensions in that case are roughly one order of magnitude smaller than that of nano-particles (despite of the term "nano").

The reason for choosing such small particles that do not absorb light (such as carbon black) is that they do not scatter light. Particles which are much less than half the size of the wavelength of visual light are transparent and do not scatter light. For clearcoats, therefore, the chosen particles have relatively low refractive indices, e.g. silica and aluminium oxide, but not titanium dioxide. Silica is preferred because there are products available which already contain an organic modification that can boost the efficiency of incorporation into the clearcoat matrix. The second reason for choosing such particles is the possibility of molecular reactions between functional groups on the surface of the particles and the surrounding matrix. The formation of covalent bonds on the surfaces of particles is usually improbable (see the description of adhesion). Chemical reactions, e.g. during the formation of molecular networks, are controlled by diffusion processes, and need a minimum of kinetic energy to be able to form chemical linkages. Normally, particles are unable to generate such energies, as they are very small and have very large surfaces areas. There are other examples in which the surfaces of solid matter reacts with other molecules of surrounding media, e.g. surface treatment of iron and steel by phosphate solutions. As the surface area of the nanoparticles is so large, there is a chance that smaller molecules from the surrounding media will react with functional groups on the surface of such particles.

Theoretically, it is possible to incorporate elastomeric behaviour into clearcoat films through judicious selection of products. For example, the aforementioned polyurethane elastomers are relatively scratch resistant. Furthermore, clearcoats containing low-branched polyesters with building blocks capable of imparting flexibility and crosslinked by aliphatic polyisocyanate adducts are scratch resistant. There are also acrylic resins available that contain monomers with long side-chains or modifications (e.g. linear polyester side-chains) which are also highly flexible and relatively scratch resistant. However, all the aforementioned examples are sensitive to other mechanical impact and are influenced by diffusion processes that lead to swelling and the possibility of damage. The chemical resistance of such clearcoats is much lower. By contrast, nanoparticles in clearcoats support chemical resistance if they are effective against diffusion processes.

The quantity of nanoparticles in clearcoats is relatively low. The first reason is that of cost. In addition, higher quantities open up the possibility of particle-particle interaction and agglomeration followed by increasing viscosity or development of structural viscosity. Agglomerates can also impair the transparency of clearcoats. This limits the amount of nanoparticles used in clearcoats. It is imperative to ensure efficient distribution of the particles throughout the clearcoat (wet phase and films). The low quantity of nanoparticles in clearcoats is compensated by the fact that

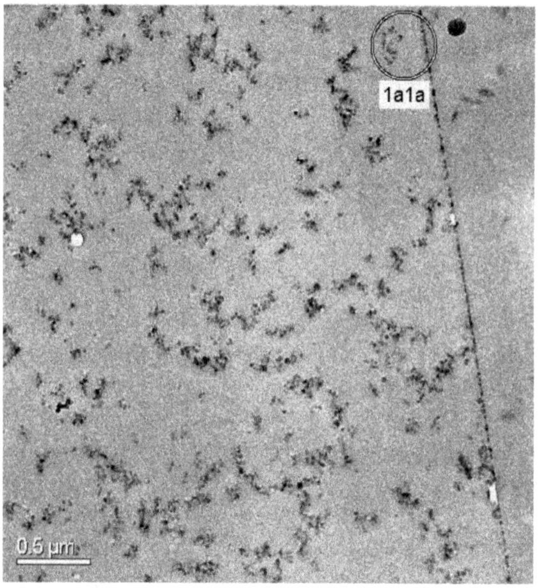

Figure 3.8.21: Distribution of nanoparticles in a clearcoat film

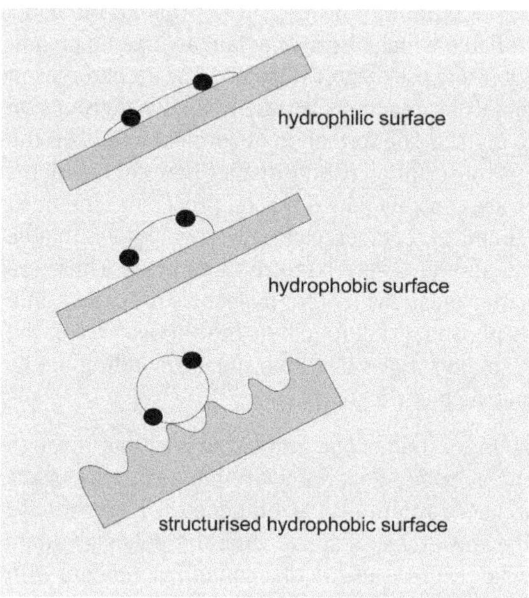

Figure 3.8.22: Principle behind the self-cleaning effect

most available particles during film forming float onto the film surface and form agglomerates. Those agglomerates support the scratch resistance very well. However, the disadvantage is that layer decomposition will occur during the lifetime of the car, through weathering, erosion, washing, and polishing. Some of the scratch resistance properties will be lost over time. Figure 3.8.21 shows the typical distribution of nanoparticles in a clearcoat film (SEM image).

3.8.3.12 Lotus effect

The phrase "Lotus" effect was mentioned a great deal in publications and press releases some years ago. After the first publications, the coatings industry was called on to make coating materials with the self-cleaning effect of the lotus plant. Automotive topcoats or clearcoats should also have this property of course. By "Lotus" effect is meant the self-cleaning effect of the leaves of lotus plants. All plants need relatively large surfaces (leaves) for absorbing light to perform photosynthesis for their metabolism and growth. It is therefore important for the plants that the surfaces for absorbing the light be free of any coating of, e.g., dust and dirt. Several plants therefore exude waxes which form a hydrophobic layer on the leaf surfaces. Not only do these layers repel dirt, but water also runs off in beads. The water may also remove dirt particles. The effect is enhanced by a structured surface. Normally, structured surfaces lead to better adhesion, as described in Chapter 3.5.2. This is true if are sufficient interactions between liquid and surface. Hydrophobic surfaces do not interact with water. Aqueous media are repelled from such surfaces. If hydrophobic surfaces are textured, the repellence is intensified. Aqueous media runs off in beads, possibly entraining dirt particles. The principle behind the self-cleaning effect is shown in Figure 3.8.22.

The leaves of the lotus plant have small conical wax structures on their surface, which makes for efficient self-cleaning. Figure 3.8.23 shows a lotus leaf repelling water and a micrograph image of the conical structures.

Various coating systems with self-cleaning effect are now available. These are mainly wall paints and facade coatings. It is very difficult to bestow this self-cleaning effect on automotive OEM clearcoats. The absence of gloss and the impossibility of recoating clearcoats which have been

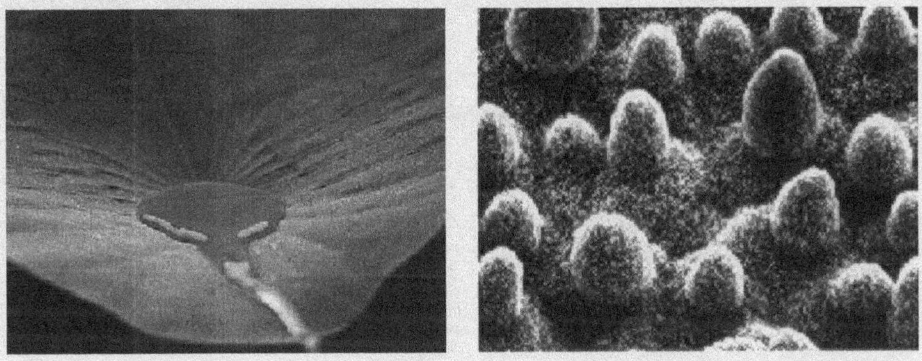

Figure 3.8.23: Lotus leaves and surface structure

modified to provide self-cleaning effects militate against the production of such clearcoats. In addition, such structures would be destroyed after a short time by weathering, washing and polishing of car body surfaces.

3.8.3.13 Sealers for clearcoats

There are other ways of protecting clearcoat surfaces against dirt and of increasing scratch resistance. One consists in covering the surface with materials that contain relatively low-molecular siloxanes to form thin layers in a sol-gel reaction (self-crosslinking reaction). This process is already used for coating glass and plastic lenses for eyeglasses to render them scratch resistant. Since glass is closely related to the siloxane materials, the application is ideal, but for plastic and paint surfaces, the coating process is more difficult. The problem lies in obtaining optimum bond. To achieve the requisite properties, it is necessary to choose specially suitable materials, specific application methods and defined crosslinking conditions. If all the conditions for homogeneous layers with good adhesion are fulfilled, the result is sealer layers which offer outstanding scratch resistance and excellent resistance to aqueous chemicals.

So far, there have been two reasons against the use of such sealers. First, there is the additional effort and cost of a separate layer that requires an additional application step and equipment. Application is difficult because the materials have a very low viscosity. Second, the product itself is rather expensive, even though only thin layers are needed. Finally, such surfaces cannot be covered by a further coating material that would not wet or adhere to the prepared surface. For re-coating, it would be necessary to sand the entire surface carefully.

3.8.4 Application

In modern application lines for automotive coatings, the clearcoats are applied first to the inner parts of the car bodies by pneumatic spray equipment, which may additionally be supported by electrostatic charging. Application of clearcoat to the outer parts of the car body is effected with electrostatic spray application equipment fitted with high-speed rotation bells. The process allows the application of relatively large quantities of clearcoat material per unit time, and additionally leads to high transfer efficiency of paint material (up to more than 80 % by weight of the wet clearcoat on the surface). The electrostatic bells are movable and arranged on both sides ("side machines") and on the top of the car body ("top machines"). The bells of the top machines follow the contour of the car body (distance and spray direction), skipping the front and rear windows. Of course, the entire process is controlled automatically. Final film forming by crosslinking of both layers (basecoat and clearcoat) is carried out in a stoving oven (combination of heat convection

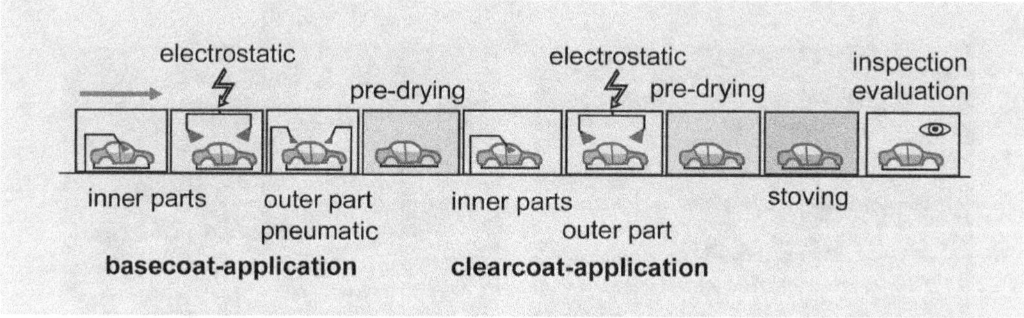

Figure 3.8.24: Topcoat segment for basecoat and clearcoat application

and radiation) at temperatures between 130 to 150 °C for 15 to 30 minutes, after the flash zone. Figure 3.8.24 illustrates the segment for the application of a topcoat system, comprising basecoat and clearcoat, in an automotive OEM coating line.

An example of the application conditions for the wet-on-wet process is as follows:

Conveyor belt speed		4 m/min
Spray basecoat	Interior parts	3 min
	Exterior parts	Electrostatic 2 min, pneumatic 3 min
Flash-off basecoat		5 min at 60 °C
Spray clearcoat	Interior parts	Pneumatic 3 min
	Exterior parts	Electrostatic 2 min
Flash-off clearcoat		6 min room temperature
Stoving	Heating phase	8 min
	Holding time	15 min at 140 °C (object temperature ± 10 °C)
Cooling		
Inspection		

The dry film thickness of clearcoats is usually 40 to 45 µm.

3.8.5 *Clearcoat with increased solid contents*

Conventional clearcoats which consist of hydroxy acrylic resins ("normal" viscosity, number average molecular weights 5,000 to 7,000 g/mol) and of highly reactive melamine resins (partly etherified) have application solids of 38 to 42 % by weight. It is possible to increase the solids content for the same application viscosity to 48 to 50 % by weight through choosing low-molecular weight acrylic resins (number average molecular weight of 1,500 to 2,500 g/mol) and relatively low-viscosity, yet still reactive melamine resins (imino types). Such clearcoats are called "medium-solids clearcoats". The use of especially-low-molecular weight acrylic resins and HMMM resins as crosslinker makes it possible to achieve 55 to 60 % by weight application solids. These clearcoats are called "high-solid clearcoats". These clearcoats must contain efficient catalysts for crosslinking, usually strong acids, e.g. sulphonic acids, to allow optimum crosslinking at the usual stoving temperatures of 130 to 145 °C. Strong acid catalysts can be disadvantageous (e.g. impair the effect substances - aluminium pigments - in the underlying basecoat). The requisite very high-solid content can be achieved by adding reactive diluents. Suitable for this are low-viscosity polyols whose hydroxyl groups may react with the melamine resins.

Specific measures are needed for making acrylic resins that have a low viscosity, yet confer nearly the same film properties as the conventional acrylic resins in the aforementioned conditions. For films of sufficient hardness and flexibility, the low-molecular acrylic resins contain high quantities of hydroxyl groups that provide maximum reactivity. However, there are monomers which generate solution viscosities lower than those of other monomers in acrylic resins of the same average molecular weight at the same solution concentration, while striking a balance of adequate hardness and flexibility. Such monomers are, for example, tertiary butyl acrylate and acrylic esters of cycloaliphatic monoalcohols [161].

The aforementioned acrylic resins modified with carbamate groups are also suitable for use in high-solid clearcoats. The use of acrylic resins containing such functional groups offsets the limitations imposed by low molecular weights, because they react more with the functional groups of melamine resins than with hydroxyl groups [162].

Combining the aforementioned low-molecular acrylic resins with aliphatic polyisocyanate adducts as crosslinker makes it possible to achieve a clearcoat solids content of 55 % by weight and more in the application state. Crosslinking with polyisocyanates generally generates better film properties than crosslinking with melamine resins. However, in that case, it makes sense to increase the hydroxyl number of the acrylic resins for the sake of optimum film properties. Of course, the use of polyisocyanate adducts requires equipment for the application of two-component materials.

As long as solventborne clearcoats are used for automotive OEM coating, the requirements and restrictions [163] concerning reductions in emissions of VOCs must be observed. The restrictions vary from region to region. This is the reason that two-component clearcoats together with water-borne basecoats are preferred in Europe. In the USA and in some Far Eastern plants, too, high-solid clearcoats together with high-solid basecoats are preferred. One reason for this difference may be the higher cost of two component clearcoats, not only of the ingredients but in addition also of the application equipment. To facilitate changing of the application process, it is necessary to install storage tanks for basecoats and hardener, and to install separate feed lines and two-component spray guns.

3.8.6 Water-borne clearcoats

Clearcoats for automotive OEM coating systems are still mostly solventborne. All other paint systems for OEM application are widely available as water-borne systems. The reason is that it is difficult to obtain clearcoat layers of the usual layer thickness that produce the familiar surface properties (appearance) of the solventborne clearcoats.

The first problem of water-borne clearcoats is to prepare films with thicknesses of 40 to 45 µm that do not sag or pop. It must be remembered that, for the sake of ensuring that such film thicknesses can be achieved, it is necessary to test the clearcoat at layer thicknesses up to 55 and 60 µm, as that will make certain that film failure does not occur during line application. Such high film thicknesses might be achieved on part of the car body, but they must not show suffer from popping or sagging.

The second problem is the nature of the polymer dispersion in aqueous phase. Colloidal particles of water-borne solutions and especially of dispersions do not fuse together as easy as particles of solutions in organic solvents. The results are poorer levelling and difficulties with achieving sufficient gloss. Since films prepared from water-borne systems are not as homogeneous as films prepared from solution, they are less resistant to chemicals and solvents. In addition, some water-borne systems have poorer storage stability than solventborne systems.

Water-borne one-component clearcoats

Water-borne one-component clearcoats contain secondary dispersions of acrylic resins and water-thinnable melamine resins. In addition, they may contain water-thinnable polyurethanes with blocked isocyanate groups for crosslinking. The clearcoats contain some cosolvents. Like, solventborne clearcoats, the additives are UV absorbers, HALS and levelling agents. The levelling agents are water-thinnable silicone oils or polyacrylates containing carboxyl groups neutralised by amines. The clearcoat formulation exhibits less application robustness than conventional clearcoats as there is a tendency to form blisters (popping) at high film thicknesses. While polyurethane supports chemical resistance, it may also promote yellowing. Such aqueous clearcoats have been introduced into the automotive OEM coatings market [164], but are not in widespread use.

Water-borne two-component clearcoats

Although it is well known that isocyanate groups react with water, it is possible to use polyisocyanates as crosslinkers for water-borne coatings. The crosslinking partners for the isocyanates are mainly secondary dispersions of acrylic resins, although water-thinnable polyesters are also suitable. Adequate weathering resistance is achieved with aliphatic and cycloaliphatic polyisocyanate adducts. Preferred are adducts of hexamethylene diisocyanate (isocyanurate trimers, allophanates) and, rarely, adducts of isophorone diisocyanate as such polyisocyanate adducts are not hydrophilic. It is believed that polyisocyanate dispersed in water will diffuse out of the first prepared droplets into the organophilic particles of the main resin, without reacting extensively with water. The combination of main resin and polyisocyanate in the colloidal particles can react and crosslink as usual. This process will be more efficient if the polyisocyanate adducts are easy to disperse. The effort involved in dispersing the polyisocyanates for effective reaction and crosslinking with the colloidal particles of the main resins is much higher than for organic solutions. The use of two-component application spray guns with a static mixing device will not lead to effective mixing.

The mixing properties of polyisocyanates can be supported in two ways. First, the polyisocyanate adducts are modified with building blocks that confer some hydrophilic behaviour. In most cases, these are ethylene glycol ethers [165]. Polyisocyanate adducts modified in this way are much more easily dispersed in water than the others. However, they also react more extensively with water to form amines that react with the remaining isocyanate groups to form ureas. Hence, the reaction with water involves two isocyanate groups that are no longer available for crosslinking reactions with hydroxyl groups. In theory, this is not disadvantageous to the final film properties since the additional urea linkages are resistant to weathering, chemicals and solvents. They confer hardness and flexibility. To allow for this side-reaction, it is necessary to increase the quantity of polyisocyanate to ensure that most hydroxyl groups of the main resin react with isocyanate so as to be part of the crosslinked network. Analysis has shown that roughly one third of the isocyanate groups react with water to form ureas. Thus the logical consequence is to use 150 % (by mol) of isocyanate, calculated on the quantity (equivalent) of hydroxyl groups on the main resin.

The second way is to employ very-low-viscosity polyisocyanate adducts. The lower the viscosity, the easier it is to disperse these in the aqueous phase. Since they are hydrophobic, they can diffuse relatively rapidly through the water phase into the colloidal particles of the resin containing hydroxyl groups.

Low-viscosity polyisocyanates can be produced in different ways: the polyisocyanate adducts are diluted in solvents which are immiscible with water. After having been admixed, the droplets of such solutions coalesce with the colloidal particles of the main resin, triggering the crosslinking reaction. The solvents may also act as levelling support. Of course, the solvent content must be calculated as VOCs. Adducts of polyisocyanates (e.g. isocyanurate trimers) with an especially narrow molecular weight distribution have significantly lower viscosities than those with broader distributions. Adducts of narrow molecular weight distribution are prepared under mild reaction conditions and the process is stopped at a very early stage. The excess, unreacted diisocyanate is distilled off. Polyisocyanate adducts that contain some allophanates have a much lower viscosity than other adducts (e.g. isocyanurate trimers). In this case, it is important to generate products with functionality far in excess of two per molecule. It is not necessary to choose an excess of isocyanate. From experience, the second method is more efficient than the first.

Films prepared by the aforementioned combination in water-borne clearcoats achieve nearly the same properties as those of solventborne two-component clearcoats. So far, such clearcoats have only been introduced into automotive OEM coating on a trial basis [166]. There are still problems: besides the difficulty of mixing both components efficiently, the products tend to form blisters (popping) in thick layers of clearcoat film.

Secondary dispersions with homogeneous particles

The experiences gained with one-component clearcoats and powder slurries (see Chapter 3.8.7) led to the preparation of secondary dispersions stabilised with suitable surfactants (emulsifiers). The products consist of acrylic resins containing hydroxyl groups and special blocked polyisocyanates [167]. Both components are mixed in the form of an organic solution and then converted into an aqueous dispersion. After that, the process solvents are stripped off. The completely solvent-free dispersions form clearcoat films with excellent weathering resistance and adequate chemical resistance. There is a much lower tendency during the application process to form blisters than is the case for the other water-borne clearcoat systems. So far, secondary dispersions containing the main resin and crosslinker homogeneously mixed in all particles have only found isolated use in automotive OEM application [168].

3.8.7 Powder clearcoats

Powder coatings are the optimum solution for reducing emissions of volatile organic compounds in coating applications. As with powder primer surfacers (see Chapter 3.5.7) and clearcoat application, a powder is converted into an aerosol (fluidised) that can be transported and sprayed. Spray application is effected with various types of electrostatic spray equipment (corona charging, tribo charging). The charged particles are fed by the electrical field and deposited on the earthed object. Film forming proceeds in a stoving oven by melting and crosslinking. So far, it takes a minimum 60 to 70 µm film thickness to achieve sufficiently smooth and glossy surfaces. Although the primary transfer efficiency of electrostatic powder application of about 50 % by weight is relatively low compared with that of electrostatically applied wet clearcoat materials with up to 80 % by weight, it is possible to recycle the overspray of powder coatings. When this recycling is factored in, the result is a transfer efficiency of about 95 % by weight. The problem with recycling powder coating material is the risk of contamination.

It is an advantage if thick coating layers do not evolve reaction products during film forming by crosslinking. Addition is therefore the chosen mechanism for crosslinking powder coating resins. For powder clearcoats, the reaction of epoxy compounds with carboxylic acids is preferred. Since conventional aromatic epoxy resins will not meet the requirement for adequate weathering resistance, epoxy acrylic resins are prepared. Epoxy acrylic resins consist of adequate quantities of a comonomer containing an epoxy group. Glycidyl methacrylate is virtually the only monomer used for preparing such acrylic resins. The acrylic resins are made by solution polymerisation. After polymerisation, the process solvent is distilled off by vacuum distillation, the resin melt is discharged, cooled and ground to a powder. The acrylic resin powder is mixed with crosslinker and suitable additives in a dry mixer. The mixture is passed through an extruder, is cooled again, milled very finely, and classified by different screening methods.

The crosslinkers for powder clearcoats are polycarboxylic acids (e.g. dodecanoic acid, azelaic acid) and polycarboxylic anhydrides. Anhydrides must be reacted first with hydroxyl groups to form free carboxyl groups that then react with the epoxy groups. The addition reaction between carboxyl groups and epoxy groups leads to the formation of β-hydroxy esters. For its part, the generated hydroxyl group can open anhydride structures to form esters and free carboxyl groups. The principle behind the addition reaction is shown in Figure 3.8.25.

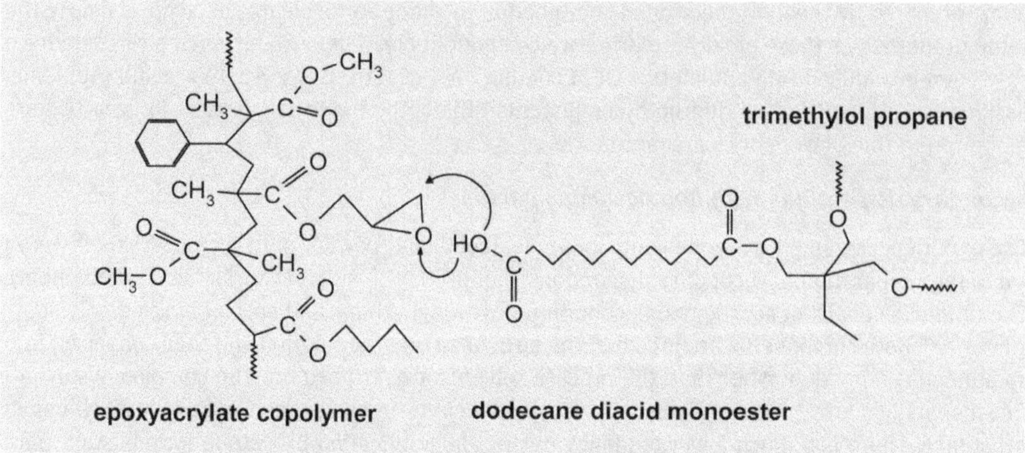

Figure 3.8.25: Principle behind the addition reaction of acrylic resins containing epoxy groups

The conditions for the reactions are 140 to 160 °C. The system is thus suitable for application as a clearcoat in automotive OEM coating lines as it meets the specified conditions. There are other powder coatings which need elevated temperatures, but they would have a negative effect on the structure of the OEM coating (over-stoving). Clearcoat films formed from the reaction products of acrylic resins containing epoxy monomer and aliphatic polycarboxylic acids or derivatives are distinguished by excellent weathering resistance, good resistance to chemicals, high hardness and gloss, and adequate yellowing resistance.

Formulation example

This example of a powder clearcoat [169] is an acrylic resin containing glycidyl methacrylate (epoxy equivalent weight of 660 g/mol), dodecane polyanhydride, with a potential carboxyl equivalent weight of 106 g/mol, and trimethylol propane as co-reactant (equivalent weight of 44.7 g/mol). The flux agent is benzoin. Other additives are a degassing additive and a combination of UV absorber (hydroxyphenyl benzotriazole) and a hindered amine light stabiliser (2,2,6,6-tetramethyl piperidine). The composition is described in Table 3.8.3.

The ingredients of the powder clearcoat are milled, blended, and then passed through an extruder (at a temperature of between 90 and 100 °C). The extruder product is cooled and milled in an impact mill and screened. The resultant powder has an average particle size of 30 μm (particle diameter). The powder clearcoat is applied to a basecoat (pre-dried for 10 min at 80 °C) by electrostatic powder spray gun and stoved for 20 min at 160 °C. The resultant clearcoat layer thickness is 60 μm. The powder clearcoat layer is notable for its optimum effect, high gloss, good fuel resistance, and intercoat adhesion. The scratch resistance and flexibility of such clearcoats still have to be improved.

In Europe, powder clearcoats have been on the automotive coatings market for several years, but are only used in few production lines. They are in competition with the water-borne systems.

Powder slurries

The possibility of obtaining high film thicknesses is often an advantage. However, for automotive clearcoat applications, the requirements are optimum properties with the thinnest possible layers, as that saves on materials, weight, and costs. There are different ways of making powder coatings which create an optimum appearance (flow, levelling, and gloss) at low film thicknesses. The first is to prepare powders with smaller particle sizes and narrower particle size distribution. Smaller powder particles can be achieved with special milling and screening equipment.

Table 3.8.3: Composition of powder clearcoat

Pos.	Components	Formulation wt-%
01	acrylic resin. containing epoxy groups. EEW 660 g/mol	73.3
02	polyanhydride of dodecanoic diacid potential COOH-EW 106 g/mol	20.3
03	trimethylol propane	1.6
04	benzoin	0.4
05	degassing agent	0.4
06	hydroxybenzotriazole	2.4
07	HALS	1.6
	sum	100.0

pos.
01 component A, example 1 patent DE 42 27 580 of methyl methacrylate, glycidyl methacrylate, styrene and n-butyl acrylate
02 component B, preparation of example 2 patent DE 42 27 580

pos.
05 Perenol F 40 (Henkel Cognis)
06 Tinuvin 900 (Ciba)
07 Tinuvin 144 (Ciba)

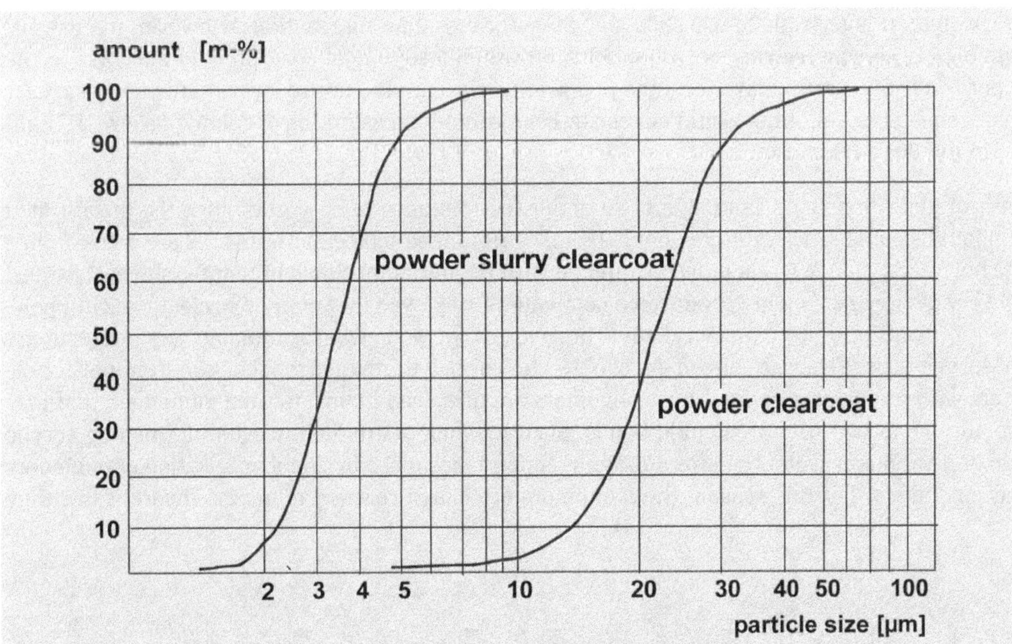

Figure 3.8.26: Comparison of the particle size distribution of a powder coating and a powder slurry

The process is very expensive. This method also has a limitation: Very small particles cannot be fluidized. In other words, they cannot form perfect aerosols for transporting or spraying.

Studies to resolve this problem yielded the process of wet milling. Powder coatings can be readily milled to much smaller particles to yield an aqueous dispersion that is called a powder slurry. The difference in particle size and particle size distribution between powder clearcoat and powder slurry is shown in Figure 3.8.26.

Application of powder slurry by conventional or electrostatic spray guns results - after film forming in the stoving oven - in clearcoat layers of optimum appearance and other properties at layer thicknesses of about 40 µm that correspond to the thicknesses of conventional clearcoats.

The general properties are comparable with those of the powder clearcoat since the same materials are used and the same chemical reactions for crosslinking take place. Besides the lower film thickness, there is another advantage relative to powder clearcoats, namely powder slurries can be cleaned more easily after contamination (filter processes). One disadvantage is that powder slurries are not as easy to recycle. Powder slurries were introduced into automotive OEM application lines some years ago [170]. Besides the aforementioned advantages, powder slurries are relatively expensive, due to the high effort required by and the cost of the two-step production process. Therefore, they have been unable to penetrate the automotive OEM coating market any further. The alternative to power slurries is clearcoats containing secondary dispersions with homogeneous particles.

3.8.8 UV clearcoats

UV light is that part of the electromagnetic spectrum with wavelengths between 100 and 400 nm. As electromagnetic radiation beyond the violet portion of the visible spectrum has shorter wavelengths, it is called ultraviolet light (UV). Electromagnetic radiation of short wavelength is energy rich; UV light has energies ranging from 12.4 to 3.1 eV. The bond energies of 6.3 eV for C=C double bonds, 4.7 eV for hydroxyl bonds, and 4.0 eV for C-H bonds lie within a range that can be activated by UV light. It is possible to use such activation for crosslinking reactions. However, the use UV light of very short wavelengths (100 to 280 nm) would lead to total absorption of that light in the upper levels of a coating layer and the penetration through the entire layer that is necessary for homogeneous crosslinking would not occur. For optimum crosslinking of coating layers, UV light of 315 to 280 nm is chosen.

Unsaturated compounds can polymerise under the influence of UV light. Since the unsaturated compounds contain two or more double bonds, crosslinked molecular networks are a possibility. The polymerisation reaction must be initiated with UV initiators. Some initiator systems also need UV-sensitising agents. The unsaturated compounds employed for automotive clearcoats are polyester acrylates and urethane acrylates. These terms are somewhat confusing. The products are not acrylic resins. They are oligomers of polyesters or polyurethanes that contain terminal acrylic esters with free double bonds. These oligomers are dissolved in unsaturated monomers that may contain one to four or more double bonds. Such monomers are mainly esters formed by acrylic acid and various polyols. The crosslinking reaction takes place by copolymerisation of oligomers and monomers. For this reason, the monomers are called reactive diluents. Theoretically, they

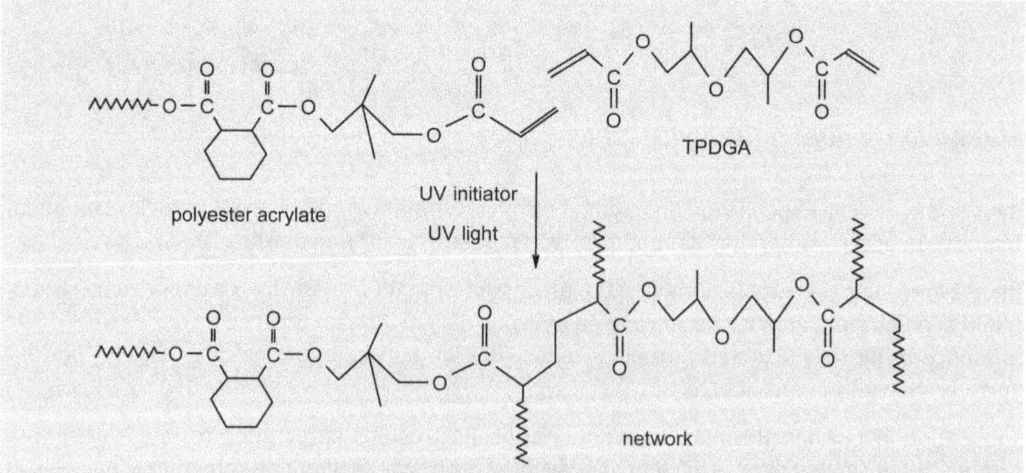

Figure 3.8.27: Principle behind UV crosslinking

do not evaporate, but are incorporated into the film matrix (molecular network). UV clearcoats are 100 % systems since they do not emit any VOCs. The principle behind the co-crosslinking of oligomer and reactive diluent is shown in Figure 3.8.27.

All initiators for UV crosslinking mainly contain the benzoyl structure. There are benzophenones, benzoin ethers, benzils, benzil ketals, α-hydroxyalkyl phenones, α-aminoalkyl phenones, and benzoin phosphine oxides [171]. Tertiary amines serve as UV sensitisers.

UV initiators and, where necessary, additional UV sensitisers are needed for starting the crosslinking reaction because the energy density of UV light is not great enough to start the polymerisation reaction directly. UV light is absorbed by the reactive groups of a UV initiator or a UV sensitiser, which become excited. The photosensitisers transfer the excited state to the molecules of the initiator partners. The excited initiator molecules generate free-radicals by decomposition or by abstraction of hydrogen atoms (hydrogen donor). The process is described in detail in the literature [172].

The initiator free-radicals initiate free-radical chain polymerisation with the unsaturated groups of the UV-clearcoat ingredients. The polymerisation reaction is comparable to that described for the preparation of acrylic resins (see Chapter 3.8.3.1). Again, the various reaction steps are initiation, chain propagation, and chain termination by recombination or chain transfer. Side-reactions which may interfere with UV crosslinking are inhibition by atmospheric oxygen, by deactivation of excited initiator molecules to generate light (fluorescence, phosphorescence), or by direct recombination of initiator free-radicals.

UV curing is very efficient. It not only affords a relatively high crosslinking density, but it is also believed that the molecular networks can be very extensive. Evidence of those network structures can be obtained by comparing the dependence of the modulus of elasticity on the temperature of conventional clearcoat (melamine resin crosslinking) with that of a UV clearcoat (see Figure 3.8.28).

Judicious choice of ingredients (oligomers, reactive diluents) yields clearcoat layers which exhibit an optimum balance of high hardness and flexibility, excellent chemical resistance, weathering resistance and very good scratch resistance. Some initiators can cause temporary yellowing. Clearcoats with such property profile would seem to be ideal for automotive OEM coating. Unfortunately, however, there are three problem areas:

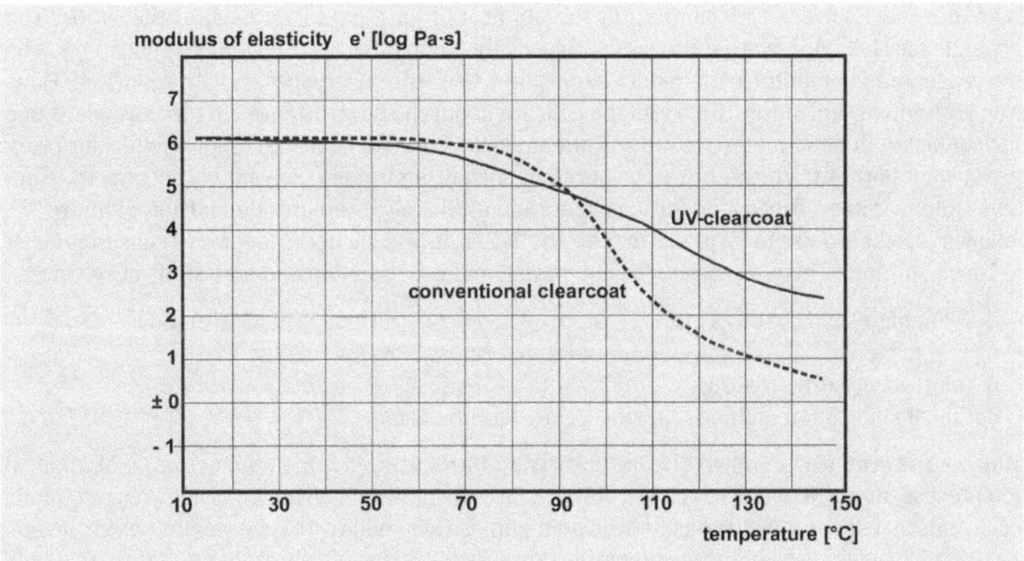

Figure 3.8.28: Modulus of elasticity as a function of temperature for a conventional clearcoat and a UV clearcoat

- inhibition by oxygen
- less crosslinking in folds, shadow zones, and on inner parts of the car body
- effort and expense involved in modifying the application line

Trials are underway to prevent inhibition by atmospheric oxygen through modifying resin components with building blocks that boost reactivity, for example modification with tertiary amines.

Other projects are seeking to replace the natural atmosphere with carbon dioxide so as to avoid the influence of oxygen. That would render crosslinking much more efficient. However, it is difficult to generate a carbon dioxide atmosphere dioxide for an application method involving a conveyor belt process and for objects of such complex dimensions as car bodies.

Interesting results have been obtained with a process that uses a plasma gas atmosphere [173]. The UV light generates free-radicals in the plasma gas that trigger the polymerisation process. There is no inhibition by oxygen. An additional advantage is that the free-radicals of plasma gas can penetrate into those parts which escape the reach of UV light. The outcome is efficient crosslinking over all parts of the car body, including shadow zones.

So far, trials have sought to combine different crosslinking methods in one coating material. Besides UV curing, conventional crosslinking reactions can generate acceptable film properties without UV light. One such example is a combination of UV crosslinking and crosslinking by polyisocyanate adducts [174]. Such products are called “dual cure systems”. The underlying assumption of dual cure systems is that it is mainly the outer part of the car body which must be resistant to weathering, chemicals, solvents and scratches and so the clearcoat is crosslinked by both reaction mechanisms. For the interior car parts, crosslinking by polyisocyanates only is assumed to be adequate. When the properties of such dual-cure clearcoats are tested, the outstanding results of clearcoats crosslinked solely by UV light are not achieved.

A key problem is that application of the UV clearcoat entails completely rebuilding the application line. Car producers are not prepared to expend time and money on that. Such systems may have a chance wherever new application lines are planned in a new plant.

3.9 Reductions in solvent emissions

Calls to reduce solvent emissions from the application of paints date back to the 1950s. The demands focused on industrial processes, especially automotive OEM coatings applications. The first reason was to minimise the risk of explosion when solvent vapours and air are mixed. However, around the same time, there was the concern about the hazards posed to humans who came into contact with paint materials and solvent emissions. At that time, the dipping tanks for primers (solvent-borne) were the primary source of solvent emissions. Solvent-borne primers were subsequently replaced by water-borne primers and ultimately by **electro deposition primers** [175]. Following the sharp rise in oil prices in the early 1970s, it became necessary to cut consumption of oil-based solvents. There are basically four ways to cut consumption and emissions of solvents:

- develop high-solid coatings
- use systems without any solvents (liquid film formers, powder coatings)
- develop water-borne coatings
- optimise application methods (increase transfer efficiency)

After water-borne electro deposition primers were introduced into the automotive OEM application process, **metallic basecoats** were now the largest source of solvent emissions. For example, a conventional silver metallic basecoat has an application solid of 13 % by weight, which means that 87 % by weight solvent must evaporate. Expressed in terms of volume fractions - with allowance for the different densities of the ingredients of such basecoats (aluminium 2.7 g/cm^3, resins

Table 3.9.1: Calculation of solvent emission of different basecoats

Metallic basecoats		Low-solid		Medium-solid		High-solid	
component	**density**	**wt-%**	**vol-%**	**wt-%**	**vol-%**	**wt-%**	**vol-%**
aluminium pigment	2.70	1.8	0.6	3.0	1.0	4.8	1.7
resin	1.20	11.2	8.3	19.0	14.6	30.2	24.3
solids		13.0		22.0		35.0	
solvent	0.85	87.0	91.1	78.0	84.4	65.0	73.9
sum		100.0	100.0	100.0	100.0	100.0	100.0
layer thickness			15 µm		15 µm		15 µm
g solid/m²			21.1		21.1		21.1
g solvent/m²			130.5		69.0		36.2

1.2 g/cm^3, and solvents 0.85 g/cm^3) - the figures are 8.9 % by volume solid and 91.1 % by volume solvents. Covering 1 m^2 surface with a basecoat 15 µm thick leads to emissions of 130.5 g solvent. Where transfer efficiencies from conventional spray guns are less than 60 %, emissions even exceed 218 g solvent per m^2. The application solid content was raised by selecting lower-molecular resins or alternative resins of optimum solubility along with suitable rheological additives. Medium-solid basecoats are the largest class of solvent-borne basecoats currently used in Europe. The application solid of a silver basecoat (comparable to the type described above) is about 22 % by weight [141]. Solvent emissions from such basecoats have fallen to 69 g/m^2, a figure which is nearly half that of low-solid basecoats (130.5 g/m^2). High-solid basecoats are primarily used in the USA, where application solid are 25 % by weight or even higher (for silver basecoats) [142]. The amount of solvent emitted from such high-solid basecoats is 36.2 g/m^2, or just 28 % of the figure for the low-solid basecoats previously employed. Table 3.9.1 lists formulations for different basecoats and the corresponding solvent emissions.

However, increasing the solid content of basecoats has given rise to some problems. Application robustness is lower than for low-solid basecoats. Optimisation of medium-solid basecoats which faithfully reproduce the colours and effects of the low-solid basecoats requires significant effort both in formulation and in the application process. High-solid basecoats offer less application robustness. It is no longer possible to duplicate the colour and effect of some very bright and light coloured, low-solid basecoats. In Europe, especially, where customers request such colours, calls to reduce solvent emissions have been increasing and have culminated in the development of water-borne basecoats [148]. Water-borne basecoats still contain some solvent, known as cosolvents, in addition to water as a liquid phase. The cosolvents boost the stability, application robustness, and levelling of such basecoats. Water-borne basecoats have a much lower solvent content than high-solid basecoats. However, the total solid content is comparable to that of medium-solid basecoats, for example 22 % by weight. The solvent content of such a water-borne basecoat (silver-grey) may be about 10 %. The calculated level of solvent emissions is 8.9 g solvent per m^2. That is just 6.8 % of the emissions from low-solid solvent-borne basecoats of the same colour. Table 3.9.2 compares the formulation for, and calculated solvent emissions, from a medium-solid basecoat and a water-borne basecoat. In addition to the optimum low-solvent content, water-borne basecoats that have the same solid content as medium-solid basecoats go a long way towards meeting the requirements on colour, effect, and application robustness.

The conventional **primer surfacers** for automotive OEM coating systems already had relatively high application solid of about 55 to 60 % by weight. The use of low-molecular resins paved the way to raising the solid content at application viscosity to 65 to 70 % by weight. The reduction in VOCs was not large. It was not until water-borne primer surfacers were developed that extensive reduc-

Table 3.9.2: Formulation and solvent emission of medium-solid and water-borne basecoat

Metallic basecoats		Medium-solid		Water-borne	
components	**density**	**wt-%**	**vol-%**	**wt-%**	**vol-%**
aluminium pigment	2.70	3.0	1.0	3.0	1.2
resins	1.20	19.0	14.6	19.0	16.4
solids		22.0		22.0	
solvents-1	0.85	78.0	84.4		
solvents-2	0.90			10.0	11.6
water	1.00			68.0	70.8
sum		100.0	100.0	100.0	100.0
layer thickness			15 µm		15 µm
g solid/m²			21.1		21.1
g solvent/m²			69.0		8.9

Table 3.9.3: Calculation of solvent emission of solvent-borne and water-borne primer surfacers

Primer surfacer		Medium-solid		Water-borne	
components	**density**	**wt-%**	**vol-%**	**wt-%**	**vol-%**
pigments	4.00	28.0	8.7	20.0	6.1
resins	1.20	32.0	33.0	25.0	25.4
solids		60.0	41.7	45.0	31.5
solvent	0.85	40.0	58.3		
co-solvent	0.90			10.0	13.6
water	1.00			45.0	54.9
sum		100.0	100.0	100.0	100.0
layer thickness			35 µm		35 µm
g solids/m²			87.7		85.6
g solvent/m²			41.6		13.5

tions were made. Although water-borne primer surfacers have only about 45 % by weight application solid, they contain only about 10 % cosolvent. Emissions drop from, for example, 41.6 g solvent per m² for a solvent-borne primer surfacer to 13.5 g/m², or to about one third. Table 3.9.3 shows the formulation of solvent-borne and water-borne primer surfacers and the calculated solvent emissions.

In the USA, the coating system for several types of passenger cars did not contain a primer surfacer layer. However, this was introduced with a view to optimising the overall film properties. The coating systems include primer surfacer layers that offer superior weathering resistance (to prevent under-chalking), and better appearance (optimum levelling and filling power). To avoid increases in VOC values, it was necessary to introduce a primer surfacer that contained no VOCs. This led to the development and application of powder primer surfacers (see Chapter 3.5.7) based on polyesters containing carboxyl groups, and crosslinked by epoxy resins [101,107]; these contain virtually no VOCs.

Conventional solid colour topcoats contain application solid of 45 to 55 % by weight, the precise figure depending on the colour shade. The solid content was increased to 55 to 65 % by weight, without any impairment of film properties, by selecting resins of lower molecular weight and optimum solubility. However, solid colour topcoat systems containing a solid colour basecoat and a clearcoat (two-layer systems) appeared that were comparable in effect to basecoat and clearcoat

Table 3.9.4: Compositions of the different clearcoats and calculation of solvent emission

OEM-clearcoat		Low-solid		Medium-solid		High-solid, 2 K	
components	density	wt-%	vol-%	wt-%	vol-%	wt-%	vol-%
resins	1.20	42.0	33.9	48.0	39.5	55.0	46.4
solvents	0.85	58.0	66.1	52.0	60.5	45.0	53.6
sum		100.0	100.0	100.0	100.0	100.0	100.0
layer thickness			40 µm		40 µm		40 µm
g solids/m²			50.4		50.4		50.4
g solvent/m²			69.6		54.6		41.2

systems. The two-layer systems offered better appearance and weathering resistance. However, introducing an additional coating layer would also increase the VOC values of the entire coating system if it contained solvent-borne materials. Therefore, it was necessary from the outset to introduce water-borne solid colour basecoats in order that the VOC requirements could be met. High-solid solid colour basecoats do not meet current requirements.

In the past, conventional automotive OEM clearcoats contained acrylic resins of average solution viscosity and high-molecular, more reactive melamine resins. The application solid were 40 to 45 % by weight. The application solid were increased to 48 to 50 % by weight through the use of low-viscosity acrylic and melamine resins. The use of HMMM resins can even push the solid content up to 55 to 60 % by weight. However, in that case, it becomes necessary to use stronger acids, e.g. sulphonic acids as crosslinking catalysts, and that has some disadvantages. These clearcoats are in competition, mainly in Europe, with two-components clearcoats that contain low-viscosity acrylic resins and aliphatic and cycloaliphatic polyisocyanate adducts as crosslinker (see Chapter 3.8.3.4). Naturally, the application of such clearcoats requires two-component spray equipment. However, these clearcoats feature greatly improved weathering and chemical resistance in comparison to clearcoats crosslinked by melamine resins. Such two-component clearcoats have application solid of 55 % by weight and more. Table 3.9.4 shows the compositions of various clearcoats and the calculated solvent emissions.

While one-component, high-solid clearcoats are preferred in the USA, water-borne clearcoats have been developed in Europe. Water-borne clearcoats offer more scope than high-solid clearcoats for reducing VOC emissions, even though they still contain some cosolvents. One-component clearcoats [164] and two-components clearcoats [165] have been developed as well. However, both technologies have so far failed to conquer the automotive coatings market. The main reasons are problems in application behaviour. They suffer from the tendency to form blisters (popping) during drying and crosslinking due to the anomalous behaviour of water as a solvent (high evaporation enthalpy). Studies on preventing popping through addition of specific additives have shown that other film properties can be adversely affected, e.g. gloss and levelling. These are the very properties that clearcoats are expected to meet very well.

If solvent emissions are to be totally prevented, one possibility is to use clearcoats containing liquid film formers. Such compositions are then UV cured. However, even UV clearcoats have some restrictions on use due to problems with application behaviour. These are:

- inhibition of effective crosslinking by atmospheric oxygen
- different radiation density on three dimensional objects
- shadow zones and interior parts are not reached by UV light

No automobile plant has so far installed UV-curing in an automotive OEM application line. However, many trials aimed at overcoming the aforementioned problems are underway [173].

Table 3.9.5: Reductions in VOC emissions from automotive OEM coating systems at the various development stages

Coatings system emission g solvent/m²	1 A-EDP LS primer surfacer LS basecoat LS clearcoat	2 C-EDP LS primer surfacer MS basecoat MS clearcoat	3 C-EDP MS primer surfacer W basecoat MS clearcoat	4 C-EDP W primer surfacer W basecoat HS clearcoat	5 C-EDP. new W primer surfacer W basecoat P/D clearcoat
primer 18 µm	10.0	2.0	2.0	2.0	0.0
powder surfacer 35 µm	50.4	50.4	33.4	14.4	14.4
basecoat 15 µm	130.5	69.0	8.9	8.9	8.9
clearcoat 40 (65) µm	69.6	54.6	54.6	41.2	0.0
sum	260.5	176.0	98.9	66.5	23.3

A-EDP = anodic electro deposition primer
C-EDP = cationic electro deposition primer
LS = low solid
MS = medium solid
HS = high solid
W = water-borne
D = dispersion, solvent-free

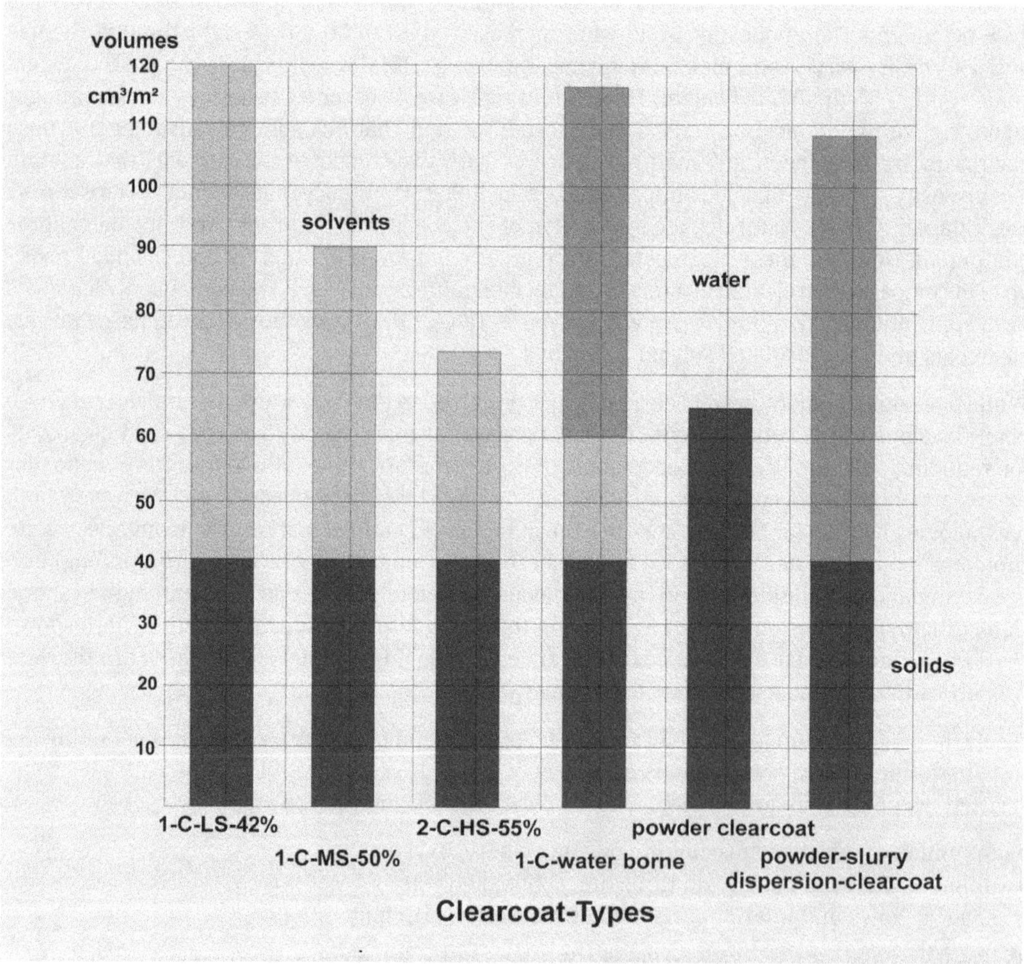

Figure 3.9.1: Composition of clearcoats and solvent emissions

There are also interesting alternatives in the guise of powder clearcoats. At least one major European car maker has introduced these into several assembly plants [176]. By contrast, powder clearcoats are not used in the USA, where they are considered to be inferior. The main problem of powder clearcoats (as described in Chapter 3.8.7) is that smooth and glossy films with optimum levelling are achieved only at relatively high layer thicknesses of more than 60 μm. This increases paint consumption and car-body weight. Trials involving complex and expensive methods for producing smaller particles yielded thinner films that were smooth and glossy, but fluidising such particles proved difficult.

The problem has been solved by preparing dispersions of powder clearcoats in water. Such dispersions make it easy to generate much lower particle sizes by milling. Powder slurries of this kind [170] yield optimum clearcoat films with a layer thickness of about 40 μm, equivalent to that of conventional clearcoats (see Chapter 3.8.8 and 3.8.7). However, the overall production process is complex and expensive.

Finally, solvent-free clearcoats based on secondary dispersions have been developed [177]. These secondary dispersions are notable for the fact that they contain homogeneous particles which contain all the film forming ingredients (resins, crosslinker, and some additives). Figure 3.9.1 shows the composition of the various clearcoat technologies, along with the respective solvent emissions at various film thicknesses (g/m^2).

Total solvent emissions from the various development stages of automotive coating systems from the 1970s to the present day are shown in Table 3.9.5.

All charts and calculations of solvent emissions presented so far are based on silver-grey basecoats (containing mainly aluminium pigments). Other colours and effects may yield other absolute figures, but the general trend is the same. They also ignore the possible complex geometry of coated objects and transfer efficiency. As already described (see Chapter 3.3), major steps have been taken to optimise spray equipment to increase transfer efficiency, and these will make an important contribution to reducing solvent emissions.

Other steps aimed at reducing VOC emissions seek to reduce layer thickness, e.g. of primer surfacers (see Chapter 3.5.7). Although there are regional differences in legislation around the world, there are various coating systems available that enable car makers to meet the different stipulations on VOC emissions [163]. Most legislative restrictions can be met with high-solid coatings. It is expected that legislative restrictions will be progressively tightened. Experience suggests that coating systems based on aqueous dispersions and systems with reactive diluents crosslinked by UV light have every chance of meeting future requirements, without loss of film properties.

4 Repair coatings

The task of repair coatings is to rectify application failures of the OEM coating process or to reproduce the coating layers after damage to car body substrates. This means reproducing on one hand the colour, effect, and appearance of the original coating layers and on the other the resistance properties of the original OEM coating. Nevertheless, because the application conditions under which repairs are performed are different, other paint materials must be chosen for repair coatings. While the primary goal of OEM coatings is to throughput high numbers of pieces in a given time and obtain high 'first-run-OK' rates, the goal of repair coatings is to faithfully reproduce the colour and effect of original coating layers. In most cases, more time is available for meeting this requirement. Unlike the OEM coating process, repair coating involves many manual interventions for meeting this demand. With the exception of manual application of repair putties, pneumatic spraying is used exclusively for all repair coating materials. The second reason for choosing other coating systems over OEM coating systems is that repair coating is performed on a fully accessorised car (fitted with interior parts, electronic equipment, engine and tyres). Exceptions here are attached car body parts, which may be coated separately. Since fully accessorised cars allow heating temperatures of 80 °C at most, it is impossible to use stoving coatings for repair coatings. Repair coatings are therefore physically drying systems or oxidatively drying paints, or they contain a resin combination which crosslinks at ambient temperatures. Currently the most important repair coating materials are two-component systems that react at ambient temperatures to form films having optimum resistance properties. The two components must be delivered separately and the pot life must be taken into consideration.

Repair coatings are mainly used in properly equipped repair shops or paint shops ("aftermarket repairs"). They also find application during car production (in-line repairs), whenever it is necessary to rectify coating flaws after OEM coating. If the car body has not been further accessorised, it can be returned to the application line and OEM-coated again with the same materials. Alternatively, it is transported to a separate special repair line. It is possible in this line to apply the OEM coating systems. However, since conventional stoving temperatures are usually unavailable, they must be catalytically crosslinked at lower temperatures (e.g. 60 to 80 °C). They may also be cured with IR radiation. The second possibility is to use the same materials that are used in the repair shops.

Repair coatings for the do-it-yourself market are also available. As home users do not have spray equipment, the DIY repair paints are supplied in aerosol cans. Manufacturers of such repair paints offer very large series of most automotive topcoat colour shades in use, ready mixed and filled into aerosol cans. These kinds of repair paints mainly contain physically drying resin systems, for example acrylic resins (TPA). Water-borne aerosol paints are now also available, mostly containing aqueous polyurethane dispersions in combination with acrylic dispersions. Some companies specialise in the production and supply of aerosol paints.

VOC regulations for repair coatings relate to the paint itself – in contrast to the regulation for OEM coating, which is related to consumption per unit time. This means that the paints may contain limited quantities of solvent in a paint system of application viscosity (grammes solvent per volume). The limitation varies from region to region. For the EU, for example, new limits came into force on 1 January 2007 [163] which mainly affect the composition of solvent-borne repair paints.

4.1 Surfaces and treatments

Optimum pretreatment is the secret to automotive repair coating. The workflow consists in:

- repairing dents by panel beating
- replacing body parts and attached parts (welding, bolting, bonding)
- cleaning damaged areas
- sanding the surfaces to be repaired
- masking the areas around the surfaces to be repaired

Beating out dents is laborious manual work. It is naturally impossible to obtain a perfectly smooth surface that way. The remaining uneven parts must smoothed with putty. On account of the huge effort involved, it is becoming more and more common to use spare parts (mudguards, bumpers, doors, bonnets, boot lids, hatchbacks) so as to circumvent the need for beating. Replacing body parts is the easier solution. Body parts made of steel are mostly welded. Other metal parts are bolted on or bonded with adhesives. In most cases, the attached body parts are already primed. However, the repair topcoat is applied after attachment of the part to the car body to ensure that the colour and effect match the surrounding surfaces.

Dents on shaped parts cannot be beaten out either and are smoothed with putties. For perform optimum adhesion of paint layers on metal body parts, the surface has to be cleaned carefully. All traces of oil and grease, rust and scale must be removed completely. There should be no economies in this regard. The best way to prepare surfaces for optimum adhesion of the repair coatings is to sand the entire area. The surfaces to which the first repair coating will be applied can vary considerably: bare metal, plastic, the primer used for attached parts, and different layers of the original coating.

Finally, the area surrounding the surface to be coated is masked. On one hand, the masking material must be resistant to the repair coating materials and process and, on the other, it must be easy to remove, without damage to the coated surface, after the repair coating has been applied. It is best to extend the area for repair coating to a joint in the structure. In that event, the very sensitive human eye does not perceive minor colour and effect differences.

4.2 Primers

Primers for repair coatings must provide optimum adhesion to the different types of surface mentioned above. In addition, they have to protect bare metal (steel) against corrosion. In the past, repair primers were based on oxidatively drying binders (e.g. linseed-modified alkyd resins). Active corrosion-protection pigments were also used. Oxidatively drying resins have very long drying times. Active corrosion-protection pigments may no longer be used in paints due to their toxicity and poor environmental compatibility. Currently, most repair primers consist of two-component coating systems containing low-viscosity **epoxy resins** (see Chapter 3.4.5.1) as resins and **polyamine adducts** as crosslinker. Primary and secondary amines react very quickly with epoxy groups to form tertiary amines at ambient temperatures. The principle behind the reaction is shown in Figure 4.1 (page 208).

The amine adducts consist of various amines: ethylenediamine, propylenediamines, diethylenetriamine, dipropylenetriamine and higher derivatives, isophoronediamine and bis-(4-amino-3-methyl-cyclohexyl) methane. Low-molecular amines have very low viscosities, but they have unpleasant odours, they absorb water during film forming, and they form carbonates by absorbing atmospheric carbon dioxide, possibly giving rise to "blooming". Therefore, like polyisocyanates, which are converted into adducts for use as hardeners, polyfunctional amines are transformed into oligomer products to circumvent these disadvantages. Amine adducts are the products of the

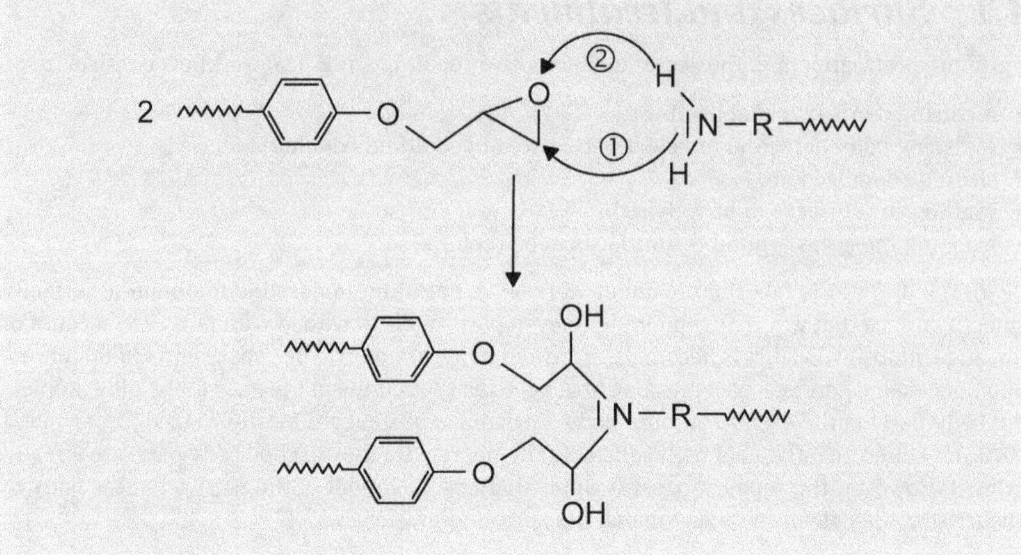

Figure 4.1: Crosslinking of epoxy resins by a primary amine

reaction between low-molecular epoxy resin and excess polyamine. One example is the product obtained from two moles of diethylenetriamine and one mole of the bis-glycidyl ether of bisphenol A. Another group of amine hardeners contains low-molecular polyamides bearing terminal and lateral free amine groups. Such polyamino amides are prepared by making polymeric fatty acids react with an excess of some of the aforementioned polyamines [178]. While the epoxy resin in repair primers bestows adhesion, hardness, and chemical resistance, the polyamino amides confer wetting, flexibility, hydrophilic behaviour, and low viscosity. The solvent for these types of primer are esters, alcohols, and aromatic hydrocarbons. Current repair primers are free of chromates. Apart from titanium dioxide as pigment and extenders, they contain zinc phosphates as active corrosion-protection pigments. Most primers are light grey and therefore contain small quantities of carbon black and yellow iron oxide. The base component and the hardener are mixed intensively just before application. The potlife is 4 to 6 hours. The resultant films are characterised by very good adhesion on different substrates, excellent corrosion resistance on steel, aluminium, galvanized steel and other metals, good chemical and solvent resistance, but poor weathering resistance.

Water-borne primers containing the same resin systems are also available.

A very special class of repair primers are the **wash primers**. These were invented in the Second World War, and were used to prime the metal superstructure of war-ships before the actual coating process. They had a very low viscosity and formed a thin layer that provided corrosion protection. They were easy to apply by long-handled scrubbing brushes used for washing, hence the name wash primer. Wash primers contain polyvinyl butyral as binder. Polyvinyl butyral is prepared from partially saponified polyvinyl acetate. The resultant hydroxyl groups are made to react with butyraldehyde to yield acetals. Figure 4.2 shows a typical structural segment of polyvinyl butyral.

The pigments in wash primers are titanium dioxide, extenders and talc (for excellent adhesion) and, most important, basic zinc chromates (zinc tetraoxy chromate, $ZnCrO_4 \cdot 4\ Zn(OH)_2$) as active corrosion-protection pigment. A specific quantity of phosphoric acid is added to the wash primer just before application. During film forming on the iron surface, mixed crystals containing zinc and iron phosphates and chromates are formed. The polyvinyl butyral participates in the forma-

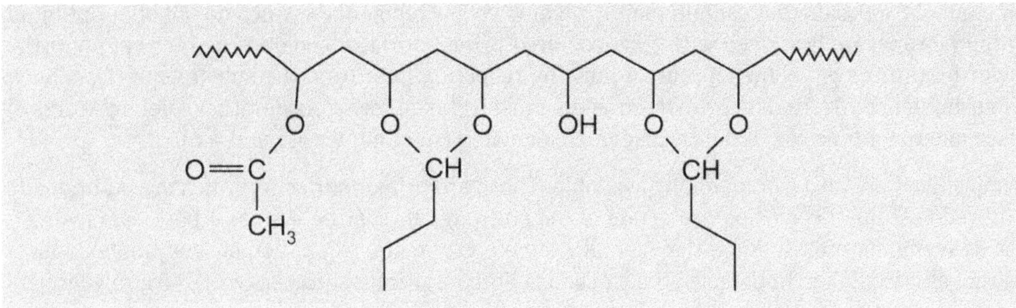

Figure 4.2: Typical structural segment of polyvinyl butyral

tion of the mixed crystals. The chemical bonding between substrate surface, pigment, and resin leads to outstanding adhesion on metal and optimum corrosion resistance, even in very thin layers. This is a precondition for optimum adhesion between subsequent layers.

However, chromate pigments have proved to be carcinogenic, especially zinc chromate pigments, which are partly water-soluble [179]. Special protection and safety measures are required when such pigments are handled. Wash primers are still used in special cases, mostly for providing optimum corrosion resistance in very thin layers. Thus, they find application as primers for aircraft fuselages, even though chromate-free products are available [180]. Because of these special applications, wash primers are still allowed to emit much higher solvent levels than other repair coating materials. Current legislation in Germany [163] stipulates 700 g/l, i.e. the application solids may be less than 30 % by weight.

4.3 Primer surfacers

If a repair coat system is applied to the sanded, original coating, there is no need to apply a primer surfacer. However, if the repair coat is to be applied to bare metal coated with primer, it makes sense to use a primer surfacer. In the past, primer surfacers also contained oxidatively drying alkyd resins. Nowadays, repair primer surfacers contain polyesters or hydroxy acrylic resins, and a polyisocyanate adduct as crosslinker, and are two-component systems. The use of polyesters (see Chapter 3.5.3.1) yields especially flexible primer surfacer layers. Acrylic resins (see Chapter 3.8.3.1) are preferred because they dry faster physically and offer scope for early recoatability. The polyisocyanate crosslinkers are based on aliphatic and cycloaliphatic polyisocyanate adducts, which have lower viscosities for application process and better weathering resistance than aromatic polyisocyanate adducts. These crosslinkers are also used in two-component clearcoats for automotive OEM clearcoats (see Chapter 3.8.3.4). Particular preference is accorded to isocyanurate adducts of hexamethylene diisocyanate. As repair primer surfacers have to form optimum film properties at ambient temperatures or at a maximum of 60 °C, catalysts must be used for optimum crosslinking. The catalysts are mainly organic tin compounds. The most common of these is dibutyltin dilaurate. On account of the long chain of the acid, the product offers ideal compatibility with resins and the film matrix. The catalyst is very effective, even in small quantities. Of course, such catalysts shorten the pot life. To strike a balance between reactivity and pot life, the optimum quantity of catalyst for all primer surfacer systems has to be tested individually. A pot life of several hours is normally required, but short recoatability times are also called for. The pigments used for repair primer surfacers are titanium dioxide and extenders. Coloured primer surfacers are used only in special cases to boost brightness and hiding power of following topcoats layers. Most repair primer surfacers come in shades of grey. Grey shades are obtained by adding small quantities of carbon black and yellow iron oxides as coloured pigments. The brightness of the subsequent topcoat systems dictates whether the dark or light grey primer surfacers are used.

For example, topcoats that contain mainly organic yellow pigments - which do not offer optimum hiding power - are best used with light-coloured primer surfacers colours, as these support the colour brightness of yellow topcoat systems by reflecting light from the substrate surface. Such colour influences exerted by primer surfacers must be taken into consideration when mixtures of base colours for topcoats and basecoats are calculated (see Chapter 4.5 and 4.6).

Primer fillers are also commercially available. These are primer surfacers which are additionally required to support corrosion protection of the entire repair coating system. Apart from hydroxy acrylic resins for crosslinking, they contain some epoxy resin. Aliphatic polyisocyanate adducts also act as crosslinker. In the past, the emissions limit for primer surfacers was 540 g solvent per l of paint; now it is 250 g/l.

4.4 Putties

Putties are used to fill out damage to metal car body parts. They must be used if the part is uneven and there is no possibility of replacing the metal part completely. Putties are paint-like fillers which can be applied in layers up to several centimetres thick. The main requirements on putties are optimum adhesion, the ready formation of smooth surfaces by manual application, sandability, and recoatability within a short time. Such requirements are not easy to fulfil with paint materials which contain solvents. Therefore, systems are chosen that contain resins and reactive diluents. During film forming and initiator-induced crosslinking, the resin reacts with the reactive diluent to form resistant film layers containing chemically bonded resin and solvent. The resins employed here are unsaturated polyesters (UP resins). The reactive diluents are vinyl compounds, mostly styrene so far.

Unsaturated polyesters

Unsaturated polyesters are members of the most important class of coating resins, polyesters. They have a similar structure and are prepared similarly to the saturated polyester resins described in Chapter 3.5.3.1. As their name suggests, they contain significant levels of unsaturated compounds, molecules which contain olefin double bonds. By this is not meant the double bond systems of monocarboxylic acids, namely fatty acids, which are suitable for preparing alkyd resins to yield oxidatively drying films (see Chapter 3.6.3.1). Rather, the suitable unsaturated building blocks are mainly polycarboxylic acids containing double bonds at the α-position of carboxyl groups or their derivatives. The most important compound is maleic acid (cis-butene diacid) or its anhydride. Other unsaturated polycarboxylic acids - which are mostly mentioned in patents - are itaconic acid, mesaconic acid and citraconic acid. There are also formulations based on fumaric acid, the *trans* isomer of maleic acid. However, fumaric acid has a very high melting temperature (287 °C under pressure) and is difficult to esterify completely. It must also be borne in mind that large quantities of the maleic acid are transformed into the isomer form (*trans* form) during the esterification process, i.e. fumaric esters are formed. Unsaturated polyester resins are made by combining the maleic anhydride with other dicarboxylic acids, for example phthalic anhydride, tetrahydrophthalic anhydride, and - for very flexible formulations - adipic acid. The other components of unsaturated polyesters are diols, e.g. ethylene glycol, propylene glycol, diethylene glycol, dipropylene glycol, and neopentyl glycol. Triols, e.g. trimethylol propane, are used only if unsaturated polyesters are modified with monocarboxylic acids. The average molecular weights of unsaturated polyesters are relatively low. They are obtained not so much by selecting a molecular excess of polyols as by limiting the degree of condensation; they therefore have high acid numbers. The reason for this is that hydroxyl groups of polyols can react with the double bonds of unsaturated compounds to yield branched molecules of high molecular weight and broad molecular weight distribution (Michael addition). This side-reaction is avoided by making the unsaturated polyester resins at relatively low temperatures, up to 180 °C, in contrast to the production conditions for saturated polyesters and alkyd resins. After the end properties have been attained, the resin melt is cooled down and dissolved in the reactive diluent. To prevent premature copolymerisation reactions during dissolution, small quantities of inhibitors, e.g. hydroquinone, methyl hydroquinone, or 4-tert.-butyl brenzcatechol, are added. The delivery forms of unsaturated polyesters are solutions in reactive diluents, mainly in styrene, with a solids content of 50 to 70 % by weight.

The putties are prepared by combining the unsaturated polyester solutions with high levels of pigments and extenders to yield a pasty product. Colouring is performed with titanium dioxide and small quantities of carbon black and iron oxides. Calcium carbonate, aluminium silicates, especially talc and kaolin, serve as extenders. The pigment and extenders are dispersed in "dissolvers" or kneaders. The type and quantity of extender varies with the rheological behaviour of the putty. The putties must be applied by knives, must spread easily and form relatively smooth surfaces, and must not run.

Chemical curing of putties consists in free-radical-initiated copolymerisation of unsaturated polyester together with reactive diluent (styrene). The initiators are peroxides, with benzoyl peroxide preferred. When used in conjunction with a redox reagent, peroxide will generate free-radicals at room temperature. Such redox reagents are mainly tertiary aromatic amines, e.g. N,N-dimethyl aniline. The generation of free-radicals is shown in Figure 4.3.

The free-radicals initiate the copolymerisation process shown in Figure 4.4 (page 212).

There is very high probability that free-radicals initiate the process at the maleic ester in the unsaturated polyester. The resultant free-radical absorbs a styrene molecule, and a new free-radical is formed. This free-radical starts a polymerisation chain by adding more styrene molecules. However, chain propagation is fairly limited, polymerisation is terminated by recombination reactions with other free-radicals. The outcome is a three dimensional molecular network containing polyester chains which are interlinked several times by short polystyrene chains.

Crosslinked UP resins are very hard, and offer resistance to chemicals and excellent resistance to solvents. As the reactive diluent is incorporated into the network structure by copolymerisation and cannot evaporate, thick layers can be effectively through-cured. There are no blisters or similar film flaws, which would occur when true solvents evaporated from film layers. The copolymerisation reaction is exothermic. There are virtually no solvent emissions, but some styrene does evaporate.

Where large areas have to be smoothed and the requisite film do not have to be very thick, spray putties (spray fillers) can be used. These contain the same ingredients as knifing putties. However, to achieve the low viscosity needed for the spraying process, they may additionally contain some real solvents. Of course, such solvents must evaporate efficiently during film forming, without causing any problems. The EU has imposed an upper limit of 250 g solvent per litre paint on such products since 1 January 2007.

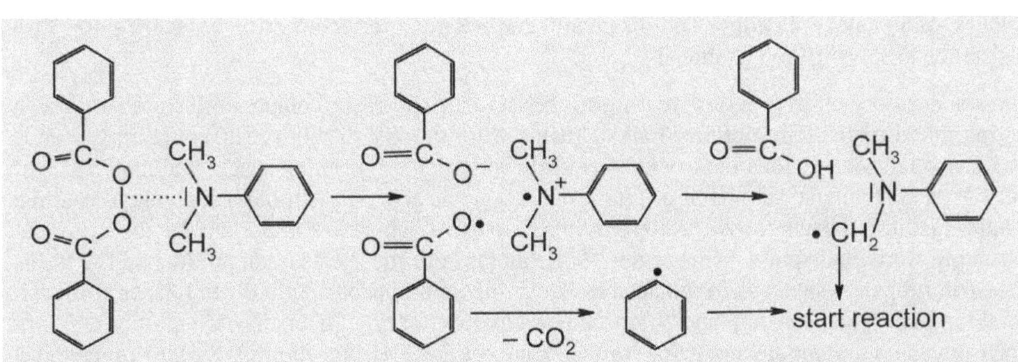

Figure 4.3: Generation of free-radicals by reaction of benzoyl peroxide with N,N-dimethyl aniline

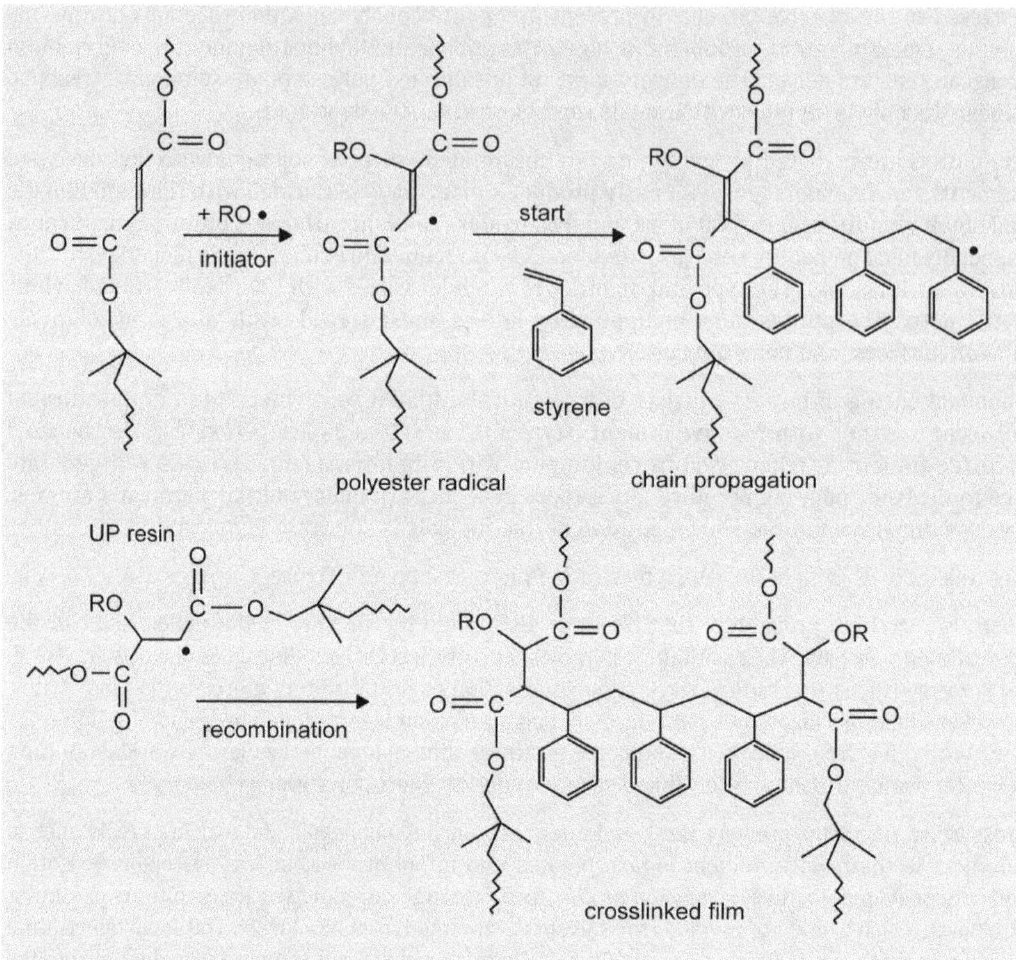

Figure 4.4: Copolymerisation of UP resin and styrene

4.5 Topcoats

Different systems were for repairing solid colour topcoats used in the past.The first were physical drying topcoats based on cellulose nitrate as main binder, combined with non-drying alkyds as plasticiser resins. In the 1920s, such systems were also used for OEM topcoats. The advantages of these 'combination enamels' were fast physical drying, high pigment loading leading to bright colours, and easy application. The disadvantages were poorer weathering resistance and little resistance to solvents and chemicals.

Their successors offered better weathering resistance. The resin components were oxidatively drying alkyd resins. The preferred alkyd resins were the fast-drying types, containing at most 50 % by weight non-yellowing fatty acids, mostly comprising 9,12-linoleic acid and modified with benzoic acids (benzoic acid itself or p-tert.-butyl benzoic acid) [86]. Although such alkyd coatings exhibit fast initial drying, the through-drying properties, which are important for recoating and assembly, were inadequate. Some regions still employ both these classes of product. In the 1970s, the earlier repair systems were replaced by two-component topcoats, initially in Europe. The resin combination of the two-components topcoats are hydroxy acrylic resins (see Chapter 3.8.3.1) and polyisocyanate adducts as crosslinkers (see Chapter 3.8.3.4). Acrylic resins were preferred to polyesters or alkyds because of their superior weathering resistance. However, the real advantage of acrylic resins lies in their far superior initial-drying properties in comparison to polyesters or alkyds. Acrylic resins also have disadvantages: the high density of colloidal particles in acrylic resin solutions confers inferior wetting properties, and coating films based on such resins have inferior filling power and poorer topcoat hold-out. These disadvantages were overcome by developing acrylic resins that behaved like alkyd resins. The most important acrylic resins based on this concept are modified with glycidyl esters of saturated fatty acids [116, 181]. The crosslinkers are low-viscosity aliphatic polyisocyanate adducts [182], mainly isocyanurate trimers of hexamethylene diisocyanate (see Chapter 3.8.3.4). Here, too, dibutyl tin dilaurate is the recommended catalyst.

The pigments used in repair topcoats are principally the same as those which are suitable for automotive OEM topcoats and basecoats. One of the preconditions for repair topcoats is faithful replication of the OEM colours. In addition, repair topcoats must provide the same weathering resistance and stability to other types of impact as OEM topcoats. However, making the right selection is not as easy as it seems: repair topcoats contain resins other than those contained in OEM topcoats and the basecoats may have different wetting properties. Repair topcoats are stored for much longer than OEM topcoats and so the pigment dispersion must be much more stable. Repair topcoats are prepared mainly from tinting base colours. Of course, when such tinting colours are mixed after being taken out of storage, they have to yield the same colour as freshly mixed material. Finally, repair topcoat systems must be capable of reproducing discontinued OEM colours. Repair topcoats require a much wider pigment range than is the case for OEM topcoats. Sometimes, it is better to choose other pigment types to meet the required colour requirement or to choose a type whose pigment particle surfaces have been specially after-treated. It may also be necessary to expend more effort on producing the optimum pigment dispersion or to use special additives for optimum pigment wetting and dispersion stability.

The same trend is to be observed for repair topcoat systems as for OEM topcoat systems. On account of the positive experiences gained with two-layer effect topcoats, solid colours for repair coating are also being applied in two layers, comprising basecoat and clearcoat. This displacement of one-layer topcoats by two-layer topcoats initiated in Europe and has spread to different extents throughout the various regions.

4.6 Basecoats

The introduction of two-layer topcoat systems comprising effect basecoat and clearcoat made it necessary to provide repair systems composed of effect basecoat and clearcoat. Repair basecoats followed OEM basecoats very closely in composition (see Chapter 3.7.3). For OEM application, the basecoat has to dry physically as fast as possible and be quickly coatable with clearcoat. Of course, repair basecoats must also have such properties. OEM basecoats can theoretically be used for repairs. Like OEM basecoats, solvent-borne repair basecoats mainly comprise the same resins as OEM basecoats, namely cellulose acetobutyrate and polyester or acrylic resins. However, they normally do not contain amino resins as it is highly unlikely that - under the repair conditions - crosslinking will occur between polyesters or acrylic resins and amino resins. However, melamine resin can support pigment wetting and increase the flexibility of basecoats. The rheological additives for repair basecoats are mainly the same as for OEM basecoats, i.e. polyethylene copolymer wax dispersions, polyamide dispersions, and low-molecular, crystalline ureas. Again, the coloured pigments and effect substances are the same as those which are suitable for OEM basecoats.

The process described above for OEM topcoats and basecoats by which series of pigment pastes or tinting colours are prepared is greatly expanded for repair basecoats. The production of coloured repair topcoats and basecoats is based on a system of tinting colours. First, the producer and supplier of repair topcoats provides different series of base colours which contain single pigments. This series covers all the pigments needed for mixing all colour shades for repair coating existing cars (both new and old). For effect basecoats, pastes containing the various effect substances must also be available. This series consists of approximately 50 base colours and several effect substance pastes. Suppliers of these topcoat and basecoat systems offer not only the various base colours, but also the mixing formulas for about 20,000 to 40,000 different topcoat and basecoat compositions. In accordance with this **colour-management concept**, suppliers of repair coats also offer electronic measuring equipment (colorimeters) for determining colour location and automatically calculating the corresponding recipe for the topcoat or basecoat. Furthermore, large suppliers of repair coatings offer their customers training programs on how to prepare and faithfully reproduce the colours of repair topcoats and basecoats with the aid of computer programs. The computer programs contain mixing formulas for all colours, listed by passenger car model and year of manufacture.

At this point, a hint on the special use of such tinting base colours should be given. The availability of such base colours has inspired persons with artistic ambitions. They blend various colours together to create fantasy hues so that they can customize their cars with figures, abstract paintings and decorations. They apply the paints with very small pneumatic spray guns called spray brushes.

Like automotive OEM basecoats, solvent-borne repair basecoats are the largest source of solvent emissions in repair coating. High-solid solvent-borne basecoats for repair coatings are particularly problematic in this regard. Since such basecoats contain lower quantities of physical drying resins and low-molecular combination resins, their drying properties are worse. And, if the repair systems are not stoved, crosslinking does not take place in the basecoats, and the resistance properties are also inadequate. This situation led to the development of water-borne repair basecoats. However, unlike the case for water-borne OEM basecoats, it took a relatively long time for water-borne basecoats to be launched onto the repair coatings market. There were different reasons for this. First of all, it is much more difficult to apply water-borne basecoats and reproduce the colour of the original coating. It is essential to perfectly maintain all the application conditions, more so than is the case for solvent-borne basecoats. Any variations in application conditions (throughput of paint material, quantity of spray air and auxiliary air, temperature, atmospheric humidity, air fall velocity) will influence the colours and effects of water-borne repair basecoats much more than for solvent-borne basecoats. However, there is more time available to apply basecoats than there is for OEM basecoats. More time can thus be allowed for basecoat spraying, this allowing a lower material feed and greater air feed (drier spraying), which contributes extensively to the desired effect.

Water-borne repair basecoats are more difficult to mix optimally and mechanical stirring equipment is needed (mechanised stirrers). It is particularly important to consider the stability of water-borne repair basecoats. The resins must be much more saponification resistant than those for water-borne OEM basecoats. The effect substances, especially aluminium pigments, must be stabilised against water (for lengthy periods of time, high pH values and probably elevated temperatures). The various suppliers of water-borne repair basecoats have met these requirements by adopting different strategies. They all offer multi-component systems. For example, there is a three-component system [183] comprising a series of aqueous tinting pastes of coloured pigments (1), a small series of water-thinnable effect substance pastes that contain only organic solvents (2), and an aqueous resin dispersion (3) for letting down all premixed formulations of the first two series.

The main resins for aqueous repair basecoats are polyurethane dispersions and secondary or primary acrylic dispersions. These are almost all exclusively aqueous systems that have been stabilised anionically or with surfactants. The rheological additives are aqueous dispersions of crosslinked acrylic resin (microgels), polyurethane thickeners, and water-swellable silicates with a layer crystal lattice.

Although the different components of the basecoat system (pigment pastes, effect substances pastes) still contain solvent, legislative restrictions on emissions reduction (VOC values) can be met [163]. On 1 January 2007, the VOC limit in the EU for repair basecoats was reduced to 420 g solvent per litre basecoat (application viscosity). As a result, water-borne basecoats - which were introduced tentatively at first - are currently spreading throughout the market in Europe and are set to become the standard systems. The VOC limit is not achievable with other systems. Solvent-borne repair basecoat are still in use in the USA and other countries.

4.7 Clearcoats

The most important demands imposed on repair clearcoats are optimum application behaviour (application robustness) and optimum resistance (durability). Repair clearcoats are applied to pre-dried basecoats and must form optimum film properties at ambient or elevated temperatures. Film forming can be forced by stoving for up to 30 minutes at 60 °C. IR emitters are also suitable for supporting film formation by repair clearcoats. Clearcoats must wet the basecoat surfaces perfectly without giving rise to any re-dissolving effects. They must provide optimum levelling and high gloss. Repair clearcoats must confer virtually the same properties as OEM clearcoats as regards weathering resistance, non-yellowing, resistance to chemicals, solvents, and mechanical impact, e.g. scratching.

In the past, repair clearcoats were used that were based only on physical drying acrylic resins (TPA, thermoplastic acrylics), but such clearcoats are unable to meet the aforementioned resistance requirements. Nowadays, nearly all repair clearcoats consist of hydroxy acrylic resins (see Chapter 3.8.3.1), which are crosslinked by polyisocyanate adducts. Repair clearcoats are two-component systems and are comparable to two-component clearcoats for automotive OEM application, which are described in Chapter 3.8.3. However, the choice of acrylic resins here is influenced by the other film forming conditions (lower crosslinking temperatures). For optimum resistance properties, the acrylic resins contain a balanced ratio of hard and plasticizing monomers. The resins have high hydroxyl numbers. The preferred crosslinkers are isocyanurate adducts of hexamethylene diisocyanate, which are more reactive than others and also generate lower solution viscosities for higher application solids. To maximise application solids, adducts of polyisocyanates are chosen that have a particularly low viscosity [182]. The low-viscosity isocyanate adducts are achieved by generating particularly narrow molecular weight distributions of isocyanurates, or by combining isocyanurate oligomerisation with the formation of allophanates.

The application viscosity of repair clearcoats is typically 60 s (ISO 4231, cup 4 mm, 23 °C). Current VOC regulations allow the mixture of base colours and hardener to emit 420 g solvent per litre coating material, i.e. the solids content must be higher than 60 % by weight. For optimum crosslinking at temperatures up to 60 °C, catalysts must be added. Until now, the preferred catalysts have again been organic tin compounds (e.g. dibutyltin dilaurate). It must be remembered that the use of tin compounds will be restricted in the near future.

As the VOC regulations can be adequately met at the moment, there has been little effort so far to develop water-borne repair clearcoats. However, there are other industrial applications which use water-borne two-component clearcoats that are based on secondary acrylic dispersions and hydrophilic polyisocyanate adducts.

Due to the fundamental concerns surrounding the use of isocyanates, attempts are underway to introduce other crosslinking mechanisms into the repair coat application (NISO or non-isocyanate technologies). Examples here are acrylic resins containing acetoacetate groups that are crosslinked by ketimimes [184] and acrylic resins crosslinked by siloxanes [185]. However, these crosslinking methods have not played any role in repair clearcoats to date.

It is likely that the market volume for repair coatings will decrease. This assumption is based on the observation that more and more car body parts are being made from plastic and that will obviate the need for repair coating. However, there is sure to be some demand for individual additional coatings and special designs in the future, as well.

5 Plastic coatings for automobiles

Plastic materials are used in both car interiors and exteriors. Interiors are usually fitted out with various plastic materials, as well as textiles (seats). Special parts are covered with leather or made of wood.

Most of the latest car models feature plastics only in specific areas. The amount of plastic used for exterior parts has been increasing continuously for several years. At the moment, plastic materials make up some 30 % of the average car exterior. There is a certain threshold to be crossed before the proportion can be increased further. From that point on, most of the car body would have be made from plastic (see Chapter 7).

Since ready-coloured plastic parts are available, it seems logical to choose parts that do not need coating. However, there are several reasons to coat plastic parts, too. Requirements on plastic parts are low cost and ease of processing, e.g. by injection moulding. In addition, plastic parts must be recycled easily. These requirements are mainly met by thermoplastics. However, most of these have poor weathering resistance. Special thermoplastics are available that offer better weathering resistance, but they are relatively expensive. Furthermore, the demand for a car of totally uniform appearance requires that coatings also be applied to plastic parts, so that they match the metal parts in terms of colour and effect. Additionally, it should be impossible to distinguish (by colour or effect) between the different types of construction materials, e.g. metal and plastic. A very important reason not to use coloured plastic parts is that the same appearances and effects of coatings cannot always be reproduced on such plastic parts.

5.1 Coatings for interior parts

Car interiors contain a lot of different plastic parts: the entire dash-board, glove compartment (front and rear), door trim, grips, winders, transmission or manual gear box covering, wheel arch covers, seat shells, frames and upholstery, overlays, steering wheel and cover, and ash-trays.

Suitable plastics for these parts are polyamides, polyurethanes, polyesters (polyethylene terephthalate), acrylonitrile-butadiene-styrene copolymers (ABS), and polycarbonates. Specialty parts consist of polyacrylates (covers for light sources and display units), polyphenylene oxide (loudspeakers, heater grilles and air conditioning), and phenolic resin (ash-trays). The parts for interior equipment are produced by subcontractors, some parts being already coloured while others are already coated. The coating materials must match the properties of the plastic parts and fulfil the requirements on coating films as well.

Requirements

The requirements on the resistance of coatings for interior parts are naturally less critical than for exterior parts. However, some special conditions have to be met. Most plastic parts have limited heat resistance. Like repair coatings, stoving coatings are unsuitable for plastic parts. Therefore, coating materials consist of physical drying resins or of resin combinations that crosslink at ambient temperatures, i.e. two-component systems.

Car interiors are usually outfitted with primarily complex three-dimensional **moulded** parts [186]. There are several methods to combine the moulding process with the coating process. For example, plastic film, artificial leather and natural leather are inserted into the mould, and then the plastic part is produced by in-mould foaming. Ready-coated films can be used, and it is also possible to apply the coating materials after the part has been removed from the mould. While there are no specific demands for weathering resistance, coatings for interior parts must be resistant to **chemicals and solvents** (e.g. food, cleaning agents, and sweat) and mechanical impact.

However, one of the key requirements is that the coating materials meet the **flexibility** properties of the plastic objects themselves. Only a few colours are chosen for interior parts, namely black, charcoal, or grey, with lighter colours used only for special items (e.g. red or white leather). The current trend is that of "**soft touch**", which closely resembles the feel of natural leather. Appealing surface structures can be obtained with special extenders or filler pigments. Although there are different possibilities, there is also a tendency to imitate the grain structure of natural leather. An additional requirement on coating materials for interior parts is that of environmental safety. The coating films must not exude any hazardous or odorous materials during the entire life-time of the products. They should not generate any allergic effects in contact with human skin. As people respond differently in this regard, it is very difficult to guarantee the harmlessness of coatings. The exclusion principle is applied here: materials about which there is any doubt whatsoever are avoided.

Products

Currently, most physical drying coating materials for interior parts consist of **polyurethanes** (see Chapter 3.5.6.1 and 3.7.5.1). Suitable polyurethanes contain flexibilising soft segments and have relatively high molecular weights (TPUs or thermoplastic polyurethanes). The latter would yield low application solids in solvent-borne coatings and thus relatively high solvent emissions. Preference is therefore given to water-borne coating systems based on aqueous polyurethane dispersions for the coating of plastic parts.

The crosslinkable coating systems for plastic parts contain polyesters, polyethers, and acrylic resins, all bearing hydroxyl groups. The crosslinkers are aliphatic polyisocyanate adducts (see Chapter 3.8.3.4). All resins must contain building blocks with plasticizing properties in order that the special requirements on flexibility in combination with plastic parts may be met.

Some prestigious cars contain interior parts of natural leather; earlier models even had leather seats. Natural leather is coated with materials which are similar to those for plastic coatings. Some car models also contain wooden parts. These are door rails, covers of glove compartments, and the upper surfaces of wooden dash-boards. They are most commonly made from wood that has been burled to create a pronounced grain structure. The wooden parts are coated with a glossy thick-layer clearcoat. Clearcoats contain unsaturated polyester resins and styrene as reactive diluents (see Chapter 4.3). Such clearcoats were also used in the past for high-gloss furniture coatings. To replace such materials with more environmentally sound products, the latest dash-board coatings are prepared from aqueous polyurethane dispersions.

5.2 Coatings for attached parts

Car exteriors made from plastic materials include bumpers, front-ends, mudguards, boot lids, sun roofs, trim, exterior mirror housings, wheel rim covers, and headlight covers. There are also plans to replace some windows with plastic parts (polycarbonate; initially for rear side windows).

Materials

Plastic materials used for exterior parts are polypropylene, polyamide, polyurethane, polycarbonate, and glassfibre-reinforced plastic (SMC, sheet moulding composites) which are made from unsaturated polyester resins [187].

Requirements

The requirements on coatings for outer plastic parts correspond to the demands on metal coatings [188]. The coating films must be resistant to weathering, chemicals, solvents, and mechanical impact (above all, scratch resistant). Since plastic materials cushion the deformation energy in a totally different way to metal objects, the coating films must be adapted accordingly. Plastic coatings are therefore submitted to special impact tests. Foremost among these are puncture tests at different temperatures, primarily at low temperatures (-20 or -40 °C). Even if the coating layers were to be damaged by high impact energy, the layers must not detach from the surface totally (loss of adhesion). The coated plastic part should have the same flexibility properties as the non-coated part - this is a very ambitious requirement. The interface between plastic part and coating layer alone creates an inhomogeneity that reduces the elastic response to mechanical impact.

Primers

Since plastic materials do not have to be protected against corrosion and it is possible to render plastic surfaces very smooth, it is not normally necessary to use primers to coat plastic parts, but they must be protected against weathering and chemicals. However, there are some important exceptions. The surface of **SMC parts** is very problematic. Reinforcing crosslinked UP resins with glass fibres leads, on one hand, to the formation of blowholes, notches and other inhomogeneities and, on the other, to the possible formation of gas bubbles after coating if air escapes from any of the voids in the material. It is therefore vital to pretreat such SMC parts with primer-sealers to fill and cover all the holes and bubbles, and prevent degassing during or after coating application. After the application of such primer-sealers, a primer surfacer is applied to smooth out all remaining unevenness to prepare the surface for the finish topcoat application.

Polypropylene is also problematic for coatings applications due to its very non-polar molecular structure. The surface of polypropylene is not easy to wet by coating materials, and there are virtually no polymers (resin) that adhere to polypropylene perfectly. The only exception is chloro-sulphinated polyethylene, which adheres to polypropylene very well. This fact is a good model for the theory of adhesion, which postulates that the widely known polar molecular interactions are accompanied by non-polar molecular interactions. Chloro-sulphinated polyethylene has a relatively high molecular weight and dries only physically. In the application state, solutions of chloro-sulphinated polyethylene have very low solids, yielding high VOC values. Suitable solvents for the polymer are chlorinated and aromatic hydrocarbons. In the past, **adhesion primers** based on the polymer were available. Many trials have been designed to replace these solvent-borne adhesion primers with more environmental friendly products, but the perfect solution is still to be found. Since polypropylene confers optimum properties, is relatively inexpensive, easy to handle (in moulding processes), and has excellent flexibility and toughness, additional effort has been expended on managing the adhesion problems. These have led to **flame-treatment** and **plasma treatment** [189]. Such methods generate oxygen-containing compounds on the surface of polypropylene (e.g. carboxyl groups), which increase the polarity of the surface significantly and offer the possibility of coating such surfaces with different materials so as to confer sufficient adhesion. The pretreated parts are coated with normal paint systems (primers and topcoats).

The third reason to apply primers to plastic parts is to offer the ability of using electrostatic application methods for topcoats. Since plastics are nearly all dielectric media, they make unsuitable conductors for electrical fields. Therefore, a **conductive primer** is applied to generate a minimum of electrical conductivity. Primers are rendered conductive by means of special pigments. The pigments chosen are metal pigments and special carbon blacks. The carbon blacks consist of nanosized particles of graphite, which is conductive. Primers containing minimum pigment volume concentrations of such carbon black pigment particles are sufficiently conductive to be suitable for subsequent electrostatic application methods. The plastic parts prepared in this way are coated with topcoats by means of electrostatic spray guns, yielding optimum transfer efficiency for coating materials.

Primer surfacers

Since plastic parts usually have a smooth surface, it is not necessary to use primer surfacers. The exception here is SMC parts, as already mentioned.

Topcoats

Topcoats for exterior plastic car parts must match the colour and effect of the topcoats on metal parts. Most plastic parts are not heat resistant and so stoving coatings cannot be applied. Therefore, it seems obvious to use the same coating materials as for repair coatings (see Chapter 4.5). In addition, the specific requirements of high flexibility must be taken into consideration. Nevertheless, the topcoats of exterior plastic parts must provide the same durability as the topcoats on metal parts: sufficient resistance to weathering, chemicals, solvents and mechanical impact, including scratching.

One-layer solid colour topcoats contain flexible polyester resins or flexibilised acrylic resins, both containing hydroxyl groups and crosslinked by aliphatic polyisocyanate adducts; they are two-component systems (see Chapter 3.8.3.4). As the film forming conditions are nearly the same as for repair coatings it makes sense to use catalysts, too.

As is the case for OEM topcoat systems on metal surfaces, the solid colour topcoats for plastic parts are currently applied as a two-layer system of basecoat and clearcoat. The formulations for basecoats are similar to the resin combinations for repair basecoats (see Chapter 4.6). Clearcoats for plastic parts must be particularly flexible. They contain the aforementioned resins. Polyesters outperform acrylic resins in terms of flexibility - but so long as they contain aromatic building blocks, they are not as weathering-resistant as acrylic resins and tend to yellow more after exposure. Flexibilised acrylic resins, which naturally do not contain quantities of aromatic monomers like styrene, are more weathering and yellowing resistant, but they are not as flexible as suitable polyesters. Particularly suitable for plastic clearcoats are acrylic resins modified by ε-caprolactone, which means they contain polyester side-chains. Two effects are at work here: the hydroxyl groups positioned on the ends of the polyester side-chain are much more reactive than the hydroxyl groups near the acrylic polymer chain, which leads to more efficient crosslinking. The segmented structure of the modified resins, containing hard acrylic main chains and flexible polyester side-chains, offers the possibility of a balance of flexibility and toughness for clearcoat films made therefrom.

In the meanwhile, plastic coatings are made not only from water-borne basecoats, but also from water-borne clearcoats. The water-borne clearcoats contain secondary acrylic dispersions; the crosslinkers are specialty polyisocyanates, more especially, either types of very low viscosity or polyisocyanates with hydrophilic modifications (see Chapter 3.8.5). Starting from Europe, water-borne two-component clearcoats for plastic coatings will replace solvent-borne clearcoats step by step in other regions as well.

A particularly big challenge is the **colour management** of plastic topcoats. As already mentioned, there must be no difference in colour or effect on the different exterior parts of car bodies. For some time, car customers prefer to have all body parts and attached parts (e.g. even mirror housings) in the same colour shade and effect as the rest of the car. This requirement is unlikely to change much in the future. Not only that but the construction joints are made as narrow as possible to convey the impression of body surface homogeneity, even for small cars. This may be the reason that small colour differences are easily noticed since the human eye is very sensitive when it comes to comparing colour shades. To achieve the goal of minimising differences in shade, it is not enough to choose the same pigmentation for all topcoat systems (basecoats for metal part and plastic parts). First of all, the two coating systems may contain the same types and quantities of pigments but different resin systems, which confer different wetting properties on pigments and effect substance, and thus lead to differences in shade. However, the most important influence is the different film-forming behaviour. Metal parts and plastic parts have totally different specific heat capacities and the film forming conditions (mainly the temperatures) are totally different. Both conditions influence the viscosity during film forming and thus the effect substance orientation, along with the re-flocculation of pigments dispersed in the paint.

There are principally two ways to manage these problems: The first is to adjust the formulation of the different systems - i.e. besides the different resin compounds, to use different recipes, quantities and, in extreme cases, different pigment types or effect substances, and production processes - with the aim of achieving the same colour and effect on all parts, even though the film forming conditions differ. These demands can be met with the help of computer-aided colour analysis and formulation guidelines for standard tinting coatings. In addition, there is the possibility of producing large series of attached parts for car bodies and to choose individual pieces from the series on the basis of best colour match. The second strategy is based on on-line application of attached parts. All attached parts, primed only, are fitted to the car body, and the entire car body is coated with the same material. Of course, the chosen coatings must confer universal properties. The must meet the relatively high demands on flexibility matching that of plastic parts, and also meet the resistance requirements imposed for OEM coatings on metal parts of car bodies. In particular, it is very difficult to balance the flexibility and solvent-resistance properties (e.g. low swellability) of such universally applied topcoat systems. Since elevated temperatures may promote crosslinking to meet the mentioned requirement combination, the plastic materials chosen are more resistant to elevated temperatures (e.g. polycarbonates, polyamides, and SMC), a fact which means that stoving coatings may be used. Unfortunately, such plastic materials are more expensive (material, production process, moulding process) than other plastic grades. On the other hand, the automotive industry has demanded for some time that the stoving temperatures of all OEM coatings be reduced with a view to saving on energy costs and to rendering the application methods more versatile (to enable earlier use of compounds and materials which are less stable). Therefore, one of the key objectives for application of OEM coatings besides reductions in VOC emission is to provide coating materials which on one hand are stable in storage, but on the other are able to crosslink efficiently at lower temperatures than is now the case, will save on energy and can be applied to all the various substrates found in car bodies. Such universal coating systems must bestow all the necessary properties (colour and effect reproduction, resistance to weathering, chemicals, solvents, water, and mechanical impact).

Radiation curable coatings are also used for plastic substrates. Until now, such coating materials have only been used for flat objects and not for pigmented systems. There are clearcoats available which crosslink in UV light or even electron beams. Important examples are clearcoats for wheel covers. In addition, front light reflectors are protected with UV coatings. This is also planned for plastic windows (polycarbonate) (see Chapter 3.8.7).

6 Coatings for heavy loader

By heavy loader here is meant buses, trucks, transporters, and those vehicles which can be classified as vans. Vans started gaining in popularity in the USA some years ago and they are increasingly to be found on the streets of Europe. In most cases, large vans are coated separately. Small vans are now fitted like normal passenger cars and also coated like passenger cars.

From the point of view of coating systems, the term heavy load vehicles extends not only to buses and trucks but also to agricultural machinery, tractors, construction machinery (e.g. cranes), municipal vehicles and, in a broader sense, rail vehicles (wagons for tram lines and trains, railcars, and other traction machines), although some of these would not be classified as automotive vehicles.

It is true, however, that all these vehicles are produced in lower numbers than passenger cars. Therefore, coating applications are quite different. Parts of the vehicles are often coated separately and then assembled. In most cases, they are at least primed. The parts to be coated are mainly made of steel, galvanised steel or aluminium. Trim and similar objects, for example, the driver's cabs, are also made from plastic.

Production and supply of coatings for heavy load vehicles are not only in the hands of producers of OEM coatings for passenger cars. There are many small companies which produce such paint materials. That is the reason why there are so many different types of coating materials. However, mostly in Europe, producers and suppliers must comply with VOC regulations [163]. Therefore, high-solid coatings, water-borne coatings and even powder coatings are available for this segment. In addition, this industrial sector is under massive cost pressure, more so than other coating technology segments. This has consequences for the formulation and composition of coatings in that application field.

Requirements

The requirements on coatings for larger vehicles concern the durability of the objects, even after particular exposure. The most important durability property is corrosion resistance. After that, the coating films must be resistant to chemicals, solvents and water. Only in a few cases is decoration a consideration.

Application

Not only must the coating materials for larger vehicles be economical, but the application process must also be simple (easily to apply), efficient and cheap. One way to meet these demands is to maximise transfer efficiencies. The coating materials are therefore applied by means of airless spray equipment that offers very high transfer efficiency (see Chapter 3.3). Pneumatic spray guns are used only decorative effects are required, for example, for coloured topcoats. Since most objects are relatively large, common stoving coatings cannot be used; however, very often film forming at elevated temperatures is possible. For this technology segment, physical drying paints, oxidatively drying paints and two-component paint systems are in use. Most of the coating systems are comparable to the systems which are used for repair coatings (see Chapter 4). Small metal parts are primed with electrodeposition primers (see Chapter 3.4.6) and then coated with powder topcoats applied by electrostatic spray guns. As already mentioned, both coating systems form films in stoving processes at relatively high temperatures.

Primers

Of course, the primary task of primers is provide corrosion protection. Substrates such as steel, galvanised steel and aluminium must be protected. In this technology field, oxidatively drying corrosion protection primers are still in use. These consist of **alkyd resins** modified with fatty acids containing two double bonds and more; the most important ones employed are linseed fatty acids. Besides coloured pigments, the primer surfacers contain extenders, which mainly confer corrosion protection by forming a barrier to water and moisture. However, active corrosion-protection pigments, e.g. zinc phosphates, are also employed. It is necessary to add siccatives to catalyse the oxidative film-forming process. Most of the products are solvent-borne. To meet the current VOC regulations, high-solids versions are available which achieve application solids of 65 % by weight and higher.

Optimum products for this application field are two-component primers containing aromatic epoxy resins (see Chapter 3.4.5.1) as binders and polyamine adducts or polyamide amines that still contain free primary or secondary amine groups as crosslinkers (hardeners). The same type of primer is used as that for repair coatings already described in Chapter 4.2. Also available are water-borne primers based on the same resin groups. These contain epoxy resin dispersion and water-tolerable amine components. Although the corrosion protection afforded by such primers is excellent, there are some disadvantages, namely the products are not easy to sand and the films are not weathering-resistant. Parts of large vehicles which are primed, assembled later and then coated with topcoats, require a minimum weathering resistance for storage and transport until assembly is started. Such elements are primed with two-component systems which contain saturated polyesters bearing hydroxyl groups and polyisocyanate adducts as crosslinkers. It is possible to choose polyesters which confer higher hardness and also a suitable pigment combination; this renders the films much easier to sand. Water-borne systems containing the aforementioned product classes are currently available.

For special parts of large vehicles, thick-layer primers are applied for optimum corrosion resistance; these parts include such building elements as frames for the vehicle floor, wheelhouses, and floor panels. In such cases, solvent-borne systems fail because they do not form homogeneous films when applied in thick layers. The plastisols suitable for thick-layer coatings of passenger cars (see Chapter 3.4.8.1) require too high a temperature for optimum film forming (gelation), which is not available for the application conditions here. Therefore, liquid 100 % systems are applied that contain liquid polyols (polyether polyols, as well as castor oil), which are crosslinked with liquid aromatic polyisocyanates, mainly MDI derivatives (highly functional isomers and derivatives of 4,4'-diisocyanatodiphenylmethane). By way of alternative, thick-layer systems can also be prepared with the aid of aqueous dispersions containing epoxy resins and water-tolerable amine adducts. All the aforementioned products are applied by airless spray methods.

It has already been mentioned that small parts for large vehicles may be coated first with electrodeposition primers. It is possible to choose the same materials as are used for the OEM process for passenger car bodies that contain amine-modified, plasticised epoxy resins and blocked polyisocyanates as crosslinker (cationic ED primers, see Chapter 3.4.5).

Finally, there are also water-borne one-component primers for small objects that contain mainly water-dilutable epoxy esters and urea resins as crosslinker and have to be stoved.

Putties

After assembly of large vehicles from primed parts, it may be necessary to smooth structures such as joints, weld seams, and bolted or riveted joins. Suitable for this purpose are the putties that have already been described for repair coatings (see Chapter 4.3).

Primer surfacers

For the objects described here, primer surfacers must be applied where high film thickness is required for obtaining optimum barrier effects to improve corrosion resistance, or if the entire coating structure has to be especially smooth, which is mostly the case for decorative effects. In all other cases, the topcoats are applied directly to the primer.

Suitable primer surfacers here are two-component systems; the compositions are comparable to those of primer surfacers for repair coatings (see Chapter 4.4). The primer surfacers contain saturated polyesters and specific quantities of epoxy resins while the crosslinkers are polyisocyanate adducts. Since such primer surfacers are occasionally also applied to bare metal parts (e.g. on through-sanded parts, joints, bolts and rivets), they must also provide some corrosion protection; that is the reason for using specific quantities of epoxy resin. They also contain some passive corrosion-protection pigments.

For some coating structures, the primer surfacers are recoated with topcoats after just a short flash-off time (wet-on-wet application). For small vehicle parts, there are water-borne stoving primer surfacers available that are based on polyesters and contain amino resins as binders (see Chapter 3.5.6).

Topcoats

For the application of topcoats to heavy vehicles, the products used are comparable to topcoat systems for repair coatings. These are two-component coating systems consisting of hydroxylic acrylic resins (see Chapter 3.8.3.1), which are crosslinked by polyisocyanate adducts (see Chapter 3.8.3.4). To comply with VOC regulations mainly in Europe, high-solids systems and also water-borne two-component paints are available. For high-solids topcoats, low-molecular weight, low-viscosity acrylic resins and also low-viscosity aliphatic polyisocyanate adducts are chosen, mostly those with a small molecular weight distribution or oligomers containing allophanates. Resins for water-borne topcoats are secondary acrylic dispersions while the crosslinkers are low-viscosity polyisocyanate adducts or polyisocyanate adducts containing hydrophilic modifications. Both must lend themselves readily to homogenisation into the entire coating system. As already described, it is necessary in these cases to bear the side-reaction of isocyanates with water in mind, by adding excess polyisocyanate (the hardener), in terms of stoichiometric ratios.

Only a few of the objects described here are coated by two-layer topcoat systems comprising basecoat and clearcoat. In those cases, water-borne basecoats and high-solid two-component clearcoats are used. Effect basecoats and clearcoats can also be applied. Such decorative considerations for heavy vehicles are needed where companies seek to identify their vehicles, for example, in their own livery colours. There are also coloured topcoats available for inscriptions and company logos. Such inscriptions are also prepared using adhesive foils. Only when large areas of coloured or effect designs are required is it necessary to use coating materials. In that event, application is carried out by spraying and masking those parts which are not to be coated.

As already mentioned, the entire coating process is arranged such that the parts are primed separately first. Then the vehicle is assembled as necessary, followed by application of putties to those parts or areas that require them. This may be followed by application of the primer surfacer and a sanding process. After that, the topcoats and, in special cases, other decorative elements are applied. For topcoats and decorative elements, the vehicle producers follow the wishes and requirements of the ordering customer. There are also vehicle parts which are coated completely before full assembly, for example with electro deposition primers, and powder topcoats applied by electrostatic powder spray equipment.

Plastic parts on heavy vehicles, for example, driver's cabs, are coated with the same paint products as are used for attached parts for passenger cars (see Chapter 5.2). In addition, it must be pointed out that many other parts and building elements on vehicles have to be coated that have not been mentioned so far.

Examples of other coatings on automobiles are those for engine housings, oil pans, axle suspension parts, carburettor lids, compressors, brake housings, battery housings, fuel tanks, exhaust pipes, cooling equipment, condensers, and heating equipment. Where aluminium or plastic is used for the construction of such parts, it is not usual to coat them. The coating of such parts is the responsibility of the parts producers. There are many products which have been formulated to meet the properties and intended usage of the parts. These coating materials are not classified as automotive coatings, but rather as general industrial coatings.

7 Alternatives and outlook

Various ideas for fundamentally changing the application process for automotive coatings have been advanced. One reason for the many studies in this regard is that, for car producers, the application of coating materials is totally different from all other assembly processes. More objective reasons for changing the application process are those seeking to make savings on materials and effort.

The coatings industry, on the other hand, has been able to comply with vehicle assembly concepts, for example by developing electrodeposition primers and the spray applications on conveyor belt processes. Most of the ideas for implementing alternative methods to replace coating applications require changes to the car assembly process. In most cases, it would be necessary to abandon the method of producing self-supporting car bodies, but that would also mean relinquishing some other advantages. Two alternative methods are still being studied and debated: the use of pre-coated panels, made by coil-coating; and an all-plastic car body from parts already coated by different application processes.

Coil coating is the coating of panels on endless metal coils by roll-coating equipment (possibly on both sides) followed by film forming in a stoving oven. The coated coil is cut and shaped into the parts. Coil coating has been implemented in the coatings industry for many years and has proved successful in different application fields. The advantages of the process are the very high transfer efficiency of the coating process itself, and the fact that the solvents evaporate in the stoving oven and are fed to a catalytic after-burning system to generate heat for the stoving oven. Available coil coating paints can also meet the requirements imposed on automotive OEM coatings, namely weathering resistance, flexibility, resistance to corrosion, chemicals, solvents, and mechanical impact. Disadvantages reside in the loss of materials due to cutting processes and the fact that the cut edges are not covered with coating initially, which may lead to corrosion. The use of pre-coated and deformed parts for car bodies would entail changes to all the technology for joining car bodies. The construction of car bodies from such pre-coated panels and parts would then take form of a space-and-frame process, which has been used in the past. Theoretically, welding, riveting and bonding would be suitable joining methods. With the exception of adhesives, these joining processes expose bare metal. These could act as the source and starting point for corrosion. Protecting the cut edges and joints with coating materials separately after car body assembly would incur additional costs and effort.

Projects and trials are underway to prepare the entire coating system for car bodies using the coil-coating process [190]. However, it is not possible to achieve all the colours and effects which are currently available. For the most part, it is the bright and light metallic effects with extensive flip-flop - and they are common in Europe - that cannot be duplicated using by coil coating methods. The idea is therefore to use coil coating only for primer. This could take place during production of the metal coil. The advantage would be elimination of the electro deposition process (see Chapter 3.4.6), which currently requires considerable outlay on equipment and process at car producers. However, again, new joining technologies would be necessary. Primer surfacers and topcoat systems would be applied by conventional spraying methods and all currently required colour shades and effects could be obtained. For a coil coating process, steel and galvanised steel make suitable substrates, but coated aluminium coils are being used more and more. Nowadays, car-making steel is already competing against plastic materials. The advantages of steel are optimum formability (for construction of car bodies, but also in the event of an accident), ease of repair, and ease of recycling. In contrast, plastic parts are not sensitive to corrosion and they are lightweight. Most plastic parts need to be protected against weathering. Since plastic parts currently account for 30 % of car exteriors, there is a specific barrier to be overcome before this proportion can be greatly increased. As already mentioned, some types of car would have to be constructed entirely from plastic parts. There are different ways of achieving this and implementing the necessary coating processes.

There are self-supporting car bodies which are made exclusively from plastic materials and which are topcoated by conventional methods. Since such car bodies need sufficient stability (rigidity), the plastic materials are chosen for their ability to meet such requirements very well. There are plastic materials, such as epoxy resins crosslinked by polyamines, which are reinforced with carbon fibres. Such materials are relatively expensive. So far, only special models have been produced in this way, and only in small numbers. These are racing cars [191] and high-class sports cars.

The second way is to construct the car body first from metal frames and then to equip the frame with plastic panels (space-frame process). This method was used in the past to produce some sports cars [192], but it is employed today for some small and less expensive cars [193]. The plastic parts are coated by conventional and common application methods (see Chapter 5.2).

One particular method is that of in-mould coating. The inside of the mould is sprayed with release agent. Then the coating material is applied by spraying or by flow coating. After a pre-drying time, the liquid plastic material is injected into the mould and cured. Upon removal from the mould, the part is already very well coated. Instead of a coating, the mould may also be lined with film or leather. This method is already widely used for interior car parts (see Chapter 4.1).

The final method would be to construct car bodies with the aid of multilayer film systems. For example, such a film system would consist of a carrier film, a film containing coloured pigment and effect substances (comparable to the basecoat layer of conventional coating methods), and a transparent film which confers optimum scratch resistance and meets other resistance requirements imposed on coating surfaces (playing the same role as clearcoats). The multilayer film system would then be shaped and cured to form body parts.

Some ideas have been proposed that seem a bit utopian. The lotus effect has already been described (see Chapter 3.8.3.12). More achievable seems the idea of generating holograms on car surfaces; this would be possible where film is used for car bodies. So far, there has been no useful solution to the question of creating surfaces that can change colour or effect.

It is anticipated that current application processes and the usage of liquid and sprayable coating materials will continue. So far, such materials have been able to meet regulations of reducing emissions of volatile organic compounds. Most of all, the use of such materials makes it possible to meet the demands and wishes of customers for custom surface designs and appearance that will surely become more widespread in the near future.

Literature

General literature

- H. Kittel: Lehrbuch der Lacke und Beschichtungen, 2nd edition, volume I-X, published by S. Hirzel, 1998-2006
- A. Goldschmidt, H.-J. Streitberger: Basics of Coating Technology, 2nd edition, published by Vincentz Network, 2007
- D. Stoye, W. Freitag: Paints, Coatings and Solvents, 2nd edition, published by Wiley-VCH, 1998
- B. Müller, U. Poth: Coatings Formulation, published by Vincentz Network, 2006
- G. Fettis (editor): Automotive Paints and Coatings, published by VCH, 1995

Notes

[1] Wikipedia, the free encyclopaedia; keyword: automotive industry

[2] Information provided by BMW

[3] List of paint producers in Farbe & Lack, 112th volume, 2006/10

[4] Example: Integrated coating concept in cooperation with a car producer (Daimler Benz), a paint producer (BASF Coatings) and a producer of application equipment (Dürr) for the assembly plant in Rastatt

[5] Bruno Vollmer: Grundriss der Makromolekularen Chemie, vol. 3, Eigenschaften des freien Makromoleküls; published by E Vollmert, Karlsruhe, 1979

[6] D. Stoye, W. Freitag: Lackharze, published by Carl Hanser, 1996

[7] H. Kittel: Lehrbuch der Lacke und Beschichtungen, 2nd edition, volume 1, Section 1.4 Naturstoffe, published by S. Hirzel, 1998

[8] Wikipedia, the free encyclopaedia; keywords: 'Light, physically' and 'Electromagnetic waves'

[9] Johann Heinrich Lambert, German physicist and philosopher, 1728-1777, described the reduction of light intensity by absorption

[10] August Beer, German physicist, chemist, and mathematician, 1825-1863, defined the law of light absorption, known as the Beer-Lambert Law

[11] Willebrord van Roijen Snell, Dutch astronomer and mathematician, 1580-1626, discovered the laws of light refraction.

[12] Augustin Jean Fresnel, 1788-1827, French physicist, discovered the wave theory of light

[13] John William Strutt, 3rd Lord of Rayleigh, English physicist, 1842-1919, investigated the theory of oscillations and interpreted the scattering of light

[14] Gustav Mie, German physicist (1868-1957) and Ludwig Lorenz, Danish physicist (1829-1891) described the scattering of light on dispersed particles, with particle sizes in the region of the wavelength of light

[15] DIN 55982, Lightening power

[16] P. Kubelka, F. Munk: Ein Beitrag zur Optik der Farbanstriche, Z. Tech. Phys. 12:593, 1931, Frédéric Cortat: The Kubelka-Munk theory, applications and modifications; 2004

[17] DIN 55982 Tinting strength

[18] Germany: „Änderung der Abwasserverordnung" (Change in Rules for Waste Water) 17. 6. 2004

[19] O. Lückert: Pigment & Füllstoff Tabellen; published by Vincentz Network, 6th edition

[20] P. Huyskens, M. C. Haulait-Pirson, X. van der Borght: Berechnung des Lösevermögens von Lösemitteln für Harze, Farbe & Lack, 92nd edition, 12/1986

[21] Calculation programs of Shell (1977), Union Carbide (1970), Eastman (1967)

[22] ISO 1523, 3673, DIN 51758, 53213, DIN/EN 456, ASTM D 93, flash point

[23] Germany: Rules for handling flammable liquids VbF and GGVSE, ADR, RID

[24] Karsten: Lackrohstoff-Tabellen, published by Vincentz-Network, Hanover, 10th edition

[25] Wikipedia, the free encyclopaedia: keywords: 'viscosity'; 'rheology', 'measurement of viscosity'

[26] B. Müller, U. Poth: Coatings Formulation, part IV, published by Vincentz Network, 2006

[27] T. Kiguchi, H. Aota, and A. Matsumoto: Approach to Ideal Simultaneous Interpenetrating Network Formation via Topological Cross-Links between Polyurethane and Polymethacrylate Network Polymer Precursors; Macromolecules, 37 (22), 8249-8255, 2004 and many articles on the Internet.

[28] Patents regarding hybrid crosslinking DE 4324947, DE 4310414, US 5780541

[29] W. Kremer, Dual Cure Process, Farbe & Lack 94 (3) 205; 1988, Vincentz Network

[30] See methods and test equipment for differential thermal mechanical analysis (DMTA)

[31] H. Kittel: Lehrbuch der Lacke und Beschichtungen, 2nd edition, volume 8, Lackherstellung (production of paints), published by Hirzel, 2002

[32] 100 Jahre Automobilbau - Automobillackierung, special edition of Kunstharznachrichten, Hoechst AG, February 1986

[33] Introduction of conveyor-belt car production in Detroit in 1913; the number of cars increased in 1914 to about 1000 units per day.

[34] Founding of the first commercial synthetic resins (named Albertol), based on phenolic resin and colophony for combination with drying oils by Ludwig Berend in the Ltd Dr. Kurt Albert GmbH 1910 and production start-up in 1917

[35] Start-up of production of cellulose nitrate suitable for automotive coatings which were used at General Motors Corporation in 1923
[36] Henry Ford answering a journalist in the 1920s
[37] R. H. Kienle: alkyd resins; AP 1 803 174 (GE, 1925), DRP 574 963 (AEG, 1926), EP 252 394 (1926), Ind. Eng. Chem. 21 (1929)
[38] First stoving enamels based on alkyd resins at General Motors in 1933
[39] First two-layers metallic coating systems, suitable for automotive OEM coatings from Celanese in the USA and the Herbol GmbH in Germany (later BASF Coatings) in 1967/1968
[40] Laws of Bernoulli; Daniel Bernoulli (1700 - 1782), a Swiss mathematician and physicist, developed equations for fluid mechanics together with Leonhard Euler
[41] Figure: Pneumatic spray gun of DeVilbis Ltd
[42] Figure: High efficiency spray gun (HLVP), of DeVilbis Ltd
[43] Figure: Electrostatic high speed rotation bell spray gun of Dürr Ltd
[44] Transfer efficiency information from Dürr Ltd
[45] Wikipedia, the free encyclopaedia; keywords: 'steel', 'production of steel'
[46] Lancia Lambda
[47] Opel-Olympia
[48] Wikipedia, the free encyclopaedia; keywords: 'corrosion', 'electrochemical model of rust formation', 'local elements'
[49] Corrosion protection, EN ISO 12944, DIN 55928
[50] Wikipedia, the free encyclopaedia; keyword: 'galvanising (with zinc)'
[51] Ullmann's Encyclopedia of Industrial Chemistry (electronic version): Pretreatment of Substrate Surfaces; published by VCH, Weilheim 1998
[52] Ullmann's Encyclopedia of Industrial Chemistry (electronic edition): Metals, Surface Treatment; published by VCH, Weilheim 1998; K. Fiedler: Die Vorbehandlung von Metallen vor der Lackierung (pretreatment of metals before coating); CHEMETALL, Chem. Lab. Oberflächentechnik
[53] Use of first dipping primers in 1945
[54] Development of first water-borne resins for stoving coatings by Herbert Hönel at Vianova: Neue Wege in der Anstrichtechnik (New ways in coatings technology); GDCh congress 1953
[55] Resin: Dow 566
[56] Gordon Fettis (editor): Automotive Paints and Coatings, section 3.4.7.1, published by VCH, 1995
[57] A. Goldschmidt, H.-J. Streitberger: BASF-Handbuch Lackiertechnik, section 7.1.2 published by Vincentz Network, 2002
[58] Basic patents US 3,169,945; US 4,419,467; US 4,170,579 of PPG
[59] Mannich base process of BASF AG
[60] Reverse-process at car producer Saab Ltd
[61] Electro Powder Coating (EPC) of Shinto for BMW and Honda
[62] Electro Slurry Coating (ESC) of ICI
[63] Application at Japanese car producers
[64] H. Schenck, H. Spoor, M. Marx: The Chemistry of Binders for Electrodeposition, Progress in Organic Coatings, (1979); A. Doroszowski: Electrodeposition of Non-Ionically Stabilized Emulsions, Colloids Surf. 17 (1986)
[65] Rotodip process applied in Japan and in the UK
[66] Erichsen test, DIN EN 13523-6, EN ISO 1520
[67] Reverse impact test DIN EN 13523-5, EN ISO 6272
[68] Rust undercutting, testing degrees of rusting, DIN 53210, ISO 4628-3
[69] D. Stoye, W. Freitag: Lackharze, 7.2.6: Aliphatische und cycloaliphatische Epoxide; ('Aliphatic and cycloaliphatic epoxies', published by Carl Hanser, 1996
[70] Gordon Fettis (editor): Automotive Paints and Coatings, section 3.4.6.2, Composition and Formula of Cathodic Primer; published by VCH, 1995
[71] U. Röckrath, K. Brockkötter, Th. Frey, U. Poth, G. Wigger: Investigations of the Crosslinking Mechanism of Etch Resistant Clearcoats; Progress in Organic Coatings, 32 (1997), 173-182
[72] Example: Bismuth compounds; patents of DuPont Performance Coatings
[73] Patent US 06 274 649 of BASF-Coatings
[74] Example 1 and 2.1; patent US 06 274 649, dispersion and crosslinker
[75] Example 1.1; patent EP 505 445 of BASF Coatings, grinding resin 1
[76] Example 1.3; patent EP 505 445 of BASF Coatings, grinding resin 3
[77] Example 4.1; patent US 06 274 649, pigment dispersion
[78] Example 5.1; patent US 06 274 649, dipping bath
[79] Gordon Fettis (editor): Automotive Paints and Coatings, section 3.4.8.1, VCH, 1995
[80] B. Müller, W. Rath: Formulierung von Kleb- und Dichtstoffe, (Formulation of adhesives and sealing compounds) part III, p. 238, published by Vincentz Network, Hanover, 2004
[81] Patents of Bollig & Kemper AG

[82] Stonechip test DIN 55996-1 multi impact, -2 single impact
[83] Stonechip test Hatchet impact at BMW
[84] Stonechip test Ball shot impact at MB
[85] Temperatures for stonechip testing at +20 °C and -20 °C
[86] U. Poth: Polyester und Alkydharze, published by Vincentz Network, Hanover, 2005
[87] H. Kittel: Lehrbuch der Lacke und Beschichtungen, 2nd edition, volume II, 2.1.5; Alkydharze und gesättigte Polyester (Alkyd resins and saturated polyesters), published by S. Hirzel, 2002
[88] Don Sanders (editor): Water-borne & Solvent Based Saturated Polyesters and their End User Applications, Surface Coatings Technology, Volume IV, John Wiley, 1999
[89] J. P. Flory: Principles of Condensation Polymerisation, Journ. Am. Chem. Soc. (1936) 58, 1877; Journ. Am. Chem. Soc. (1941) 63, 3083
[90] Resamin HF 450, Cytec Surface Specialties (Vianova)
[91] Powderlink C 1174; Cytec Surface Specialties (Dyno-Cytec)
[92] Oil absorption value, DIN 53199
[93] Akzo-Nuplex: REC 96110 in 'Specialty Resins'
[94] Recommendation of AMOCO
[95] EP 0 339 433 (BASF-Coatings), example 1
[96] EP 0 339 433 (BASF-Coatings), polyester 1
[97] EP 0 339 433 (BASF-Coatings), polyurethane 1
[98] Cymel 301, Cytec Surface Specialties (Dyno-Cytec)
[99] Chrysler assembly plant in St. Louis and Toledo
[100] Application at PSA, Mulhouse
[101] Patent example 8 in DE 02 163 962 of Unilever
[102] Dow, technical data sheet of D.E.R. 663 U
[103] B. Müller, U. Poth: Coatings Formulation, part IV, chapter 3.2, published by Vincentz Network, Hanover 2006
[104] Eurostar, Graz, until 2000, currently still in the USA
[105] Ulrich Poth, Heinrich Wonnemann: Properties of Epoxy-Polyester-Powder-Primer Surfacers for Automotive OEM; congress IBEC, 1993, Stuttgart
[106] Coating of the metal frame of the Smart
[107] DIN 55677, sagging limit
[108] Test method of BASF Coatings: application of a coating wedge, short flash-off, and horizontal position during baking process.
[109] Method of rolling ball; DIN EN 1939; DIN ISO 1628
[110] Perthometer, DIN EN ISO 4288; DIN ISO 12085
[111] Reflectometer, EN ISO 2813
[112] DOI, ASTM D 5767
[113] Adhesion, cross-cut test, EN ISO 2409
[114] Definition of alkyd resins, DIN 55958, DIN 53183
[115] Technical data sheets: Hexion (Shell-Resolution) - Cardura; Exxon-Glydex
[116] U.Poth: Polyester und Alkydharze, chapter 4.4.5.1; published by Vincentz Network, Hanover, 2005
[117] Luwipal 018, BASF AG
[118] Ullmann's Encyclopedia of Industrial Chemistry (electronic version): Titanium dioxide; Synthesis; VCH, Weilheim 1998
[119] Carbon black TEM photography of DEGUSSA
[120] Colour Index, C. I.
[121] Aerosol; technical data sheets of DEGUSSA
[122] SCA-Resins, technical data sheet of Duplex (earlier AKZO Resins)
[123] AKZO Resins, (now Duplex) REC 96061
[124] Cellulose acetobutyrate EAB 381-2BP of Eastman-Kodak
[125] Technical data sheets of Eastman-Kodak
[126] Patent example B3 in EP 0 137 256 of BASF Coatings
[127] Ullmann's Encyclopedia of Industrial Chemistry (electronic version): Aluminium, Physical Properties; VCH, Weilheim 1998
[128] SEM-pictures of Eckardt Ltd (Altana Group)
[129] H. Jaquin described fish silver in 1655
[130] F. J. Maile, G. Pfaff, P. Reynders: Effect pigments - past, present and future; Prog. in Org. Coat.; Elsevier, 2005
[131] Patents US 2,278,970 (1942), US 3,087,828 (1963)
[132] Ullmann's Encyclopedia of Industrial Chemistry (electronic version): Pigments, inorganic; VCH, Weilheim 1998
[133] Paliochrome pigments of BASF AG
[134] Variochrome pigments of BASF AG, EP 0708154; EP 0753545
[135] Chromaflair pigments of Flex Products Inc. (USA), USP 4,434,010, USP 5,029,245
[136] Colorstream pigments of Merck KGaA, EP 0608388

[137] Paliocolor pigments of BASF AG, EP 0601483 (LC polyester)
[138] Heliocone pigments of Wacker Chemie, EP 0799877 (LC polysiloxane)
[139] Dr. Frank J. Maile, Merck KGaA Darmstadt, Euroforum-conference (20059: Farben und Lacke für das Automobil/ Die Zukunft der Effektgebung (future of effect forming)
[140] Duplex (earlier AKZO-Resins) Specialty Resins (2001) REC 98072
[141] Patent examples for medium-solids basecoats EP 0052776, EP 0137256
[142] Patent examples for high-solids basecoats containing microgels, EP 0480959
[143] Bruno Vollmert: Grundriss der Makromolekularen Chemie, Bd. 1, S. 181 Emulsionspolymerisation; E Vollmert Verlag, Karlsruhe, 1979
[144] Patent DE 3825278
[145] Patent examples EP 0521928, EP 0593454
[146] Patent example polyester 3 in EP 0239149
[147] Examples of pretreated aluminium pigments / Eckardt Ltd (Altana Group): Hydrolux, Hydrolan, Hydrolac
[148] Patent examples EP 0228003, EP 3825278
[149] R. Renz, H. Töpfel, W. Vesper; Einfluss des UV-Lichts auf die Zerstörung von Lackfilmen; Farbe + Lack, 80, 6 (1974) S. 513-517, Vincentz Network
[150] Orange Peel measurement on high to semi gloss surfaces; Byk-Gardner; Geretsried, December 6th, 2005 DOI and Wavescan ASTM E 430
[151] PRA: Testing Paints for Resistance to Weathering; ASTM G154; EN ISO 11507, EN ISO 11341, ISO 12040
[152] Example: Testing of chemical resistance using the gradient bar oven (Daimler-Benz method)
[153] B. Müller, W. Rath: Formulierung von Kleb- und Dichtstoffe, (Formulation of adhesives and sealing compounds) part III, p. 246, Vincentz Network, Hanover, 2004
[154] Patent DE 4324917, example 1
[155] Cylink 2000 of Cytec, Laro-TACT of BASF AG
[156] Molecular structure and proposed reaction mechanism of free-radical quencher by Ciba Specialty Chemicals
[157] SCA-modified resins of Duplex Resins (earlier AKZO Resins)
[158] Example: Setalux 81198 of Duplex Resins (earlier AKZO Resins)
[159] Example of Duplex Resins (earlier AKZO Resins), Specialty Resins REC 95041
[160] Wolfgang Bremser, University of Paderborn, Lecture in seminar: chemistry and physics of automotive coatings, Vincentz Network, Cologne 2006
[161] TCD Acrylate of Röhm
[162] Patents EP 0710676, EP 0710707, EP 0915113 of BASF Corporation
[163] Guideline 2004/42/EG of the European Parliament and Council regarding the limitation of emission of volatile organic compounds using organic solvents in defined paints and coatings and in products for automotive repair coatings as well, and variation of the former guideline 1993/13/EC of April 21st 2004, second step, valid from January 1st 2007
[164] Water-borne clearcoat of DuPont Performance Coatings for GM-Opel, assembly plant in Eisenach; patent example DE 19623371
[165] Examples: Product group Bayhydur of Bayer Material Science
[166] Line trials of DuPont Performance Coatings and Bayer AG at Audi in the Ingolstadt assembly plant
[167] Patent example DE 19841842
[168] Solvent-free water-borne clearcoat for Daimler Benz, type A production in Rastatt
[169] Powder clearcoat, patent example in DE 42 27 580
[170] Powder Slurry, Patents EP 0652264, DE 19540977, DE 19617086
[171] UV initiators of Ciba Specialty Chemicals
[172] Peter G. Garratt: Strahlenhärtung (Radiation Curing), Vincentz-Network, Hanover, 1996
[173] Dürr Ltd constructed in 2006 an application device for UV curing in a plasma atmosphere
[174] David James, Henrik Bernquist, Pia Appelkvist, Peter Sandell, Kent Sörensen (Perstorp Specialty Chemicals AB): Dual functionality monomers for photo crosslinking open up new opportunities
[175] Changeover to anodic electrodeposition primers based on maleinised linseed oil in 1965
[176] Powder clearcoat at BMW in assembly plants Ingolstadt, Regensburg and Leipzig
[177] Dispersion clearcoat, patent DE 19841842
[178] Polyamidoamines of Huntsman (Eurolon) and Cognis (Versamid)
[179] Römpp, Lexicon of Chemistry; keyword: Zinc chromate
[180] Report in Farbe & Lack, 2005, October; Vincentz Network
[181] Resins of Cytec Surface Specialties (formerly Vianova): Macrynal
[182] Low viscous polyisocyanate adducts of Bayer Material Science and Rhodia
[183] Three-component system; described in patent DE 195 42 626 of BASF Coatings
[184] Products of Duplex Resins (earlier AKZO Resins): SETALUX 17-7101; Setalux 10-1440; Setalux 7001 SS-45 acetoacetate-ketimine systems
[185] Gommans, Finish Coatings for the New Millennium, Minutes of: Corrosion Protection, 1999, Australia

[186] Wikipedia, the free encyclopaedia; keywords: injection moulding, in-mould process
[187] Wikipedia, the free encyclopaedia; keyword: Sheet moulding composition (SMC)
[188] Gordon Fettis (editor): Automotive Paints and Coatings, Chapter 6, VCH, 1995
[189] Wikipedia, the free encyclopaedia; keyword: plasma, low-pressure plasma
[190] R. Drufke: Coil-coating; Internet
[191] Car body based on carbon fibre reinforced plastic materials at BMW
[192] The legendary Fierro produced in the USA was one of the first sport cars to be made by the space-frame process
[193] Body parts for the Smart car

Acknowledgements

Thirty years in research and development and the production of resins and automotive coatings were followed by optimisation work that required a great many conversations, interviews and discussions. The best results were always achieved by combining research ideas and models with the results of practical trials. It is impossible to name all the partners with whom I was involved in this field over the years. However, some names stick in the mind and shall be mentioned here on behalf of the others. In Würzburg: my colleagues Werner Gehringer, Adolf Machholz, Willy Vesper and Dr. Bodo Müller, who were involved in the development of solvent-borne and water-borne basecoats, and solvent-borne clearcoats. In Münster: Helga Middendorf, who worked on primer surfacers, and Dr. Rainer Henning on plastics coatings. Heinz Töpfel in Würzburg was a vital source of information on the influences exerted by various application processes.

More recent information was kindly provided by my former colleagues in Münster, namely Dr. Günther Ott (electro deposition primers), Dr. Peter Mayenfels (OEM clearcoats), Dr. Bernhard Lettmann (automotive repair coatings), and Dr. Lutz-Werner Groß (coatings for heavy loaders).

Index

Symbole

A

B

C

D

E

F

G

H

Q

R

S

T

U

V

W

Z

www.ingramcontent.com/pod-product-compliance
Lightning Source LLC
LaVergne TN
LVHW081322300926
849161LV00008B/1853
* 9 7 8 3 8 6 6 3 0 9 0 4 3 *